# The Physiology of Flowering Plants: Their Growth and Development

*Third Edition*

## H. E. Street
D.Sc., F.I.Biol.

The Late
Professor of Botany
University of Leicester

## Helgi Öpik
Ph.D.

Senior Lecturer in Plant Physiology
University College of Swansea

Edward Arnold

© Helgi Öpik and P. Street 1984

First published in Great Britain 1970
by Edward Arnold (Publishers) Ltd
41 Bedford Square
London WC1B 3DQ

Reprinted 1971, 1973
Second Edition 1976
Reprinted 1977, 1979
Third Edition, 1984
Reprinted 1986

Edward Arnold (Australia) Pty Ltd
80 Waverley Road
Caulfield East 3145
PO Box 234
Melbourne

First Published in United States of America 1984
By Edward Arnold
3 East Read Street
Baltimore
Maryland 21202

**British Library Cataloguing in Publication Data**

Street, H. E.
    The physiology of flowering plants.—3rd ed.—
(Contemporary biology)
    1. Flowers    2. Plant physiology
    I. Title    II. Öpik, Helgi    III. Series
    582.13′041    QK711.2

ISBN 0-7131-2883-6

Text set in 10/11pt Times New Roman
by Castlefield Press, Northampton
Printed and Bound by Thomson Litho Ltd, East Kilbride, Scotland

# Preface to the First Edition

The present volume is an introduction to the many problems posed by the growth and development of flowering plants. We hope that such an introduction will meet the needs especially of first and second year University students of biology and those concerned with the teaching of biology in our schools.

Chapters dealing with the water relations, solute movement, nutrition and energy balances of the developing plant lead on to those describing growth, growth movements and morphogenesis. Such an approach inevitably omits consideration of major aspects of plant metabolism, more appropriately treated in an introduction to cellular physiology and biochemistry. This restriction of the scope of the present volume has been imposed, not only by considerations of space, but because we feel it corresponds to a widely-adopted pattern in modern university teaching.

Text figures and plates are particularly valuable to the teaching of our subject matter and we are deeply appreciative to authors and publishers for permission to reproduce many of our illustrations.

Throughout the writing and particularly in the final stages of preparing the manuscript, we have benefited from the expert and painstaking help of our editor, Professor A. J. Willis.

Leicester and Swansea           H.E.S.
1970           H.Ö.

# Preface to the Second Edition

The revisions introduced for this new edition have, we hope, appropriately updated but not altered the standard and general theme of our 1970 text. Those aspects of plant physiology which feature prominently in current research have needed most extensive revision; these include the metabolic aspect of germination (Chapter 1), phloem and hormone transport (Chapter 6), growth movements (Chapter 10) and many facets of developmental physiology (Chapters 11 and 12 replacing our original Chapter 11). The treatment of the water relations of plant cells has been rewritten in water potential terminology (Chapter 4). This introduction of new material (with appropriate text figures) has been achieved without substantial increase in size of the volume by deleting (somewhat regretfully) certain sections which seemed less central to our main themes.

We have again benefited from the expert help of our editor, Professor A. J. Willis.

Leicester and Swansea           H.E.S.
1975           H.Ö.

# Preface to the Third Edition

Following the untimely death of Professor Street in December 1977, the task of preparing the third edition of this book has fallen to me in its entirety. I have sadly missed his presence, but I have still endeavoured to work within the framework of his original concept of the book. I can only hope that I have not proved too unequal to the undertaking.

Plant physiology is a rapidly developing science and progress made in the eight years since we completed the second edition has again required considerable updating of the physiology of germination (Chapter 2), phloem transport (Chapter 6) and tropic movements (Chapter 10). The emergence of new knowledge and ideas about plant growth hormones has led to the revision of numerous passages; in particular Chapter 9 – Cell Growth and Differentiation – has been re-organized to include a section on the mode of hormone action, as well as a consideration of modern methods of hormone analysis. A general treatment of the physiology of photosynthesis has been included in Chapter 3; the account of mineral nutrition (Chapter 5) has been made more comprehensive, and discussions of salinity stress and high temperature resistance have been incorporated in Chapter 7. It is hoped that these addenda will increase the usefulness of the text. The topics of photomorphogenesis and seed dormancy also have been given somewhat fuller treatment.

The advice of Professor A. J. Willis, editor of this series, is gratefully acknowledged. I should also like to thank my colleague, Dr J. Hayward, for much support during the preparation of the manuscript.

Swansea, 1984                                                                                    H.Ö.

# Contents

Preface to the first edition      v

Preface to the second edition      v

Preface to the third edition      vi

1. **Introduction**      1

2. **Germination**      4
Seed structure, chemical composition and metabolism
Water imbibition: the hydration phase of germination
Phase of activated metabolism and growth
Interrelationships between the growing axis and the storage tissues

3. **Energy Flow and Carbon Turnover**      26
Introduction
Photosynthesis: the overall process
The efficiency of photosynthesis
Utilization of photosynthate: biosynthesis and respiration

4. **Water Relations**      47
Water relations of plant cells
Water relations of whole plants and organs

5. **Mineral Nutrition**      74
Essential mineral elements
Mineral nutrients in the soil
The uptake of minerals by the root
Physiological roles of the mineral elements

6. **Transport of Metabolites**      88
The transporting system
The composition of phloem and xylem sap
The mechanism of phloem translocation
The transport of hormones and stimuli

7. **Resistance to Stress**      110
Resistance to water stress (drought resistance, drought hardiness)
Resistance to low temperature stress (cold hardiness)
Resistance to high temperature stress

Resistance to salinity stress
Resistance to heavy metals
The human viewpoint: breeding plants for stress resistance

8. **Growth: Progress and Pattern**                    135
The definition and measurement of growth
Localization of growth in space and time
Conditions necessary for growth
Growth rates
Morphogenesis

9. **Cell Growth and Differentiation**                 157
Plant growth hormones: a brief survey
Cell division
Cell expansion
Cell differentiation
Patterns of differentiation
The problem of the mode of hormone action

10. **Growth Movements**                               189
Tropisms
Gravitropism
Phototropism

11. **Vegetative Development**                         211
Morphogenesis
Initiation of lateral roots and buds
Leaf initiation and growth
Apical dominance
Seasonal bud dormancy
Leaf senescence and leaf abscission
Photomorphogenesis

12. **Reproductive Development**                       232
Flowering
Fertilization
Fruit development
Embryo development
Seed dormancy

Units of measurement – conversion table               267
Index                                                 268

# 1
# Introduction

*'For this appointment preference will be given to a whole plant physiologist'*

The quotation at the head of this chapter, taken from an advertisement for a university post, is meant to attract a particular kind of plant physiologist, one interested in the physiology of the whole organism rather than a student of cellular physiology! The distinction implied is sometimes reflected in the university teaching of plant physiology in separate courses, one termed 'metabolism' or 'cellular physiology', and a second termed 'growth and development' or 'whole plant physiology'. The present book attempts to give an account of the physiology of flowering plants from the viewpoint of the whole plant, with particular emphasis on processes of growth and development.

When plant physiology teaching is so divided into two complementary courses, the distinction between them is not only one of subject matter but of emphasis. The course in metabolism, by having the cell as its primary territory, emphasizes the basic similarity of metabolic patterns in different organisms and develops the theme of the unity of living organisms rather than that of their diversity. Indeed many of the experimental data come from work with unicellular organisms, because of their suitability as experimental material. The subject matter of the second course, though also concerned with basic phenomena, must to a far greater extent be concerned with diversity. In the development of a discussion of photoperiodism (see Chapter 12), the differences in response of different species are an essential part of the story. Similarly in considering the initiation of leaves, buds and roots it is essential to emphasize that different species react in different ways to experimental treatment with growth active substances.

The same broad topic may be treated in both courses, but with a different emphasis. Thus the mechanism of water uptake by individual plant cells may be treated in the first course and water uptake by the growing root, water transport to the shoot, and water loss by transpiration in the second course. Similarly such processes as respiration and photosynthesis may be treated as metabolic pathways in one course and as aspects of the overall energy economy in another. The account of photosynthesis developed in a course on metabolism draws upon work with unicellular algae such as *Chlorella* as well as upon work with higher plants and its emphasis is upon the mechanisms of conversion of radiant energy into utilizable chemical energy and with the biochemical pathways of carbon assimilation. It is in the second course that attention may however be directed to such aspects as the intercellular diffusion pathway of gases in multicellular plants, the regulating activity of stomata and the movement of metabolites into and away from the photosynthetic cells. Lectures on salt accumulation at the cellular level in the first course may be complemented by lectures on mineral nutrition in the second.

There are as many ways of teaching plant physiology as there are teachers of this subject. Even when a broad distinction is drawn between 'metabolism' and 'growth and development' or between 'cellular physiology' and the 'physiology of the whole organism' the exact scope of the two aspects will reflect the individual judgements of the teacher concerned. The scope of the present book represents such an individual judgement by the authors. Although we have placed our emphasis on the physiology of the whole organism, various aspects of cell structure and physiology are discussed in so far as they seemed essential to the proper consideration of the major aspects of growth and development which form the subject matter of the Chapters 2, 8, 9, 10, 11 and 12. Within these chapters consideration is not confined to the growth and development of the plant and its separate organs, but extends to include processes of cell division, cell enlargement and cell differentiation, cellular processes involved in the initiation of organs, in organ growth and in overall plant development. Similarly a brief account of the water relations of plant cells was felt to be necessary before the water relations of the whole plant could be considered.

The development of the text can be briefly summarized as follows: a chapter on germination (Chapter 2) introduces a number of fundamental processes which are then treated in more detail in the immediately following chapters (Chapters 3 to 6). The background of general physiology thereby established forms a basis for the chapters relating to growth and development and to the influence on these processes of the natural environment.

The chapters follow a logical sequence and should normally be read in that order. Nevertheless it is recognized that a different, though equally defensible, order of presentation may be followed in the student's course at his/her university, college or school. With this in mind we have endeavoured, even to the extent of introducing a very limited element of repetition, to make each chapter as self-contained as possible and therefore readable on its own. Further each chapter is linked to other relevant parts of the book by cross references.

It is always possible to criticize a work of this kind on the grounds that it states what is known without indicating in sufficient detail how that knowledge has been obtained by observation and experiment, and further, that it favours this or that hypothesis or interpretation of the experimental data without a really critical evaluation of the theoretical and technical considerations involved. Shortcomings of this kind are inevitable when a large body of knowledge is surveyed in a book of limited size and when it is regarded as important that basic concepts and interrelationships should not be obscured by excessive facts, figures and references. With this in mind we have, at the conclusion of each chapter, referred the student and teacher not only to *Further Reading* but to *Selected References*. The latter are particularly important to those seeking the 'nature of the evidence' and details of important techniques. For advanced students each of our chapters can thus be regarded as the essential baokground to contact with primary data through the discussion of these and other research publications in tutorials and seminars.

We need not emphasize that the student reader should pursue a parallel course of practical work. The challenge of the study of plant physiology can be appreciated only by handling the research material, using techniques of measurement and interpreting one's own experimental data. The organization of such a course is, however, the province of the teacher working with the particular laboratory facilities available. Our text will, we hope, be of assistance to those concerned with developing such practical courses in so far as, at many points, it indicates useful plants for experimentation and outlines experimental approaches to the study of particular problems. Rigorous control of the physical and nutritional environment is the essential basis for the experimental study of many of the physiological problems raised in this book. Growth rooms represent, in this context, an important technical facility for which many of the design problems have now been satisfactorily solved. Growth room construction and details of other physical techniques applicable to the study of plant physiology are discussed in *Physics in Botany* by J. A. Richardson (Pitman and Sons, London,). The problems of lighting growth rooms are treated in *Lighting for Plant Growth* by E. D. Bickford and S. Dunn (Kent State University Press, 1972). It will also become clear, particularly in Chapters 9, 11 and 12, that our understanding of a number of aspects of plant physiology has, in recent years, been advanced by using the techniques of organ, tissue and cell culture. Certain of these techniques are valuable in developing an interesting practical course on plant growth and development. An introduction to plant tissue culture work can be found in *Plant Tissue Culture* by D. N. Butcher and D. S. Ingram (Studies in Biology No. 65, Edward Arnold, London, 1976). Comprehensive accounts of the techniques and their application to the study of a wide range of problems will be found in *Cells and Tissues in Culture* Vol. 3 (ed. E. N. Willmer, Academic Press, New York, 1966) and *Plant Tissue and Cell Culture* (ed. H. E. Street, Blackwell Scientific Publications, Oxford, 1973).

COMPLEMENTARY READING TO THIS VOLUME

CUTTER, E. G. *Plant Anatomy. Part 1, Cells and Tissues* (1978); *Part 2, Organs* (1971). Contemporary Biology Series, Edward Arnold, London.

GUNNING, B. E. S. and STEER, M. W. (1975). *Ultrastructure and Biology of Plant Cells.* Edward Arnold, London.

STREET, H. E. and COCKBURN, W. (1972). *Plant Metabolism,* 2nd edition, Pergamon Press, Oxford.

TROUGHTON, J. and DONALDSON, L. A. (1972). *Probing Plant Structure.* Chapman and Hall, London.

# 2
# Germination

## Seed structure, chemical composition and metabolism

The life of a flowering plant normally begins with a double fertilization within the embryo sac of the female parent. The egg nucleus fuses with one of the male nuclei contributed by the germinating pollen grain to form the zygote, while the two polar nuclei of the embryo sac and the second male nucleus fuse to give the triploid endosperm nucleus. Sometimes this sexual fusion is by-passed (apomixis) and an embryo develops from a diploid cell of the ovule. In either case, the embryo usually passes rapidly and without interruption through its early embryology (see Chapter 12). Then growth stops, water content falls and metabolic activity slows down prior to seed dispersal.

The degree of development of the embryo when the seed is shed varies; in the orchids (Orchidaceae) the embryo consists only of a small group of undifferentiated cells; in the grasses (Gramineae) several internodes with leaves and several embryonic roots are already distinguishable within the grain. Usually at least a radicle and plumule, each with an apical meristem, are differentiated, and one or two first leaves, the cotyledons. The triploid endosperm nucleus gives rise to the endosperm which may be a transient nutritive tissue or persist; the nucellus may also persist as perisperm, but more usually it disappears. The seed coat or testa is derived from the integument(s) of the ovule; sometimes, as in cereal grains, the testa and the ovary wall (pericarp) fuse to form the protective coat. Seeds present an immense variety of size, shape and structure, largely associated with modes of dispersal and with the conditions encountered in the natural habitat of the species. Some examples of seed structure are given in Fig. 2.1. The mature seed when released from the parent plant nearly always contains the embryo in a metabolically inactive, dormant state, capable of withstanding adverse environmental conditions. The success of the flowering plants is in a great measure due to the effectiveness of the seed as a perennating and dispersal organ.

Nearly all seeds contain some reserve nutrient; in some cases this may make up 85–90% of the seed by weight. Even in small seeds, such as those of lettuce (*Lactuca sativa*), weighing only a few mg, the reserves can support embryo growth for several days. In seeds like peas and beans, weighing up to

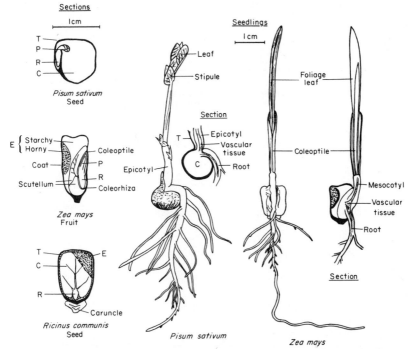

**Fig. 2.1** Some examples of seeds and seedlings: the pea, *Pisum sativum,* a dicotyledon with cotyledonary reserves; the castor bean, *Ricinus communis,* a dicotyledon with endospermic reserves; and maize, *Zea mays,* a monocotyledon with endospermic reserves. In the *Ricinus* seed section, part of the cotyledon is cut away to show the endosperm (stippled). C, cotyledon; E, endosperm; P, plumule; R, radicle; T, testa.

1.5 g, the stores suffice for several weeks, while coconut (*Cocos nucifera*) seedlings have been found to deplete only one-half of their reserve during 15 months of growth in darkness. The reserves may be found (as is common in the dicotyledons) in the cotyledons, in the endosperm (as in the Gramineae), or more rarely in the perisperm (e.g. in some Liliaceae and Piperaceae). Usually one type of storage tissue is present in any one seed, but this is not universally so. Both perisperm and endosperm are for instance present in the nutmeg (*Myristica fragrans*).

The chemical composition of some seeds is given in Table 2.1. The analysis reflects mainly the composition of storage tissues. The high lipid contents are noteworthy; no other plant organs achieve such high lipid levels. Lipids provide the highest amount of potential energy per unit weight. All the basic protoplasmic components are present in seeds, including nucleic acids, amino acids, vitamins, coenzymes and minerals. The phosphorus content is often high; the element occurs most commonly as phytin (salts of inositol hexaphosphate). All seeds contain protein as a protoplasmic component, but in some cases, e.g. in legumes, large quantities of special reserve proteins, with a distinctive chemical composition, are also stored. Chlorophyll is usually absent.

The reserves are found in the storage cells mostly as insoluble compounds.

**Table 2.1** The chemical composition of some seeds of economic importance. The percentages are based on the fresh (air-dry) weights of the seeds, except for Date palm, where the percentages are expressed on a dry weight basis.

| Species | Family | Nature of reserve tissue | Per cent content Carbo-hydrate | Protein | Lipid |
|---|---|---|---|---|---|
| Maize *(Zea mays)* | Gramineae | Endosperm | 51–74 | 10 | 5 |
| Wheat *(Triticum vulgare)* | Gramineae | Endosperm | 60–75 | 13 | 2 |
| Pea *(Pisum sativum)* | Leguminosae | Cotyledons | 34–46 | 20 | 2 |
| Peanut *(Arachis hypogaea)* | Leguminosae | Cotyledons | 12–33 | 20–30 | 40–50 |
| Soybean *(Glycine* sp.*)* | Leguminosae | Cotyledons | 14 | 37 | 17 |
| Brazil nut *(Bertholletia excelsa)* | Lecythidaceae | Hypocotyl | 4 | 14 | 62 |
| Castor bean *(Ricinus communis)* | Euphorbiaceae | Endosperm | 0 | 18 | 64 |
| Date palm *(Phoenix dactylifera)* | Palmae | Endosperm | 57 | 6 | 10 |
| Sunflower *(Helianthus annuus)* | Compositae | Cotyledons | 2 | 25 | 45–50 |
| Oak *(Quercus robur)* | Fagaceae | Cotyledons | 47 | 3 | 3 |

Small amounts of soluble sugars are usually present, but are concentrated mainly in the growing parts of the embryo. Carbohydrate occurs chiefly as starch or hemicelluloses. Starch is stored as grains up to 50 $\mu$m in diameter, formed in amyloplasts. Hemicelluloses are normal cell wall components in all tissues and where they form a major seed reserve they are laid down as heavy cell wall thickenings which almost fill the cell lumen, as in seeds of the date (*Phoenix dactylifera*) and ivory nut palm (*Phytelephas macrocarpa*). Reserve lipids are deposited as droplets of varying size enclosed within a limiting membrane though this does not show the usual unit membrane structure. The storage proteins occur as membrane-enclosed protein bodies (aleurone grains) or more rarely as crystals. The cells of reserve tissues are packed full of storage materials, and even the embryonic tissues of seeds are rich in reserve granules (Figs 2.2 and 2.3).

Seeds often contain unusual amino acids which are not constituents of proteins and these can act as a nitrogen store, being present to the extent of several per cent of the seed dry weight, and often containing a high proportion of nitrogen in the molecule. Pyrazol–l–yl–alanine, found in seeds of some members of the Cucurbitaceae, has a nitrogen content of 26%; the nitrogen content of canavanine, a constituent of seeds of the legume *Canavalia ensiformis,* amounts to 32%. (An average protein contains about 16% nitrogen by weight, though in seed reserve proteins the value may reach 19%.) The non-protein amino acids are often toxic to animals and are thought to fulfil also the role of repellents against seed-eating animals. Caterpillars and mice have been noted to refuse a diet

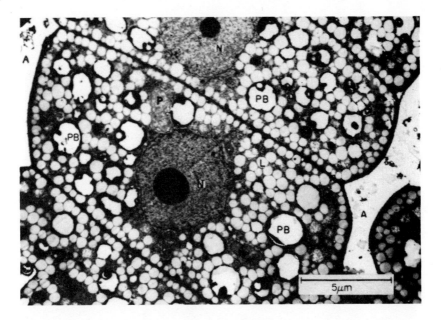

**Fig. 2.2** Electron micrograph of cells in the embryonic coleoptile of an imbibed but ungerminated grain of rice *(Oryza sativa* L.). The cells contain many lipid vesicles (L, small, electron-transparent) and fewer but larger protein bodies, PB. The concentration of protein body contents around their peripheries may be a fixation artifact. A, air space; N, nucleus; P, plastid. Fixation in glutaraldehyde and osmium tetroxide. × 5000.

containing 5% canavanine, the percentage normally present in *Canavalia* seeds, and similar experimental results have been obtained with other non-protein amino acids of seeds. Moreover it has been observed that, in nature, leguminous seeds with toxic amino acids are avoided by insects and other seed-eating animals.

The metabolic rate of ungerminated seeds is usually extremely low. The factor responsible for this is the low water content, 5–20% on a fresh weight basis, compared with 80–95% for most plant tissues in an active state. Most of the water in 'dry' seeds is moreover firmly bound to colloids, inaccessible for hydrolytic reactions, unfreezable, and removable only by temperatures approaching 100°C or by storage in high vacuum. Due to this dehydration, the cells and subcellular organelles in dry seeds are often shrunken and angular.

Germination is the resumption of metabolic activity and growth by the seed tissues, involving rehydration, utilization of nutrient reserves, and the gradual development of synthetic systems which enable the young plant to assume an autotrophic existence.

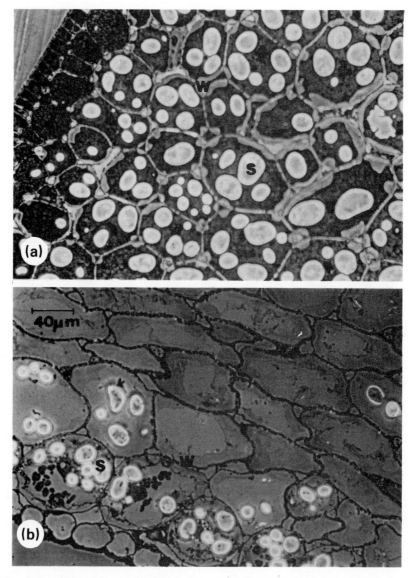

**Fig. 2.3** (**a**) Photomicrograph of part of a cotyledon of bean *(Phaseolus vulgaris)* prior to germination; the cells are packed with starch (S) and protein bodies; cell walls are thickened, especially at corners (W); epidermis at top left. (**b**) The same tissue after 8 days of germination at 25°C; the cells are almost empty and have shrunk, except beneath the epidermis at bottom left. × 270.

## Water imbibition: the hydration phase of germination

The first process that occurs in germination is water uptake, involving both *imbibition* and *osmosis*. These are physical processes not dependent on metabolic activity – seeds become hydrated under anaerobic conditions and, albeit slower, at temperatures too low to permit subsequent development; even dead seeds take up water. Colloidal imbibition is dominant in the initial phase of water uptake, the magnitude of the force of colloidal imbibition being at first very high; 'dry' seeds can absorb water from a saturated LiCl solution with a water potential of − 96.5 MPa (megapascals; see Chapter 4 and table of units, page 267; the seed can be regarded as absorbing water *against* this force opposing water entry). Imbibitional forces of more than −100 MPa have been claimed. As the water content rises, the imbibitional force rapidly decreases so that the rate of water uptake slows down, and osmotic forces become relatively more important and determine the final water content reached in the hydration phase.

During this hydration phase, the testa is frequently the limiting factor and in such cases removal or puncturing of the testa significantly speeds up the rate of water uptake. Imbibition causes swelling of the seed and this is often greater in the living cells than in the dead coats, so that the coats are consequently ruptured during the hydration phase. The micropylar pore and the hilum may be the chief areas of water entry; for example in beans (*Vicia* and *Phaseolus*) the micropyle is responsible for 20% of the water entry in the first 24 hours. In the seed of horse chestnut (*Aesculus hippocastanum*) and the peanut (*Arachis hypogaea*), the testa is more permeable to water from the outside inwards than in the opposite direction, thus showing some adaptation to water uptake. The testa of the lima bean (*Phaseolus limensis*) is impermeable to water from the outside until it has been wetted from the inside by water entering through the hilum. The limitation of water uptake by the testa may be a protective mechanism. A number of leguminous seeds when decoated and immersed in water suffer damage, some surface cells rupturing. The testa also protects to some extent against the leakage of solutes during imbibition.

Inside the seed, hydration normally proceeds inwards, cell layer by cell layer, so that in a partially imbibed seed the hydration is not even. Water movement is faster in the embryonic tissues than in the storage regions. In the castor bean (*Ricinus communis*), the cotyledons which are embedded in the middle of the endosperm are quickly hydrated and conduct water to the endosperm, which thus gets hydrated from two sides. The initial hydration phase brings the water content of the seeds to 50–60% of their fresh weight, a value determined by the presence of a high proportion of storage cells which do not necessarily become vacuolated at this stage. This is in consequence lower than the water content of 80–95% generally characteristic of mature tissues not packed with storage material and of the embryo axis itself at the completion of the hydration phase.

During hydration, the cells and subcellular organelles regain the size and shape which they had before the drying out which occurred during the ripening of the seed, and protoplasm resumes its normal submicroscopic structure. Significant changes may occur in cellular membranes. During the first few minutes of imbibition, decoated seeds undergo a rapid leakage of solutes – ions, sugars, and even proteins: this can be interpreted to indicate that the plasmalemma is at that moment highly permeable to compounds which it usually retains. While the precise events at the molecular level are not known with certainty, it is clear that a minimal level of hydration of c.17% (per fresh weight) is critical for the maintenance of normal membrane structure and semipermeability. Seed hydration therefore involves some reconstruction of cellular membranes at the molecular level.

As the protoplasm becomes hydrated, it resumes physiological activity. Diverse enzyme activities have been detected in extracts prepared from air-dry seeds, and these enzymes are activated by hydration. The activation of pre-existing enzymes is, however, soon followed by synthesis of more enzymes and of new enzymes undetectable in the dry material, as described in the next section.

Under certain conditions, the seed's development may be arrested at the stage of completed hydration. The light-sensitive seeds of foxglove (*Digitalis purpurea*), of certain varieties of lettuce (*Lactuca sativa*), and of tobacco (*Nicotiana tabacum*) will in the dark undergo hydration associated with a limited activation of their metabolism. However, only when the seeds are illuminated does their metabolism become further activated and their germination proceed. Seeds can also be maintained in the hydrated state without embryo growth and development at low temperature (this is one technique of vernalization, see Chapter 12).

## Phase of activated metabolism and growth

Hydration is followed by a stage of intense metabolic activity. Development follows a different course in the two functional regions of the seed. In the embryo axis (radicle plus plumule) cells begin to elongate and divide; according to species, either of these processes can start first, or both simultaneously. The metabolism of the growing regions of the embryo is directed towards the synthesis of new cell components and structures. Growth is usually visible in the radicle before the plumule; the emergence of the radicle is frequently taken as the criterion of germination. Indeed some authors insist that 'germination' is completed with the emergence of the radicle. There is, however, no sudden change in the physiology of the seed at that moment, and a wider definition of germination is employed here, including a consideration of what according to the stricter view is the early 'post-germinative' period, when the embryo axis is growing at the expense of the nutrients stored in the reserve tissues.

It is not known whether any specific stimulus is required, during germination, to induce the embryo axis to embark upon synthetic activity and growth. Cell division, cell elongation and the increase in dry weight begin very suddenly, and not simultaneously in the radicle and plumule. The suddenness with which these processes take place has suggested to various workers the operation of a specific 'trigger' substance, but it is equally possible that hydration initiates a predetermined programme of reactions, leading the cells to start division and elongation when some sequence of reactions is completed.

The metabolism of the storage tissues is directed to a hydrolysis of the storage reserves, and translocation of the resulting soluble products to the growing regions. There is as a rule no cell division in the storage tissues, and cell expansion is limited to that associated with rehydration. The activity of the storage tissues during germination is thus the reverse of that which occurred during ripening, when they received soluble compounds by translocation and converted them into insoluble reserve materials. Storage tissues of parent origin, endosperms and perisperms, die when their reserves are exhausted; their metabolic activity during germination is high but brief. Reserve-carrying cotyledons, too, may die quickly, or they may become green and persist for some time as the first photosynthetic leaves of the plant.

### Respiration during germination

The intense metabolism of germinating seeds is accompanied by high rates of respiration per unit weight of tissue both in the embryonic and the storage regions. Respiratory enzymes are already present in the dry seeds and hydration leads to a steep rise in respiration rate, once a critical water content has been passed (Fig. 2.4). The concentration of ATP rises rapidly in imbibing seeds. After the hydration phase, further increases in respiration rate are associated with increases in the amounts of respiratory enzymes.

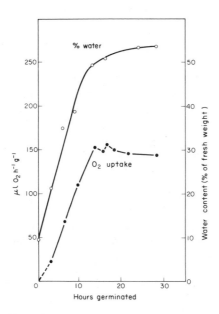

**Fig. 2.4** The activation of respiration, measured as oxygen uptake, in cotyledons of bean *(Phaseolus vulgaris)* during germination. After the initial hydration is completed (at about 50% water content), the respiration rate remains constant for some hours.

Mitochondria in cells of dry seeds are relatively poor in cristae, and increases in crista density subsequent to imbibition have been observed in numerous tissues – e.g. embryo axes of rye (*Secale cereale*), pea (*Pisum sativum*) and rice (*Oryza sativa*); and also storage organs such as cotyledons of bean and peanut; and in maize scutellum. Maize scutellum and castor bean endosperm have yielded some of the most active mitochondrial preparations isolated from plant tissues. Since the mitochondria carry out the main energy-transducing reactions of respiration, the storage tissues develop a high potential capacity for ATP production. The respiration rate of storage organs destined to die during germination typically rises to a peak and then falls as their senescence sets in. This rise to a peak is observed whether measurements are made on a basis of organ, cell, or unit weight. The overall course of development and metabolism in the growing tissues is more complex. For the growing axis as a whole, the total respiration continues to increase throughout germination. The axis, however, soon comes to contain regions of different structure, function and age, within each of which the respiration per cell or unit weight reaches a peak and then decreases again to a more or less steady value characteristic of maturity.

*Respiratory quotient during germination*

The respiratory quotient (RQ) of germinating seeds has been studied extensively. The RQ is defined as the fraction:

$$\frac{\text{Volume of } CO_2 \text{ evolved}}{\text{Volume of } O_2 \text{ absorbed}}$$

For the aerobic oxidation of carbohydrate, the RQ = 1.0; for lipids, 0.7–0.8, according to the chemical structure of the lipid molecules; for proteins, 0.8–1.0, according to the degree of completeness of amino acid oxidation. During anaerobic respiration, $CO_2$ evolution takes place without $O_2$ uptake, giving an RQ of infinity; hence the simultaneous occurrence of aerobic and anaerobic respiration will give RQs above unity. Mature plant tissues respire aerobically on a carbohydrate substrate and with an RQ of about 1. During germination however the RQ deviates widely from unity and changes rapidly. There are several reasons for this.

In the first hours of imbibition, very many seeds exhibit RQ values above unity, which fall quickly once hydration has proceeded to a critical extent. Reports of very high RQ values *at the very beginning of imbibition* should be accepted with caution. Considerable quantities of adsorbed air, which may be enriched with carbon dioxide (presumably accumulated during the slow respiration of the air-dry seeds) can be released very rapidly when dry seeds and any other dry plant materials, living or dead, are wetted. Therefore if water is added to dry seeds and changes in gas pressure determined immediately, the release of adsorbed gas could be mistaken for a large respiratory output of carbon dioxide. There is also a possibility of carbon dioxide evolution from metabolic reactions distinct from respiration. Glutamic decarboxylase, an enzyme which splits off carbon dioxide from glutamic acid, is activated at lower water contents than many respiratory enzymes. However, there is no doubt that in many instances high RQ values at the beginning of germination result from anaerobic respiration. The testa can severely limit oxygen access. For intact seeds of pea (*Pisum sativum*), RQ values of up to 8 have been reported in early germination, but, with removal of the testa, the RQ drops to 1.5. In two species of pine, the oxygen uptake of decoated seeds was found to be 3 to 4 times higher than in seeds with intact coats. (Not all seeds have testas highly impermeable to oxygen; e.g. in seeds of dwarf French bean (*Phaseolus vulgaris*), the RQ is unaffected by removal of the testa and oxygen uptake is increased by only 10%.) A second factor restricting oxygen access to the cells is the diffusion resistance offered by incompletely hydrated tissues, which lack large air spaces until the cells swell and become rounded. Dry tissue also does not wet easily, and water films form on the surfaces of seeds hindering gaseous exchange.

Once hydration is complete and seed coats are split, it is common for the RQ to drop below unity; after a minimum lasting some days it may rise again. An explanation of these RQ changes in terms of underlying physiological processes is not easy. Where products of fermentation such as alcohol and lactic acid have accumulated during a period of oxygen shortage, oxidation of these can lower the RQ. In pea seeds, for instance, alcohol and lactic acid do accumulate during soaking and disappear when the testa is ruptured, the RQ during the oxidation of these compounds being close to 0.53. The soluble sugars of the seed which serve as substrates in early germination are soon used up and respiration then depends on the main reserves. The oxidation of fats and proteins produces RQ values below unity, and the usual and probably generally valid explanation put forward is that, in germinating seedlings, these compounds rather than or in addition to carbohydrates serve as respiratory substrates. The occurrence of RQ values as low as 0.3 has been explained as due to a conversion of lipid to carbohydrate proceeding concurrently with lipid and carbohydrate oxidation. This conversion is an oxidative process not accompanied by carbon dioxide release and thus has an RQ of zero. Lipid-rich seeds generally have RQs lower than those of carbohydrate-storing seeds. Yet there are instances of seeds with high lipid contents in which the RQ never falls far below unity, and seeds low in lipid may have very low RQs. Again there is the possibility of carbon dioxide fixation reactions lowering the RQ; such 'dark' fixation of carbon dioxide has been demonstrated in germinating seeds. The RQ is therefore clearly not a value determined by one single process, but the resultant of a number of biological reactions involving oxygen and carbon dioxide.

In the embryo axis, the RQ value deviates much less from unity than in the storage tissues. In the lipid-storing castor bean seed, the RQ of the embryo, after a brief minimum immediately after the hydration phase, stabilizes at 0.8–1.0 in contrast to the endosperm which has an RQ of 0.4–0.5. It may be concluded that the RQs of storage tissues reflect their special type of metabolism, concerned with the hydrolysis and further chemical modification of reserves, while the growing regions receive their respiratory substrate from the storage cells largely in the form of soluble carbohydrates.

## Activation of protein synthesis

The synthesis of protein is a key process in growth and development. It commences rapidly in imbibed seed tissues: the incorporation of radioactive amino acids into protein can be detected within 30–60 min after the addition of water to isolated embryos of rice, rye and wheat, and isolated embryo axes of the bean species *Phaseolus lunatus* and *P. vulgaris*. Protein synthesis is assumed to be a prerequisite for 'germination proper', i.e. radicle emergence.

The metabolic apparatus required for protein synthesis is complex. The cellular sites of protein synthesis are the ***ribosomes***, particles composed of ribosomal RNA and proteins. Amino acids are transported to the ribosomes by ***transfer RNA*** (tRNA) molecules in the form of aminoacyl–tRNA complexes, individual amino acids reacting with specific tRNA molecules. The synthesis of each individual kind of protein is finally directed by a specific ***messenger RNA*** (mRNA), which attaches to the ribosomes, binding them into functional units, the ***polysomes***. The primary structure (amino acid sequence) of each protein is coded in the base triplets of the mRNA and is 'translated' into the protein amino acid sequence during synthesis on the ribosomes. The master code for the protein resides in the nuclear DNA from which it is 'transcribed' into the mRNA. It is the presence of a particular mRNA that enables a particular protein to be synthesized in a given cell at a given time, for only a limited number of messengers is transcribed at any moment, and mRNA molecules have a limited lifespan. A considerable number of enzymes is also required for protein synthesis. The question is therefore posed: do dormant seed tissues possess a complete protein synthesizing system needing only hydration for its activation, or must some components of the system itself first be synthesized? Particular interest attaches to whether mRNA is present in dormant tissues, or whether the messengers active during ripening all break down on dehydration, to be replaced during germination – perhaps by a different set.

The existence of ribosomes in both embryonic and storage regions of dry seeds has been clearly revealed by electron microscopy but they do not seem to be joined to polysomes. Ribosomes have been isolated from ungerminated seeds, and the density of ribosomes per unit volume of cytoplasm may be very high in embryonic regions. The presence in dormant tissues of a complement of tRNA molecules, and of the enzymes necessary for protein synthesis, also has been demonstrated. The situation with respect to mRNA has proved much more difficult to elucidate. Messenger makes up only 1–2% of total cellular RNA and the isolation and even partial purification of mRNA present formidable problems. Good evidence for the existence of mRNA in dry tissues has been obtained for embryos of wheat, rye, radish and castor bean, and for castor bean endosperm. The rapidity with which protein synthesis can commmence on imbibition can be quoted in support of the presence of preformed mRNA in the dormant tissues. The application of actinomycin D, an inhibitor of mRNA transcription, fails to affect protein synthesis in barley embryos during the first 4 hours of imbibition; this is again evidence for the presence of mRNA in the dormant embryo.

But although some mRNA is present in dormant tissues, the synthesis of mRNA may be detected very soon after the start of imbibition by the addition of labelled precursors of RNA: the radioactive label rapidly finds its way into the mRNA which is isolated. Experiments with wheat embryos have shown that messenger activity increases measurably in 2 h at 30°, while original messenger is gradually degraded. Work with castor bean embryo and endosperm has led to the same conclusion: while mRNA was present prior to hydration and was used at the beginning of germination, synthesis of new mRNA occurred rapidly. Ribosomal and tRNA are also synthesized early in germination.

If it is accepted that mRNA is present in dormant seeds, codes for some protein synthesis, but is replaced over a matter of hours by newly transcribed messages, the question arises as to whether the new messenger molecules are the same as, or different from, those stored in the dormant seed. Attempts to answer this question have been made by examining the patterns of proteins synthesized coded by mRNA extracted from seed tissues at different stages of development. The extracted mRNA fractions were allowed to direct protein synthesis *in vitro;* the resulting protein mixture was then subjected to gel electrophoresis, which causes proteins to separate into different bands along a strip of gel, according to their molecular weights, i.e. each band on the gel corresponds to protein(s) of a certain molecular size. By this method it has been shown that with mRNA from wheat embryos there was no qualitative change in the protein pattern detectable over 8 h of germination, though some proteins became relatively more abundant. At least 35 bands were resolvable on the gels. In whole seeds of castor bean, the protein pattern did change during 8 h of imbibition, but much more information is needed on this subject.

It is not known which kinds of protein are the first to be synthesized at the onset of germination. None of the protein bands visible on gels has been identified. While much emphasis has been placed on enzyme synthesis during germination, more recently it has been suggested that the first proteins to be formed might be structural proteins of membranes; some membrane increases occur very rapidly. In the embryo of rye (*Secale cereale*), increases in mitochondrial cristae are apparent within two hours of the start of imbibition and endoplasmic reticulum has proliferated extensively within five hours; in pea radicles, endoplasmic reticulum shows an appreciable increase by eight hours. Golgi apparatus, a site of membrane formation, is also absent or sparsely represented in dry seeds, becoming apparent during early germination.

### Mobilization of food reserves

The first stage in the utilization of the nutrient reserves is their hydrolysis to soluble products. Table 2.2 lists the hydrolases (hydrolytic enzymes) responsible for the most universal classes of reserves and the primary products of their hydrolysis. RNA is listed in the table though it is not a major reserve; however, the hydrolysis of RNA does commonly occur in reserve tissues. The cellulose cell walls of storage cells often remain undigested, but cellulose is degraded in, e.g. cotyledons of *Lupinus*, and enzymes promoting cellulose breakdown have been detected in germinating barley. Several enzymes may be needed to break down one type of compound. Only $\alpha$-amylase can attack intact starch grains; the partly-cleaved molecules then become subject to hydrolysis by $\beta$-amylase and dextrinases. In protein hydrolysis, too, numerous enzymes are implicated, attacking peptide links in various positions of the protein molecule; in barley grains there are at least six proteinases and two peptidases.

**Table 2.2**   The main classes of hydrolytic enzymes involved in the mobilization of the food reserves of seeds.

| Reserve compound | Enzymes | Products of hydrolysis |
|---|---|---|
| Starch | 1) $\alpha$- and $\beta$-amylase<br>2) Dextrinase<br>3) $\alpha$-glucosidase<br>Phosphorylase | Maltose, glucose, dextrins<br>Maltose, glucose<br>Glucose<br>Glucose-1-phosphate |
| Hemicellulose | Mannanase<br>$\alpha$-galactosidase | Pentoses, mannose, galactose |
| Lipid | Lipases (esterases) | Fatty acids and glycerol |
| Protein | 1) Proteinases<br>2) Peptidases | Amino acids, peptides<br>Amino acids |
| Phytin | Phytase | Inorganic phosphate and inositol |
| Ribonucleic acid | Ribonuclease | Ribonucleotides |
| Cellulose | 1) Cellulase<br>2) Cellobiase | Cellobiose<br>Glucose |

In a number of instances it has been noted that hydrolase enzymes are not soluble in the cytoplasm, but hydrolase activity is associated with the storage organelles – proteinase activity in protein bodies (e.g. mung bean), lipase in oil bodies (e.g. castor bean). This would serve not only to facilitate digestion within the storage bodies, but would protect the rest of the cell from hydrolysis. It is interesting to note that while amylases and lipases are not species-specific, i.e. the amylase of one species will hydrolyse the starch from another, the seed proteinases show considerable specificity, being much more efficient in hydrolysing the proteins from the same species. A proteinase preparation from seeds of cabbage (*Brassica*), for instance, has been found to be completely inactive towards the proteins from the seeds of bean (*Phaseolus*).

Enzymes indispensable to basic metabolism, such as respiratory enzymes, are present in the dry seed at quite high levels and need only hydration to become active (even though later their activity may still rise). By contrast, the activity of a hydrolase is in many cases low or undetectable in a dry or just-hydrated seed, the appearance of activity, or increase in the initially low activity, beginning hours or even days after the hydration phase, and continuing for several days (Fig. 2.5). Radicle emergence precedes appreciable reserve hydrolysis. Increases in hydrolase activity in storage tissues may come about either by activation of an enzyme already present in an inactive ('zymogen') form, or by enzyme synthesis. In cereals, appreciable amounts of $\beta$-amylase are already present in ungerminated grain in zymogen form, and in extracts of ungerminated grain this enzyme can be activated by treatment with a proteinase; *in vivo,* the increase in activity may be presumed to be elicited by cellular proteinases. On the other hand, increases

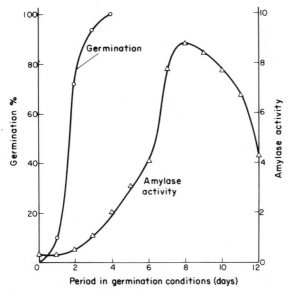

**Fig. 2.5** The increase in amylase activity in germinating grains of barley *(Hordeum vulgare).* The activity is expressed as mg starch digested in 2 minutes by a volume of extract equivalent to half a grain. The emergence of the radicle was taken as the criterion of germination. (From Drennan and Berrie, 1962, *New Phytologist,* **61**, 1–9.)

in $\alpha$-amylase in the endosperm of barley (*Hordeum*), wheat, rice and oat (*Avena*) result from enzyme synthesis. This has been shown by observations of the incorporation of radioactively-labelled amino acids into the enzyme protein, and by noting inhibition of the rise in enzyme activity when grains were germinated in the presence of inhibitors of protein synthesis. Yet the cereal sorghum (*Sorghum vulgare*) stores $\alpha$-amylase, together with numerous other hydrolases, in the dry grain. Thus even for such closely-related plants as cereals, the same enzyme, $\alpha$-amylase, is stored in some and not in others. It has been suggested that the conditions to which the ripening seeds are subjected may influence the enzyme complement laid down in the dormant tissues. In numerous instances it is not known with certainty whether the rise in an enzymic activity results from activation of a zymogen or from *de novo* protein synthesis. Both processes may contribute: RNAase activity in cotyledons of field pea (*Pisum arvense*) is stated to increase first due to activation of stored enzyme; later a second rise owing to enzyme synthesis follows.

Whether the mRNA for the hydrolases which are newly synthesized is already present in the dormant tissues, or is transcribed during germination, has been reliably investigated in only a few tissues. There is fairly conclusive evidence for the transcription of the mRNA coding for α-amylase in the aleurone layer of the endosperm of barley, wheat and oats (see p. 23), before the enzyme is produced. Claims for the presence of stored mRNA for hydrolases have been based on observations that a rise in enzyme activity has taken place in the presence of actinomycin D, an inhibitor of mRNA synthesis. But this is not conclusive, for complete inhibition of all mRNA synthesis was not demonstrated, and in plant tissues actinomycin D is not always very effective.

The physiological activity of seed storage tissues during germination has been described as being catabolic, or degradative, in contrast to the anabolic or synthetic activity of the growing regions of the embryo. The foregoing discussion, however, emphasizes that synthesis of protein and RNA occurs in the storage tissues; rises in mitochondrial activity and respiration rate have already been mentioned. Xylem and phloem elements differentiate in cotyledons; even short-lived cotyledons may become green and differentiate chloroplasts. The mobilization of food reserves is the consequence not of a purely degradative self-digestion of the storage cell contents, but a process involving complex energy-dependent syntheses.

The hydrolysis of reserves does not proceed uniformly throughout the storage tissues and, correlated with this, certain enzymes have been shown to have characteristic distribution patterns. In the cotyledons of the bean (*Phaseolus vulgaris*) digestion begins in the centre, but avoiding the vascular bundles. The cells where hydrolysis first starts are dead before digestion becomes evident in the sub-epidermal layers and next to the bundles (Fig. 2.3). Lipid hydrolysis in castor bean endosperm proceeds simultaneously from the outside inwards and from the cotyledons outwards. Such patterns of digestion complicate studies in that any assay of, say, an enzyme activity carried out on whole cotyledons or endosperm gives an average value only and increases in some cells might be masked by decreases in dying cells. In cereals, the endosperm is differentiated into the metabolically-active aleurone layer cells and the endosperm cells which are dead or of very low metabolic activity. In resting cereal grains, proteinase, peptidase, esterase and phytase activities are confined to the aleurone layer and the embryo. The aleurone cells contain the protein and phytin reserves. However β-amylase is found in the endosperm and is absent from the aleurone layer, being at its highest concentration just outside this layer. During germination, the aleurone layer secretes other enzymes, e.g. α-amylase, into the endosperm and digestion in the endosperm starts in its vicinity. This digestion is also promoted by the secretion of enzymes by the scutellum Fig. 2.1). In the date seed, the tip of the cotyledon burrows into the endosperm, secreting enzymes which digest the hemicellulose reserves.

The secretion of enzymes from cell to cell requires respiratory energy and what little evidence there is (from barley aleurone) suggests that enzyme is packaged in vesicles which are discharged through the plasma membrane of the aleurone cells. It would then have to diffuse through cell walls to the starchy endosperm and, if cells in the latter still retain membrane semipermeability, would have to be ingested by formation of vesicles again ('reverse pinocytosis'), by infoldings of plasma membrane. Proteins are too large to diffuse through an undamaged plasmalemma.

There are a number of puzzling features concerning the activities of hydrolytic enzymes in germinating seeds. There is often no obvious proportionality between the level of activity of a particular enzyme and the quantity of its substrate. For example, soybean seeds are rich in $\beta$-amylase, but contain little or no starch; seeds with high lipase activity are not necessarily rich in lipid. Activities of particular enzymes frequently continue to rise after their substrates are depleted. This suggests that the complex control mechanisms which operate in growing cells may be impaired in the senescent storage cells.

**Interconversions and translocation of products of hydrolysis**

Hydrolyses convert insoluble reserves to soluble derivatives. In maize, the proportion of soluble compounds rises from 2% to 25% of the dry weight during the first 5 days of germination. A fraction of the soluble products is used in respiration and for syntheses in the storage tissues themselves, but by far the greater part is transported to the growing parts. The dry weight of the storage parts decreases and that of the growing parts increases (Fig. 2.6). As their reserves are exhausted, storage cells collapse and die (Fig. 2.3b).

The transport is assumed to proceed by diffusion from cell to cell and by phloem translocation. Endosperm does not develop vascular tissue, and reserves from it are absorbed where it makes close contact with embryonic tissues. In cereals, the scutellum acts as an absorbing organ; it absorbs the glucose arising from starch breakdown in the endosperm and converts the glucose into sucrose, which is the form of sugar involved in phloem transport. In date and coconut the cotyledons grow deep into the endosperm and act as absorbing organs.

The utilization of the products of hydrolysis for synthetic reactions in the embryo necessitates extensive interconversions of metabolites since the chemical composition of the new cells formed in growth is very different from that of the storage cells. In particular most of the stored lipid is converted to carbohydrate. Table 2.3 shows an analysis of the lipid and carbohydrate content of germinating castor bean seeds, which at the beginning of germination contain 60–70% lipid. The conversion occurs in the storage tissue and the sucrose formed is translocated to the embryo. The biochemical pathway for this conversion is the glyoxylate cycle and the two enzymes unique to this cycle, isocitratase and malate synthetase, are found in high activity only in the tissues of fatty seeds during the period of lipid

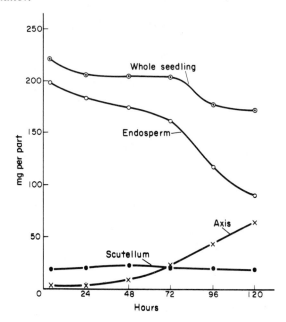

**Fig. 2.6** Dry weight changes in the different parts of the seedling of maize *(Zea mays)* germinating at 25°C in the dark; the weights are given as mg per seedling or mg per seedling part. (From Ingle, Beevers and Hageman, 1964, *Plant Physiology, Lancaster,* **39**, 735–40.)

breakdown. They appear on germination and disappear again when the lipid is used up. The glyoxylate cycle proceeds in small organelles, the *glyoxysomes*, which proliferate in lipid-metabolizing storage tissues. The protein metabolism of seedlings includes interconversions between amino acids, the amino acid composition of the storage proteins differing from that of the proteins of the growing embryo. The amino acid interconversions, or transaminations, proceed in the tissues of the embryo, catalysed by aminotransferases.

**Table 2.3** Changes in lipid and carbohydrate content in castor beans during germination at 25°C in the dark. (Data of Desveaux, R. and Kogane-Charles, M., 1952, *Annales de l'Institut national de la Recherche Agronomique, Paris,* **3**, 385–416.)
The increase in total dry weight results from the incorporation of water and oxygen during the conversion of lipid to carbohydrate.

| Days germinated | Weight per 100 seedlings (g) | | |
| :---: | :---: | :---: | :---: |
| | Lipid | Carbohydrate | Total dry weight |
| 0 | 26.2 | 1.51 | 37.6 |
| 4 | 25.0 | 5.10 | 39.0 |
| 6 | 10.8 | 18.2 | 45.1 |
| 8 | 5.40 | 23.3 | 43.9 |
| 11 | 1.78 | 17.7 | 38.4 |

## Interrelationships between the growing axis and the storage tissues

The embryo axis exerts some control over the activity of the storage tissues. In pea seedlings, mitochondria in the cotyledons soon lose their activity if the embryo axis is removed, and axis removal has been claimed to prevent hydrolysis of the starch and protein reserves (although this has been disputed; see below). In bean cotyledons, a lower level of $\alpha$-amylase activity is attained on excision of the axis. Promotion of reserve hydrolysis, and/or stimulation of the activity of hydrolases by the axis has been reported also for seeds of cotton (*Gossypium*), sunflower (*Helianthus annuus*), and pine (*Pinus* spp.).

The system in which the interrelations between the embryo axis and the storage tissues have been studied most thoroughly is the grain of barley (*Hordeum vulgare* and *H. distichon*), particularly with respect to the synthesis of $\alpha$-amylase, which initiates starch breakdown in the endosperm. The barley grain owes its popularity partly to its importance in the brewing industry; since malting involves starch hydrolysis, the control of starch hydrolysis in barley is of economic value. Moreover the structure of the grain permits dissection of the various regions which is highly advantageous for experimental purposes. The barley grain has the typical structure of a cereal grain (Fig. 2.7): on one side lies the embryo consisting of radicle, plumule and scutellum; the rest of the grain is taken up by the storage tissue, the endosperm, the entire structure being enclosed in the fused seed and fruit coats. The bulk of the endosperm consists of starch-packed cells which are no longer physiologically fully functional. The three outermost layers of the endosperm are differentiated into the aleurone, smaller fully functional cells, which surround the starchy endosperm on all sides except next to the scutellum. During germination, the aleurone cells secrete a number of hydrolytic enzymes into the endosperm, $\alpha$-amylase being the most thoroughly studied; the products of hydrolysis are absorbed by the scutellum for translocation to the growing axis.

For a number of years the view has prevailed that the synthesis of enzymes by the aleurone is stimulated by the secretion of the hormones, **gibberellins**, by the embryo axis. (For a description of gibberellins, see Chapter 9; the major ones involved in the barley grain are $GA_1$ and $GA_3$.) The evidence for this is as follows:

(*i*)   Embryo-less barley grain halves fail to develop hydrolase activity; control by the axis is thereby indicated

(*ii*)  Embryo-less endosperm halves do synthesize hydrolases when supplied with gibberellic acid

(*iii*) Isolated aleurone layers respond to gibberellins by secretion of hydrolases

(*iv*)  Isolated embryo axes can secrete gibberellins.

The sequence is summarized in Fig. 2.8. The timing of events is dependent on conditions such as temperature; at *c.* 25°C, gibberellin from the embryo

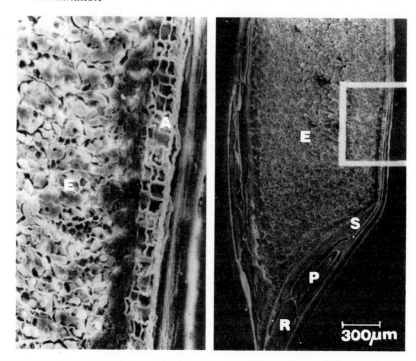

**Fig. 2.7** Scanning electron micrograph of an ungerminated barley grain in longitudinal section. The low power image on the right (× 32) shows the embryo and part of the endosperm (E); plumule (P), radicle (R), scutellum (S). On the left, the area enclosed by the white lines is magnified (× 180) to show the three-layered aleurone (A); the starch grains in the endosperm are clearly visible, but the aleurone cells have lost contents during specimen processing. The assistance of Mr M. Williams, Swansea, with specimen preparation is gratefully acknowledged.

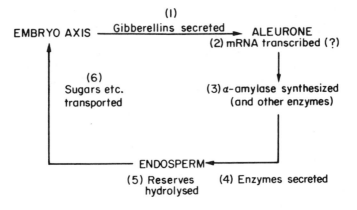

**Fig. 2.8** A summary of the postulated sequence of main events in the mobilization of endosperm reserves in the barley grain, numbered in chronological order.

axis is stated to reach the aleurone after 17–18 h, and enzymes begin to be secreted some 5–8 h later. During this lag period the activity of the mRNA coding for $\alpha$-amylase increases and, while unequivocal proof is lacking, the data are consistent with the transcription of the messenger being induced by gibberellin – a point used to support the argument that plant hormones act at the genetic level (see Chapter 9, p. 184). Similar evidence for a gibberellin-mediated control of endosperm digestion has been obtained also for some other cereals – wheat, rice and oats.

Nevertheless doubts are now being raised about the action of gibberellins as specific enzyme-inducing stimuli from the embryo axis. Examination of the timing of events in intact grains (as opposed to dissected parts) has shown that, in certain circumstances, amylase synthesis can *precede* the increase in gibberellin content. Moreover the native level of gibberellin attained in the seeds seems to be too low to account for the level of enzyme synthesis observed. While gibberellins are undoubtedly involved in the germination physiology of cereal grains, their precise role is not clear. Similar difficulties arise in interpreting the roles of plant hormones in other developmental processes (Chapter 9, 10); a detailed evaluation of the cereal aleurone problem is given by Trewavas, 1982 (in *Selected References*).

The control mechanisms and hormonal relations of other species are even less clearly understood. For pea seeds (and for some other species, too) an aqueous extract of the embryo axes can substitute for the presence of the axis, suggesting that the embryo produces a chemical factor which controls cotyledon activity. Applications of gibberellins and/or cytokinins (another class of hormone: see Chapter 9, p. 163) have been reported to enhance the development of hydrolase activity in a number of storage organs; but it has not been shown that the axis under normal circumstances produces the hormone and the observed stimulation might be a fortuitous effect. Reports are conflicting; e.g. in peas, some workers have claimed that reserve mobilization is dependent on the axis and others have found it to be independent; both positive and negative results with cytokinins on peas have been published. The precise experimental conditions are probably of much significance here. For the cotyledons of pumpkin (*Cucurbita maxima*) and cucumber (*Cucumis sativus*), it was initially reported that the presence of the embryo axis enhanced the development of enzyme activity and greening. A later study showed these results to be spurious: under the conditions of the original experiment, the embryo-less cotyledons remained submerged in the liquid medium while in whole seeds the growing axis lifted the cotyledons up; the submerged cotyledons suffered a shortage of oxygen which retarded their development. When adequate aeration of the embryo-less cotyledons was ensured, they developed just as fast as cotyledons with axis attached! This example shows the importance of having all other conditions equal for whole seeds and axis-less storage organs – a situation not always easy to achieve, and a criterion which may be overlooked.

A negative feedback control via end-product inhibition sometimes operates. The increase in proteinase activity of pea seeds is inhibited when the seeds are germinated in a mixture of amino acids, the products of proteinase activity. In other seeds, too, inhibition of storage protein breakdown by amino acids and of starch hydrolysis by sugars has been noted, though the activities of the hydrolytic enzymes were not measured. The embryo axis clearly depends on the storage tissues for its nutrient supply. The developmental changes in the storage tissues which result in this supply becoming available may be dependent on stimuli of hormonal nature from the embryo axis, and may also be promoted by the embryo axis acting as a 'sink' for nutrients. Many aspects of the interrelationship between the growing regions and storage tissue have yet to be elucidated.

Gradually, the root system of the seedling becomes able to absorb and supply mineral ions to the developing plant; the photosynthetic apparatus develops and takes over the production of organic compounds. The heterotrophic seedling becomes an independent autotrophic plant and germination is complete.

### FURTHER READING

BEEVERS, L. (1975). *Nitrogen Metabolism in Plants.* Contemporary Biology Series, Edward Arnold, London.

BEWLEY, J. D. and BLACK, M. (1978). *Physiology and Biochemistry of Seeds in Relation to Germination. 1. Development, Germination and Growth.* Springer-Verlag, Berlin.

HAYWARD, H. E. (1938). *The Structure of Economic Plants.* Macmillan, New York.

MAYER, A. M. and POLJAKOFF-MAYBER, A. (1982). *The Germination of Seeds* (3rd Edition). Pergamon Press, Oxford.

### SELECTED REFERENCES

CAERS, T., PEUMANS, W. J. and CARLIER, A. R. (1979). Preformed and newly synthesized messenger RNA of germinating wheat embryos. *Planta,* **144**, 491–6.

DUKE, S. H. and KAKEFUDA, G. (1981). Role of the testa in preventing cellular rupture during imbibition of legume seeds. *Plant Physiology, Lancaster,* **67**, 449–56.

FORD, M. J., SLACK, P., BLACK, M. and CHAPMAN, J. M. (1976). A re-examination of the reputed control of cotyledonary metabolism by the axis. *Planta,* **132**, 205–8.

HALLAM, N. D., ROBERTS, B. E. and OSBORNE, D. J. (1972). Embryogenesis and germination in rye (*Secale cereale* L.). II. Biochemical and fine structural changes during germination. *Planta,* **105**, 293–309.

INGLE, J., BEEVERS, L. and HAGEMAN, R. H. (1964). Metabolic changes associated with the germination of corn. I. Changes in weight and metabolites and their redistribution in the embryo axis, scutellum and endosperm. *Plant Physiology, Lancaster,* **39**, 735–40.

MARTIN, C. and NORTHCOTE, D. H. (1981). Qualitative and quantitative changes in mRNA of castor beans during the initial stages of germination. *Planta,* **151**, 189–97.

NISHIMURA, M. and BEEVERS, H. (1978). Hydrolases in vacuoles from castor bean endosperm. *Plant Physiology, Lancaster,* **62**, 44–8.

SEEWALDT. V., PRIESTLEY, D. A., LEOPOLD, A.C., FEIGENSON, G. W. and GOODSAID-ZALDUNO, F. (1981) Membrane organization in soybean seeds during hydration. *Planta*, **152**, 19–23.

TREWAVAS, A. J. (1982). Growth substance sensitivity: the limiting factor in plant development. *Physiologia Plantarum*, **55**, 60–72.

VAN DER WILDEN, W., HERMAN, E. M. and CHRISPEELS, M. J. (1980). Protein bodies of mung bean cotyledons as autophagic organelles. *Proceedings of the National Academy of Science, U.S.A.*, **77**, 428–32.

# 3
# Energy Flow and Carbon Turnover

## Introduction

The photosynthesis of green plants is the process by which the thermonuclear energy of the sun, transmitted to the earth as electromagnetic radiation, is transformed into the chemical bond energy of organic substances. It is a process of energy transduction and the pathway by which the inorganic molecules $CO_2$ and water are combined to organic form. In flowering plants the first stable photosynthetic products are predominantly sugars. The sugars form the starting point for the synthesis of further organic plant constituents including macromolecules of proteins, polysaccharides, nucleic acids and compound lipids, which are the units of cellular architecture. The sugars and sugar-derived polysaccharides also form a store of potential energy for plant cells. This store is continuously drawn upon during respiration: the sugars are oxidized to $CO_2$ and water again, while some of the potential energy of the sugar molecules is transferred to molecules of ATP. This compound has been termed the energy currency molecule of living cells and its potential energy can in turn be utilized for cellular work such as growth (biosynthesis of macromolecules and their organization into protoplasm), membrane transport of solutes, and movements such as protoplasmic streaming. The ATP is broken down in the course of these activities to ADP or AMP and inorganic phosphate, from which it is resynthesized with an energy input. There is a constant flow of energy and material through a plant (Fig. 3.1), with photosynthesis providing the primary energy input. The energy converted to heat in the processes illustrated is irradiated back into space as infra-red radiation, low energy quanta that cannot be used for photosynthesis again. The $CO_2$, water and mineral elements flow through to be recycled, although the recycling may take a long time. A living tree may retain organic material in its body for several thousand years, and we are currently releasing to the atmosphere the $CO_2$ from coal derived from the photosynthetic activities of the Carboniferous forests of over 300 million years ago. The material and energy flow patterns within the plant and between the plant and its environment must proceed in a highly co-ordinated way if they are to be compatible with the normal growth and development of the organism.

The rate and efficiency of photosynthesis are central to any consideration

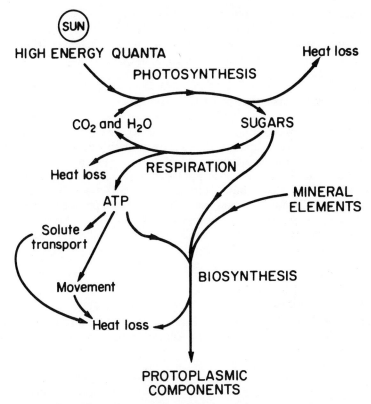

**Fig. 3.1** The flow of energy and metabolites through the plant. The protoplasmic components will eventually be returned to inorganic form through microbial decay.

of the energy relations of plants. Moreover photosynthesis is, directly or indirectly, the source of energy and organic material not only for plants but almost all other terrestrial organisms. An exception is provided by chemosynthetic bacteria; also water movement through plants is largely energized by solar radiation directly without the intervention of photosynthesis (see Chapter 4). The efficiency of photosynthesis can be considered at many levels. Photosynthetic efficiency can be visualized in terms of the whole earth, of a population of plants, of a whole plant, of the individual photosynthetic organ, of the chloroplast, or of the individual reactions, photochemical and thermochemical, that go to make up the process. At one end of a complex chain we have the total available energy utilizable in photosynthesis, and a supply of inorganic material; at the other end, either an energy gain in the form of living matter, or, at the level of human economy, a harvestable plant product. At each step in this chain some energy loss will occur and the ultimate efficiency will be compounded of the efficiencies of the component processes. In this chapter we shall touch on some of the levels at which photosynthetic efficiency can be studied, and shall examine some of the factors that affect the rate and efficiency of photosynthesis. An understanding of the energy relations at various levels of the process, and of the factors controlling photosynthetic rate, may help us to recognize where and how improvements may be achieved in the energy conversion of the green plant, on which all lives depend.

## Photosynthesis: the overall process

The raw materials required for photosynthesis are $CO_2$, water, and quanta of light in the visible range of the spectrum, with wavelengths $c$. 370–760 nm. Water is ubiquitous in living cells; of the water taken up by plants, only a very small fraction is utilized in photosynthesis. The process is extremely sensitive to water stress, but this is an indirect effect (see below, effects on stomata, and Chapter 7). Plants show no special adaptations which ensure a water supply specifically for photosynthesis. The need to obtain light and $CO_2$ has, however, profoundly conditioned the evolution of higher plants both as regards structural organization and with respect to physiology and biochemistry.

### The capture of light

The photosynthetic organ *par excellence* is the leaf, a thin flat surface which catches the light effectively. The photosynthetic pigments have their

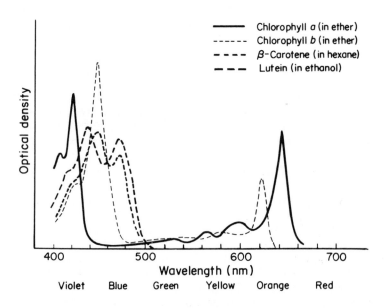

**Fig. 3.2** Absorption spectra of chlorophyll *a*, chlorophyll *b*, β-carotene and lutein (a xanthophyll). (Adapted from Goodwin and Mercer, 1972, *Introduction to Plant Biochemistry*, Pergamon Press, Oxford.)

absorption peaks in the blue and red wavelengths (chlorophylls *a* and *b*), or in the blue (carotenoids and xanthophylls), i.e. at the two ends of the visible spectrum (Fig. 3.2). Since sunlight that has penetrated the atmosphere has its maximum energy output almost in the middle of the visible spectrum, in the green and blue-green, the photosynthetic pigments may seem at first sight to be somewhat poorly adapted to capture solar energy. When, however, the action spectrum of photosynthesis for a whole leaf is considered, i.e. the rate of photosynthesis for the same number of incident quanta is plotted against wavelength, it is found that photosynthesis shows only a moderate dip in the green region of the spectrum, where none of the isolated pigments absorbs strongly (Fig. 3.3). A quantum of green light which is not absorbed immediately may not pass straight through a leaf, but is likely to be reflected and scattered between many internal surfaces. If the quantum spends long enough inside the leaf it may be absorbed by a photosynthetic pigment molecule in spite of the low absorptivity of pigments in the green region. Carotenoid absorption is shifted towards

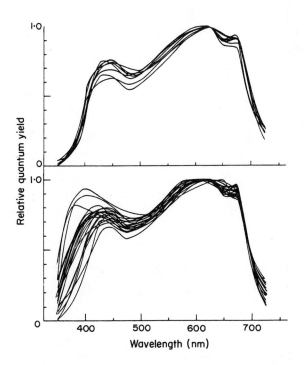

**Fig. 3.3** Relative quantum yields of photosynthesis at different wavelengths in leaves of 8 species of crop plants grown in the field (upper graph) or 20 species grown in a growth chamber (lower graph). (From McCree, 1972, *Agricultural Meteorology*, **9,** 191–216.)

green wavelengths through association with chloroplast membranes, and this gives some increase in absorption in the green wavelengths.

As the level of irradiance on a photosynthetic organ is increased, the rate of photosynthesis at first rises linearly, then levels off to a steady rate as light saturation is reached (Fig. 3.4). The irradiance level at which saturation occurs depends on a number of factors. If the temperature is very low, light saturation is attained at low levels of irradiance: the rate of thermochemical reactions soon becomes limiting; similarly at low $CO_2$ concentrations, light saturation is reached once $CO_2$ has become limiting. Conversely at high temperatures and high $CO_2$ concentrations, light saturation is reached only at high levels of irradiance. Other conditions being equal, significant differences in light saturation values are exhibited by individual photosynthesizing systems. Shade plants, which normally grow beneath other species, are light-saturated at much lower levels of irradiance than

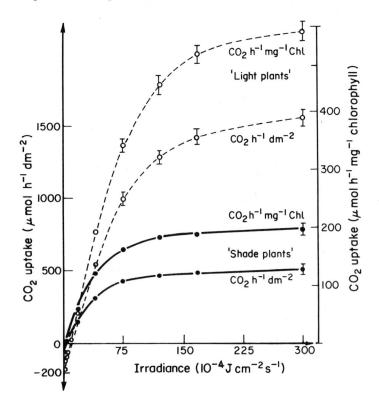

**Fig. 3.4** Light saturation curves of photosynthesis for plants of *Sinapis alba* grown under strong illumination, 'light plants' (o – – – o) and weak illumination, 'shade plants' (● ——— ●). Rate of photosynthesis is expressed per unit leaf area and per unit of chlorophyll. Where the curves cut the x axis is the light compensation point, below which respiration exceeds photosynthesis; this lies at lower irradiance for the shade-grown plants. (From Grahl and Wild, 1972, *Zeitschrift für Pflanzenphysiologie*, **67**, 443–53.)

species of open habitats. It is common for trees to produce *sun leaves* on the outside of the canopy and *shade leaves* within the canopy. The shade leaves become light-saturated at lower levels of irradiance than sun leaves, and at low levels of irradiance the shade leaves have higher rates of photosynthesis than sun leaves, both per unit leaf area and per unit weight of chlorophyll. The *light compensation point*, the level of irradiance at which photosynthesis exactly equals respiration (and below which respiration exceeds photosynthesis leading to a net loss of organic matter) (Fig. 3.4), is low in shade plants and shade leaves. Individuals of the same species vary in their light saturation levels according to the levels of irradiance at which they were grown (Fig. 3.4). Plants and leaves are adapted to the irradiance levels of their environment.

For whole plants, light saturation requires much higher levels of irradiance than for single leaves because leaves shade each other in an intact plant. Whereas a single leaf may be light-saturated with *c.* 25% of full sunlight, an entire plant may not reach light saturation even with irradiances equivalent to full midsummer sun. Heavy clouding may bring a plant to its light compensation point. Light can thus often become the limiting factor for photosynthesis in the field.

## The absorption of carbon dioxide

$CO_2$ is present in the atmosphere at an average concentration of 0.032% by volume, or 320 ppm ($6.47 \times 10^{-1}$ g m$^{-3}$). The large surface:volume ratio of a leaf, which is favourable for light absorption, is also favourable for the diffusion of $CO_2$ into the leaf. Inside the leaf, abundant intercellular spaces permit gaseous diffusion between the cells; the internal surface of a leaf, i.e. the total cell surface area in contact with the intercellular air spaces, is much larger than the external leaf surface area (Chapter 4, Table 4.5). The fixation of $CO_2$ in photosynthesis maintains a gradient of $CO_2$ between the sites of fixation and the external atmosphere. This concentration gradient, $\Delta CO_2$, is equivalent to a gradient of free energy, and is the driving force for the inward movement of $CO_2$; the value of $\Delta CO_2$ is given by the difference between the $CO_2$ concentration in the external air and the concentration of $CO_2$ at the reaction sites:

$$\Delta CO_2 = [CO_2] \text{ external} - [CO_2] \text{ internal} \qquad (3.1)$$

The rate of $CO_2$ diffusion is directly proportional to $\Delta CO_2$. During diffusion to the photosynthetic reaction sites, however, the $CO_2$ molecules encounter resistance at various points, and the rate of diffusion is inversely proportional to the resistance, R. If we denote the rate of entry of $CO_2$, which equals the rate of net photosynthesis, by P, then:

$$P = \frac{\Delta CO_2}{R} \qquad (3.2)$$

R comprises various components contributed respectively by stomata, cuticle, a boundary air layer, and by the tissue traversed after the $CO_2$ has entered the leaf. These factors will be considered in turn. (It may be noted that some authors refer to conductances rather than resistances; conductance is the reciprocal of resistance, i.e. $1/R$.)

Most of the entry of $CO_2$ into leaves occurs through the **stomata** (singular: **stoma**). These are minute structures in the epidermis, consisting of two highly specialized elongate cells, the guard cells, enclosing a pore between them (Fig. 3.5). The guard cells are surrounded by a few subsidiary cells,

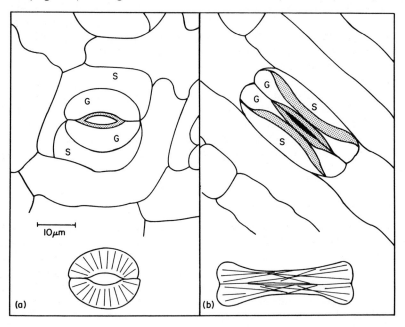

**Fig. 3.5** The structure of stomata as seen in surface view. Guard cell (G), subsidiary cell (S), thickened parts of guard cell walls stippled. The diagrams at the bottom show the directions of cellulose microfibrils in the guard cell walls. (**a**) Stoma in young leaf of mung bean *(Phaseolus aureus)*, pore open. When the sausage-shaped guard cells increase in turgor, the microfibrils resist longitudinal stretching and the thick parts of the walls cannot expand; hence the cells must bulge outwards and open the pore. (**b**) Stoma in grass leaf, pore closed. The guard cells are dumb-bell-shaped and here the microfibril arrangement and wall thickenings are such that appreciable cell expansion can occur only at the bulbous ends; thus with increasing turgidity, the ends bulge and pull the central parts apart, opening the pore.

different in morphology from the typical epidermal cells. The shape of the guard cells, and the arrangement of their cell wall thickenings, ensure that when the guard cells are more turgid than the subsidiary cells, the guard cells bulge outwards into the subsidiary cells and separate in the middle, opening the pore. When the guard cell turgor equals, or is less than, that of the adjacent cells, the guard cells shrink together and the pore closes. All intermediate stages between maximal opening and complete closure are possible. The pore sizes vary according to species; at full opening, the stomatal apertures of *Phaseolus vulgaris* measure only $3 \times 7 \ \mu m$, while fully open stomata of *Zebrina pendula* reach pore sizes of $12 \times 31$ $\mu$m. In grasses, the stomata are very elongate, e.g. a fully open stomatal pore of *Avena sativa* measures $8 \times 38$ $\mu$m. The stomatal frequency per cm$^2$ of leaf surface usually ranges from *c*. 1000 to 200 000. The apertures are, however, so small that at the most 3% of the total leaf area is occupied by the pores. Yet an illuminated leaf can absorb $CO_2$ from the atmosphere with great efficiency. A leaf can maintain a steep diffusion gradient for $CO_2$ and many small pores have a large amount of edge in relation to their surface area; gas diffusion through a hole is more rapid round the edges, where the molecules can fan out into the region of lower concentration. The stomatal pores nevertheless offer an appreciable resistance to $CO_2$ diffusion, denoted by $R_s$, which depends on the degree of opening of the stomata: the larger the pore diameter, the lower the value of $R_s$.

The cuticle which forms a continuous layer over the epidermis presents a very high resistance to $CO_2$ diffusion. Paradoxically for that very reason cuticular resistance, $R_c$, can often be ignored in considerations of $CO_2$ diffusion into leaves: as long as the stomata are open at all, the fraction of $CO_2$ passing through the cuticle is very small. In mathematical terms, since the stomatal and cuticular resistances act in parallel rather than in series, they are not additive; the relationship is:

$$\frac{1}{R} = \frac{1}{R_s} + \frac{1}{R_c} \qquad (3.3)$$

As values of $R_c$ are 500–1000 times higher than $R_s$, with open stomata, it can be seen that $1/R_c$ is negligible compared with $1/R_s$.

The boundary layer resistance, $R_a$, is the result of a layer of still air (unstirred layer) immediately adjacent to the outside of the leaf. In this layer the $CO_2$ concentration is lower than in the bulk atmosphere which is the source of the $CO_2$ diffusing in; the boundary layer may be regarded as insulating the leaf from the general atmosphere.

After a $CO_2$ molecule has entered a leaf it has still some distance to travel before it reaches the carboxylation sites. All the internal resistances encountered are collectively termed the mesophyll resistance, $R_m$. The boundary layer, stomatal and mesophyll resistances all act in series, i.e. they are additive, so that we can now write, expanding *equation 3.2*;

$$P = \frac{\Delta CO_2}{R_a + R_s + R_m} \qquad (3.4)$$

The $CO_2$ concentration at the reaction centres during photosynthesis is taken to be virtually zero, so that the value of $\Delta CO_2$ is determined by the external $CO_2$ concentration. Under field conditions, photosynthesizing plants do not greatly deplete the $CO_2$ supply in their vicinity, because of the rapid mixing of air, so that $\Delta CO_2$ remains fairly constant. During the day, the $CO_2$ concentration within a crop may fall, for example, to about 0.027%. But artificial enrichment of the ambient $CO_2$ concentration above normal atmospheric level leads to an increased rate of photosynthesis (unless light or temperature are severely limiting), and it may be deduced that $CO_2$ supply is in nature an important limiting factor in photosynthesis.

$R_a$ is a component dependent on wind speed. In still air, a relatively thick boundary layer builds up. This layer is disturbed by wind, and $R_a$ diminishes with increasing wind velocity. Under field conditions there is usually sufficient air movement to keep $R_a$ low relative to $R_s$ and $R_m$, and variation in wind speed does not have much effect on the rate of $CO_2$ uptake.

The mesophyll resistance, $R_m$, is a comprehensive term covering the resistance met by the $CO_2$ in traversing intercellular spaces, the cell wall, the plasmalemma, the cytoplasm and the chloroplast envelope. The biochemical reactions involved in $CO_2$ assimilation also contribute to $R_m$. The $R_m$ makes the greatest total contribution to total resistance (when stomata are open), but it cannot be studied or measured directly; the value of $R_m$ is obtained by difference when R, $R_s$ and $R_a$ have been determined.

Of all the factors controlling photosynthetic $CO_2$ uptake, stomata have received most attention. Stomata provide a variable resistance to $CO_2$ diffusion according to their degree of opening and the rate of photosynthesis depends on stomatal aperture; stomatal closure can reduce photosynthesis almost to zero. Nevertheless the controlling effect of stomata is not expressed independently of other factors such as irradiance (Fig. 3.6). It is generally agreed that stomata have evolved in response to the need of the plant to permit the entry of $CO_2$ and to prevent excessive water loss. The high surface: volume ratio of a leaf which makes it an efficient $CO_2$ absorber is also very conducive to the loss of water vapour. The cuticle affords protection against water loss from the epidermal surface, but the cuticle is also highly impermeable to $CO_2$; hence there has been the necessity for the development of pores for $CO_2$ entry. The stomata obviously act as channels for the exit of water vapour, but their opening and closure are under physiological control, in response to environmental stimuli. In many species stomata show a diurnal rhythm, opening by day and closing by night, light stimulating stomatal opening. Accordingly the pores are open during the period when light is available for photosynthesis; in the dark when $CO_2$ cannot be assimilated, water loss is avoided. Stomatal opening is also promoted by low concentrations of $CO_2$ within the leaf; this has been interpreted as a feedback type of control, a low internal $CO_2$ level signalling a depletion of $CO_2$ in the leaf, requiring stomatal opening if photosynthesis is to proceed. Water stress, however, has an overriding effect over other stimuli, causing stomatal closure. This occurs at quite moderate levels of water stress, before the leaf wilts, i.e. the stomata do not close because of an overall lack of water, but the decline in plant water status acts as a stimulus for closure. A hormonal signal appears to be involved: the concentration of the hormone ABA (abscisic acid) rises rapidly in response to water stress, and ABA induces stomata to close. The closing of stomata reduces water loss and is a vital mechanism for preventing desiccation. Photosynthesis is, however, drastically inhibited as a consequence. The physiological mechanisms underlying the changes in guard cell turgor have been the subject of much speculation. The current view is that metabolically

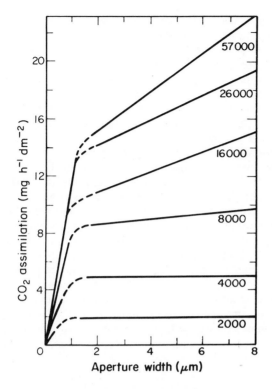

**Fig. 3.6** Effect of stomatal aperture on $CO_2$ assimilation of leaves of oat *(Avena sativa)* at six levels of irradiance. The illumination is given on each curve in lux, an old photometric unit; very approximately, 1 lux (white incadescent source) = $4.7 \times 10^{-3}$ J $s^{-1}$ $m^{-2}$. Carbon dioxide uptake rises with increasing stomatal width, but, at low levels of irradiance, increased aperture ceases to have an effect once a minimal width has been reached, light then becoming limiting. (From Stålfelt, 1935, *Planta, **23***, 715–59.)

controlled increases in guard cell turgidity follow an active pumping of $K^+$ ions into the guard cells, water entering by osmosis; decrease of guard cell turgor is consequent on an outward leakage of $K^+$ ions into subsidiary cells, because of increased membrane permeability to $K^+$. Chloride and malate ions accompany the $K^+$ ions so that ionic balance is maintained. Passive changes in guard cell turgor can follow changes in leaf water potential.

### Temperature effects: general comments on limiting factors

The effects of temperature on photosynthesis are complex. Temperature has no effect on light absorption and the primary photochemical reaction, and it has little effect on the physical process of $CO_2$ diffusion. But photosynthesis involves many ordinary thermochemical reactions and therefore the overall process would be expected to show a $Q_{10}$ of 2–3. If, however, irradiance is very low, raising the temperature over a certain minimum will not increase the rate of photosynthesis, this being limited by the light absorption with a $Q_{10}$ of 1. Even with adequate light and $CO_2$ supply, the temperature optimum of photosynthesis in temperate zone plants is quite low, 20–30°C. For total (gross) photosynthesis the value would be a few degrees higher; respiration rate rises more rapidly with rising temperature than the rate of photosynthesis, and this lowers the apparent optimum of photosynthesis. For plants native to warm climates, the temperature optimum is higher. Temperature can have indirect effects on photosynthesis, high temperatures promoting water loss, with potential stomatal closure. Extremes of temperature moreover promote stomatal closure independently of effects on water content.

Clearly in the preceding discussion it has scarcely been possible to evaluate the effect of any one factor on photosynthesis without specifying the status of other factors. The rate of photosynthesis is affected by irradiance, $CO_2$ concentration, temperature, plant water status, wind speed, and the degree of stomatal opening. The effect of varying any one of these factors depends on the value of the others. An increase in the rate of photosynthesis can be obtained by optimizing the factor – and only that factor – which happens to be limiting the process in the given situation. If photosynthesis is light-limited, an improvement in rate can be achieved only by increasing the irradiance; the system may then reach a state where $CO_2$ becomes limiting, and so on. The final limit would be set by the capacity of the plant's photosynthetic apparatus. In the field, a plant would very seldom, if ever, be working at its theoretical maximal photosynthetic capacity.

## The efficiency of photosynthesis

### Quantum efficiency

The overall process of photosynthesis can be expressed by the Van Niel equation:

$$CO_2 + 2H_2O \rightarrow (CH_2O) + H_2O + O_2 \qquad (3.5)$$

This reaction in the direction of the arrow involves a gain in free energy ($\Delta G$ has a positive value); $\Delta G$ can be calculated for *equation 3.5*, making reasonable assumptions for the concentration of the reactants, to be *c.* 490 kJ mol$^{-1}$ of $CO_2$. However, the radiant energy used in photosynthesis is involved in photosynthetic phosphorylation reactions which can be summarized thus:

$$2H_2O + 3ADP + 2NADP^+ + 3P_i \rightarrow 3ATP + O_2 + 2NADPH + 2H^+$$
$$\Delta G = 565 \text{ kJ} \qquad (3.6)$$

(where ADP = adenosine diphosphate, NADP = nicotinamide adenine dinucleotide phosphate, $P_i$ = inorganic phosphate, ATP = adenosine triphosphate.) Subsequently the 3 molecules of ATP and the 2 molecules of reduced NADP (NADPH) effect the reduction of one molecule of $CO_2$.

Hence a more realistic value for the gain in free energy per mole $CO_2$ fixed into carbohydrate is that calculated for *equation 3.6*, i.e. 565 kJ.

Although we can think of light as a continuous wave motion, the energy of light comes in discrete units or packets termed *photons,* each associated with a quantum of energy. Quanta are indivisible and photochemical reactions involve molecules or atoms activated by the absorption of a single quantum or, in successive steps, of several quanta. The quantum is not a constant; its energy value is inversely related to the wavelength of the radiation. The magnitude of the quantum ($E'$) is given by the equation.

$$E' = h\nu \text{ joules} \qquad (3.7)$$

where $\nu$ is the frequency of the radiation ($\nu = c/\lambda$ where c = the velocity of light, $3 \times 10^{10}$ cm s$^{-1}$ and $\lambda$ = wavelength in cm) and $h$ is Planck's constant ($= 6.62 \times 10^{-34}$ Js).

The quantum, $E'$, is the unit of energy absorbed by an individual molecule or atom. To obtain the amount of energy acquired by 1 mole when each individual molecule absorbs one quantum, we must multiply $E'$ by Avogadro's number $N$ (the number of molecules per mole) = $6.02 \times 10^{23}$. The energy per mole quanta, $NE'$, is sometimes called the Einstein. For red light of wavelength 670 nm ($= 6.7 \times 10^{-5}$ cm), the energy per mole quanta is:

$$NE' = Nh\nu = \frac{(6.02 \times 10^{23}) \times (6.62 \times 10^{-34}) \times (3 \times 10^{10})}{6.7 \times 10^{-5}} \text{ J}$$
$$= 179 \times 10^3 \text{ J} = 179 \text{ kJ} \qquad (3.8)$$

For blue light of wavelength 400 nm, the value is 299 kJ.

The smallest whole number of moles quanta of red light which can supply enough energy to effect the fixation of 1 mole $CO_2$ according to *equation 3.6* is 4 (equivalent to $4 \times 179 = 716$ kJ). If only 4 moles quanta were used per mole $CO_2$, this aspect of photosynthesis would be proceeding with 100% efficiency (although an energy loss is involved, since the free energy gain for *equation 3.6* is only 565 kJ mol$^{-1}$), and the quantum yield, moles $CO_2$ fixed per mole quanta used, would be 0.25. In so far as more than 4 mole quanta are required the efficiency is below 100%. Measurements have indicated that 8–12 moles quanta are needed per mole $CO_2$, giving quantum yields of 0.125–0.084. Against a theoretical requirement of 4 moles quanta per mole of $CO_2$ this implies a quantum efficiency of 33–50%. Further it has been shown that the quantum efficiency is more or less constant over the range of wavelengths 400–670 nm so that, in terms of joules energy, red light is most efficient. If a quantum requirement of 10 moles quanta per mole $CO_2$ is assumed, the efficiency of energy conversion is (from *equation 3.6* and *3.8*), $565/(10 \times 179) = 32\%$; for blue light of 400 nm the efficiency of energy conversion would be 19%. The extra energy of the short wavelength quanta is dissipated as heat because, as a first approximation, the energy level of the light-excited chlorophyll molecules reacting in photosynthesis corresponds to the quantum energy of *red* light.

The efficiency of quantum utilization and energy conversion calculated above applies, however, only for the quanta actually absorbed by photosynthetic pigments. A proportion of the radiation absorbed by a photosynthetic organ will be absorbed by other molecules and be transformed to kinetic (heat) energy and cause simply a rise in temperature and loss of long wavelength radiation. The quantum efficiency for total incident light intercepted by a leaf will therefore be less than the values just quoted. The efficiency in terms of total incident radiation falling on an area of vegetation will be much lower still.

## Crop efficiency

In estimates of the photosynthetic efficiency of whole plants or of a crop it is very important to consider the time scale. Values may be based upon a period of photosynthesis on a bright summer day, a whole 24-hour period, a growing season or over a year. Photosynthesis may be measured directly (e.g. by $CO_2$ uptake or $O_2$ evolution) or in terms of dry weight gain or crop yield.

The radiation reaching the surface of the earth from the sun has a maximum intensity of about 6.7 J cm$^{-2}$ per minute. Of this energy about 47% is in the wavelength range utilizable in photosynthesis.

In a study of photosynthesis of maize (*Zea mays*) under field conditions, undertaken at the time of maximum coverage at Ellis Hollow, New York, it was found that, with incident solar radiation of 4.2 J cm$^{-2}$ min$^{-1}$, the recorded photosynthetic $CO_2$ uptake was $4.8 \times 10^{-7}$ mol $CO_2$ cm$^{-2}$ min$^{-1}$. From these data the quantum yield, moles $CO_2$ fixed per moles incident quanta, can be calculated:

$$\frac{\text{moles } CO_2 \text{ fixed}}{\text{moles quanta supplied}} = \frac{4.8 \times 10^{-7} \text{ mol cm}^{-2} \text{ min}^{-1}}{(0.47 \times 4.2 \text{ J cm}^{-2} \text{ min}^{-1})(3.8 \times 10^{-6} \text{ J}^{-1})}$$

$$= 0.064 \tag{3.9}$$

where 0.47 = the fraction of incident solar energy in the utilizable wavelength range, and $3.8 \times 10^{-6}$ = moles quanta per joule of incident energy. The value of 0.064 may be compared with the best laboratory values for quantum yield of 0.125, i.e. it is about 50% of the maximum efficiency achieved in laboratory determinations of quantum yield. The efficiency of energy conversion corresponds to a utilization of about 13.5% of the energy in the visible range, or some 6.4% of the total incident radiation. This however is the photosynthetic efficiency and, to work out the efficiency of net fixation of $CO_2$ per day, we have to reduce the figure by the respiratory loss of $CO_2$ throughout the 24 hour period. This reduces the figure by 30–50% even under conditions of a high photosynthetic rate during the day. Thus at the time when the crop is at maximum efficiency we can expect only 3–4.5% efficiency of carbon gain relative to incident utilizable energy falling on unit crop area per day. Using the data for the maize crop quoted above and with total radiation per day taken as 2090 J cm$^{-2}$ (this would be just over

8 hours at 4.2 J cm$^{-2}$ min$^{-1}$), and assuming a respiratory loss of carbon equivalent to 33% of that fixed by photosynthesis, the calculated net gain is just over 20 g C m$^{-2}$ day$^{-1}$. This value can be compared with actual values averaging the carbon gains per day over the growing season for various 'crops' (Table 3.1). Lowered photosynthetic efficiencies due to such factors as water stress, limiting temperatures, mineral deficiency, disease and limiting $CO_2$ availability are involved in these actual seasonal averages. However of great importance in actual yields are: (1) the fact that over much of the growing season energy strikes bare ground, not the leaves of the crop, and that at the height of growth (time of maximum leaf area index, i.e. ratio of total leaf area to the ground area occupied) many leaves may be so shaded that they have very low net photosynthesis; (2) the seasonal energy input which at many times may be limiting carbon fixation; a cloudy sky reduces the mid-day irradiance to between 5–15% of that from a cloudless sky.

**Table 3.1**  Carbon gains per day averaged over the year or growing season for some natural, semi-natural and cultivated plant communities. (Data quoted by Westlake, D. F., 1963, Comparisons of plant productivity, *Biological Reviews*, **38**, 385–425.)

| Vegetation | g C m$^{-2}$ day$^{-1}$ (mean for growing season) |
|---|---|
| Marine phytoplankton, Denmark | 0.32 |
| Birch forest, England | 2.2 |
| Tropical rain forest | 7.6 |
| Maize crop, Minnesota | 8.1 |
| Sugar cane, Java | 11.1 |

Latitude and season limit the maximum input at different points on the earth's surface. The maximum value at the latitude of London varies from 4800 J cm$^{-2}$ day$^{-1}$ in summer, down to 830 J cm$^{-2}$ day$^{-1}$ in winter.

When all the actual losses are taken into account we begin to be able to predict the finding that, even under conditions of intensive agriculture, crop yields expressed as dry matter production relative to total incident energy work out at less than a 1% conversion of total available radiant into chemical energy. When it is further recognized that only a fraction of the dry weight yield is utilizable as food or in manufacture, the energy of the harvest may be as low as 0.2–0.5% of the solar energy input. Averaged out over the year the conversion will normally have a still lower value.

It can be seen that the (quantum) efficiency of the process of energy transduction, which can be of the order of 33–50%, is not the limiting factor in the energy conversion efficiency of a crop; the limiting factors must be sought at the whole plant, and environmental (climatic and geographical) levels.

Studies in crop growth in relation to plant habit, climate and soil factors are now being carried out by techniques which reveal the factors limiting energy conversion in the growing crop. Thus in studies on the corn crop in the United States it has been shown that the leaf area index reaches a value of 4 but computer calculations indicate that a value of 8 is needed for maximum production. However to be effective this increased leaf area index will require plants to have near vertical leaves and not horizontal or drooping leaves. The plants with vertical leaves, if the plant breeder can provide them, will need to be planted at a higher density than now used and to be planted equidistant from one another rather than in rows. Again yield in corn is limited by the 35 days during which the grain grows; if this could be extended grain yield would be increased (a greater proportion of the dry matter would appear in the harvested grain). An effective crop plant should be rapidly growing and capable of close planting so that an effective leaf area index is rapidly established and the period reduced when the radiation strikes bare earth. Avoidance of water stress is important not only for rapid growth but to ensure that diffusion of carbon dioxide is not restricted by stomatal aperture. Under high irradiance fully open stomata are needed for maximum photosynthesis. From appropriate monitoring of such moisture stress, transpiration loss and irradiance, irrigation can be programmed to be most effective. At high irradiance the natural carbon dioxide level is often a limiting factor in photosynthesis and there are many instances reported of large yield increases of glasshouse crops from increasing the carbon dioxide level often up to values of $c.$ 0.5%, but $CO_2$ enrichment is not feasible out of doors. Plants do, however, vary in the efficiency with which they use $CO_2$, so that there may be hope of selection of plants for efficient utilization of $CO_2$.

**Variations in physiological efficiency**

In the preceding paragraphs the influence of a number of external conditions on photosynthetic rate has been considered. There are also substantial intrinsic differences in the rates of photosynthesis of different species when measured under similar external conditions. Leaves of maize, sugar-cane (*Saccharum officinarum*), sorghum (*Sorghum vulgare*) and numerous other tropical grasses, have at high levels of irradiance rates of net photosynthesis 2–3 times greater than leaves of wheat, sugar beet (*Beta vulgaris*), tobacco (*Nicotiana tabacum*) and most other temperate zone crop plants. These differences are shown whether the rate is expressed per unit leaf area, leaf weight, or unit chlorophyll, indicating that variations in the amount of photosynthetic apparatus per unit area or mass of leaf do not account for the variations in rate. Interspecific differences exist in the biochemistry of $CO_2$ fixation and in the ancillary reactions involving newly-fixed carbon. The differences in photosynthetic rate between the 'high photosynthesis' and 'low photosynthesis' species, mentioned above, can be attributed at least partly to differences in their respiratory physiology in the light, and to variations in their $CO_2$ fixation strategies.

**Photorespiration**

The net rate (or amount) of $CO_2$ fixation is given by the difference in photosynthetic fixation and respiratory loss:

net photosynthesis = total photosynthesis − respiration

The net rate of photosynthesis, and hence of dry matter production, therefore depends not only on the rate of photosynthesis but also on respiration rate.

When leaf respiration rates are measured in the dark, they appear low compared with the net rate of photosynthesis under favourable conditions of irradiance, $CO_2$ concentration and temperature. Accurate values for the respiration rate of a photosynthesizing organ in the light are very difficult to obtain, for while respiration utilizes $O_2$ and evolves $CO_2$, photosynthesis gives off $O_2$ and takes up $CO_2$. Estimates can, however, be made, e.g. by the use of isotopes; radioactive $^{14}CO_2$ can be supplied for photosynthesis while respiration is evolving unlabelled $CO_2$. It has been found that in the low photosynthesis plants, respiration is strongly stimulated by light. This light-stimulated respiration does not follow the same biochemical pathways as the 'dark' respiration proceeding continuously in all organs (in darkness and light). There is a biochemically distinct *photorespiration*, involving a special organelle, the *peroxisome*, in addition to chloroplasts, mitochondria and cytosol. In the high photosynthesis plants such as maize, photorespiration is undetectable or very low. Where photorespiration occurs it can reduce net photosynthesis to from 85% to 50% of gross photosynthesis. Photorespiration is stimulated by high irradiance, high $O_2$ concentration, low $CO_2$ concentration, and high temperature: a $Q_{10}$ of 8 between 20°C and 30°C has been claimed. It seems to be a wasteful process dissipating newly-fixed carbon to $CO_2$ again, and while some ATP may be produced there is no net ATP gain.

Photorespiration depends on the potential for dual action by the $CO_2$-fixing enzyme of the $C_3$ cycle (Calvin cycle), ribulose-1,5-bisphosphate (RuBP) carboxylase. The enzyme carboxylates RuBP with $CO_2$ to give two molecules of 3-phosphoglycerate (PGA), which are reduced to sugar by further reactions of the $C_3$ cycle. The enzyme can, however, also oxygenate RuBP to yield PGA and phosphoglycolate, and the phosphoglycolate becomes the substrate for photorespiration with eventual $CO_2$ loss (Fig. 3.7). Since $CO_2$ and $O_2$ compete in the reaction, high levels of $O_2$ favour photorespiration and high $CO_2$ concentrations suppress it. High

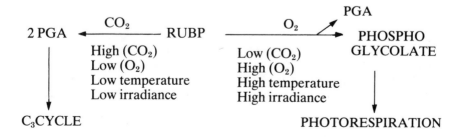

**Fig. 3.7** The initiation of photorespiration.

temperatures may similarly create a $CO_2$ shortage. With rising temperatures the affinity of RuBP carboxylase for $O_2$ relative to $CO_2$ becomes greater. Moreover the solubility of $CO_2$ relative to that of $O_2$ decreases with rising temperature; between 10°C and 30°C the ratio of their solubilities, $\alpha CO_2/\alpha O_2$, falls from 31.5 to 25.5. Accordingly it is at high irradiance levels and high temperatures that photorespiratory losses are most serious, resulting both from respiratory release of newly-fixed carbon, and from competition for RuBP by $O_2$. When photorespiration is diminished by $\alpha$-hydroxysulphonates which inhibit glycolate oxidation, or by low $O_2$ concentration, the rates of net photosynthesis in low photosynthesis plants are much enhanced and can reach values characteristic of high photosynthesis plants (Table 3.2). Experimental or genetic suppression of photorespiration has therefore been suggested as a possible means of

**Table 3.2**    $CO_2$ budget for a tobacco leaf at 25° and 35°C in high illumination and 300 ppm $CO_2$. (Data from Zelitch, I., 1971, *Photosynthesis, Photorespiration and Plant Productivity*, Academic Press, New York.)

|  | mg $CO_2$ dm$^{-2}$ hour$^{-1}$ | |
|---|---|---|
|  | 25°C | 35°C |
| Gross photosynthetic $CO_2$ uptake | −25.0 | −47.0 |
| $CO_2$ output from dark respiration | + 3,0 | + 6.0 |
| $CO_2$ output from photorespiration | + 7.0 | +26.0 |
| Net $CO_2$ uptake observed | −15.0 | −15.0 |
| When photorespiration inhibited net $CO_2$ uptake observed | −22.0 | −41.0 |

improving the photosynthetic yield of crop plants. This presupposes that photorespiration serves no vital physiological function. One view does regard photorespiration as an evolutionary relic. Photosynthesis evolved under an atmosphere rich in $CO_2$ but devoid of $O_2$, so that there would have been no selection pressure against RuBP evolving with a potential for oxygenase activity. However, photorespiration has also been assigned the function of protecting photosynthetic pigments by permitting a dissipation of surplus energy and a recycling of $CO_2$. Otherwise, it is surmised, when $CO_2$ runs short in strong light, the unstable light-excited pigments might accumulate and undergo breakdown.

### $C_3$, $C_4$ and CAM plants

The 'high' and 'low' photosynthesis plants differ with respect to their pathways of $CO_2$ fixation. The $C_3$ pathway, in which $CO_2$ is fixed by RuBP carboxylase to yield PGA, is universal. The PGA is reduced to carbohydrate level by the expenditure of three ATP molecules and two molecules of NADPH per molecule of $CO_2$ fixed (*equation 3.6*). The plants in which the $C_3$ cycle is the sole photosynthetic $CO_2$-fixation pathway are the low photosynthesis plants with high rates of photorespiration. The high photosynthesis plants, including tropical grasses and also a number of dicotyledons, possess in addition to the $C_3$ cycle a second $CO_2$-fixing pathway. This is the $C_4$ cycle (Hatch–Slack cycle, dicarboxylic acid cycle), and plants which possess it are now generally known as $C_4$ plants. In the $C_4$ pathway $CO_2$ reacts with phosphoenolpyruvate (PEP), catalysed by the enzyme PEP carboxylase, to give the dicarboxylic, 4-carbon acid oxaloacetate; this is then converted very quickly to either malate or aspartate depending on the species of plant. The malate and aspartate are in due course decarboxylated to yield $CO_2$ which is refixed in the $C_3$ cycle to produce carbohydrate, whilst the remaining 3-carbon moiety is processed to form PEP again. The significance of the $C_4$ cycle lies in the high $CO_2$ affinity of PEP carboxylase, which enables this enzyme to act at lower $CO_2$ concentrations than RuBP carboxylase. When $C_4$ plants are illuminated in an airtight container, they can lower the external $CO_2$ level to 0–5 ppm, while $C_3$ plants can lower it only to 50–70 ppm. Because of the high $CO_2$ affinity of PEP carboxylase, $C_4$ plants can maintain a steeper diffusion gradient for $CO_2$ than can $C_3$ plants. The efficiency of the $C_4$ species is further aided by their structural organization. Many $C_4$ plants have a distinctive leaf structure called Kranz anatomy. The $C_3$ cycle is concentrated in (or confined to) a layer of bundle sheath cells tightly packed around the vascular bundles so that sugar can be easily translocated away; $C_4$ plants have high rates of translocation of photosynthate out of leaves. The PEP carboxylase is concentrated in (or confined to) mesophyll cells arranged around the outside of the bundle sheath. The PEP carboxylase is thus positioned to receive the $CO_2$ diffusing inwards, and to prevent the escape of respiratory $CO_2$ from inner tissues. Malate and aspartate pass from the mesophyll to the bundle sheath where they are decarboxylated, in the case of malate actually inside the bundle sheath chloroplasts. The PEP carboxylase collects $CO_2$ for RuBP carboxylase and the Calvin cycle. Considerable traffic of metabolites between bundle sheath and mesophyll cells is necessitated by this division of labour, but the cells are amply provided with plasmodesmata, and it has been calculated that, over the distances in question, diffusion should be adequate to account for the intercellular exchanges.

Some photorespiration does occur in $C_4$ plants; biochemical analysis has revealed the presence of photorespiratory enzymes in their bundle sheath cells (where the RuBP carboxylase is located), and peroxisomes are present there. The enzyme levels, and the numbers of peroxisomes are, however, smaller than in $C_3$ plants, and the evolution of photorespiratory $CO_2$ is almost undetectable presumably because it is recaptured by PEP carboxylase in the mesophyll cells.

The conversion of $CO_2$ to carbohydrate by the $C_4$ pathway in combination with the $C_3$ cycle does demand the expenditure of two extra molecules of ATP per molecule of $CO_2$ over and above the ATP required in the $C_3$ cycle alone. The $C_4$ cycle is advantageous under conditions where $CO_2$ is limiting, rather than light, and the $C_4$ species are native to tropical and subtropical habitats, where high irradiance and high temperature would tend to bring about $CO_2$ limitation. The temperature optima of photosynthesis are relatively high in $C_4$ species. Another characteristic of $C_4$ plants is that they are efficient in their use of water, being able to synthesize 2–4 times more dry matter per unit mass of water transpired than $C_3$ plants. $C_4$ metabolism may be regarded as an adaptation to warm climates.

Interest has been expressed in the possibility of breeding the $C_4$ cycle into crop plants which lack it, in the hope of improving their productivity. The $C_4$ system is complex and as such must be under polygenic control; it would therefore be no simple matter to introduce it into a plant. It has also been queried whether the possession of $C_4$ photosynthesis would confer much benefit to plants under temperate climatic conditions.

A third group of plants, the CAM plants, fix $CO_2$ by CAM, *crassulacean acid metabolism*. The CAM plants are mainly succulents of hot arid climates; the name is derived from the family Crassulaceae, which contains many succulent members, although CAM is by no means confined to this family; e.g. cacti (Family Cactaceae) are CAM plants. In CAM plants stomata open at night and close in the day. During darkness $CO_2$ is fixed by PEP carboxylase, as in $C_4$ plants, but the malate is stored in vacuoles. The photosynthetic cells of CAM plants typically possess large vacuoles and the vacuolar pH falls steeply at night ('dark acidification'). During the day, the malate is decarboxylated and the $CO_2$ serves as substrate for RuBP carboxylase. Biochemically CAM resembles $C_4$ photosynthesis, but whereas in $C_4$ plants $CO_2$ fixation by PEP carboxylase and RuBP carboxylase occurs simultaneously but separated in space, in CAM plants the two enzymes act in the same cells but sequentially, separated in time. CAM is obviously an adaptation to hot and dry conditions; it enables the plants to conserve water by keeping stomata closed in the heat of the day when transpiration would be most rapid, and CAM plants are in fact extremely economical on water. Their rate of net photosynthesis per unit area of plant or ground, and their growth rate, are, however, very low. The small surface to volume ratios of the succulent organs are unfavourable for gas exchange.

## Utilization of photosynthate: biosynthesis and respiration

Photosynthetic productivity is measured in terms of a mass (dry weight) gain. Some of that dry mass gain is in the form of stored photosynthate. Much of the dry mass increase is, however, in the form of a multitude of compounds derived from the primary photosynthate by biosyntheses. Growth means synthesis of protoplasmic components. When the starting materials and the final end products are considered, such biosyntheses are energetically (thermodynamically) unfavourable: there must be inter-mediate stages with an energy input additional to the primary input in photosynthesis. Metabolites must moreover be moved to the sites of growth, sometimes over many metres, and translocation of organic material requires energy (Chapter 6). Mineral ions are taken up by cells against a free energy gradient. The driving force for this metabolic work is provided by *respiration*, utilizing primary or stored photosynthate as substrate.

Respiration is an exergonic process; starting with glucose as substrate, respiration can be represented by the equation:

$$C_6H_{12}O_6 + 6O_2 \rightarrow 6CO_2 + 6H_2O \qquad \Delta G = -2880 \text{ kJ} \qquad (3.10)$$

The oxidation of 1 mole glucose yields 2880 kJ of free energy. During combustion, all this energy is released as heat; during respiration (in this section meaning 'dark' respiration), coupling of the oxidation to ATP synthesis results in conservation of part of the energy in ATP bonds, and up to 38 molecules of ATP can be gained per molecule of glucose respired. This corresponds to the conservation of about 55% of the potential energy of the glucose. (The figure of 55% is derived by taking the molar free energy of hydrolysis of ATP as equivalent to the energy conserved in ATP, and multiplying by 38. The free energy of a reaction depends on the concentrations of the reactants; slightly different figures are therefore quoted according to the value assumed for cellular ATP concentration. Here the molar free energy of ATP hydrolysis is taken as 42 kJ $mol^{-1}$.) The ATP is used as the energy source for cellular work.

If the symbol E is used for the energy produced in respiration,

$$E = W + S + H \qquad (3.11)$$

where W = physical work, S = synthetic work, and H = heat released. From measurement of the respiration rate of a plant tissue, it is simple to calculate the value of E, knowing the molar free energy of glucose oxidation (*equation 3.10*); H is measurable calorimetrically. Such measurements have shown that 90–99% of the potential energy of respiration is released as heat in plant tissues and only 1–10% seems to be utilized – a very low value compared with the theoretically possible conservation of some 55% in ATP.

If, however, the way in which ATP is utilized in biosynthesis is considered, it is seen that the energy released as heat during metabolism does not necessarily represent a waste. The high reactivity of ATP is utilized in cells to make phosphorylated or nucleotide-linked derivatives of metabolites such as sugars and amino acids. This synthesis already entails the release of some energy as heat. The 'activated' derivatives can now participate in reactions not possible energetically to the parent compounds. Thus when macromolecules are synthesized, phosphorylated or nucleotide-linked precursors are involved and for each bond between the subunits at least 1 ATP molecule is used up, while most of the ATP energy is released as heat. In the extreme case of protein synthesis, there is an expenditure of at least 5 molecules of ATP per peptide bond formed, equivalent to a conservation of ATP energy of no more than 1%. Yet the 99% of energy released as heat cannot be regarded as uselessly wasted: it is released in reactions without which cellular protein synthesis cannot take place. The transport of numerous chemicals across cellular membranes is linked to hydrolysis of ATP; again energy would be released as heat, while concomitantly with the hydrolysis a proton gradient is built across the membrane and used to drive solute transport. A proportion of respiratory energy is used for maintenance processes, i.e. repair and resynthesis of protoplasmic components which are labile and in a state of continuous turnover. The energy output from maintenance respiration would be expected to appear entirely as heat, there being no *net* work or synthesis accomplished.

Nevertheless it should not be assumed that the coupling of respiration to ATP synthesis is 100% efficient, so that some of the heat release observed in plant tissues may represent a waste. This cannot easily be ascertained experimentally.

According to the laws of thermodynamics, any system should assume the state of maximum *entropy*, i.e. maximum randomness, disorganization, and minimum free energy. A growing plant increases in complexity and organization both at the chemical level and at the structural level. This is possible only at the expense of entropy increasing somewhere else, that is, in the plant's environment. The high energy quanta from the sun are transduced during photosynthesis to chemical bond energy, but (as shown above) always with some energy loss as heat, low energy quanta, of higher entropy. Then as the products of photosynthesis are incorporated into cells, more heat energy is released during respiration and the subsequent utilization of ATP. All this heat loss represents an increase in entropy, and dissipation of energy, which compensates for the increased complexity and free energy content of the living plant.

## FURTHER READING

BLACK, C. C. (1973). Photosynthetic carbon fixation in relation to net $CO_2$ uptake. *Annual Review of Plant Physiology*, **24**, 253–86.

CHOLLET, R. and OGREN, W. L. (1975). Regulation of photorespiration in $C_3$ and $C_4$ species. *Botanical Review*, **41**, 137–79.

GIFFORD, R. M. and EVANS, L. T. (1981). Photosynthesis, carbon partitioning, and yield. *Annual Review of Plant Physiology*, **32**, 485–509.

JONES, G. (1979). *Vegetation Productivity*. Longman, London.

SAN PIETRO, A., GREEN, F. A. and ARMY, T. J. (eds.) (1967). *Harvesting the Sun*. Academic Press, New York.

ZELITCH, I. (1971). *Photosynthesis, Photorespiration and Plant Productivity*. Academic Press, New York.

## SELECTED REFERENCES

BERRY, J. and BJÖRKMAN, O. (1980). Photosynthetic response and adaptation to temperature in higher plants. *Annual Review of Plant Physiology*, **31**, 491–543.

LONG, S. P., INCOLL, L. D. and WOOLHOUSE, H. W. (1975). $C_4$ photosynthesis in plants from cool temperate regions, with particular reference to *Spartina townsendii*. *Nature, London*, **257**, 622–4.

MANSFIELD, T. A., TRAVIS, A. J. and JARVIS, R. G. (1981). Responses to light and carbon dioxide. In *Stomatal Physiology*, ed. Jarvis P. G. and Mansfield T. A., 119–135. Cambridge University Press, Cambridge.

PALEVITZ, B. A. (1981). The structure and development of stomatal cells. *Ibid.*, 1–23.

PENNING DE VRIES, F. W. T. (1975). The cost of maintenance processes in plant cells. *Annals of Botany N.S.*, **39**, 77–92.

RAVEN, J. A. (1976). The quantitative role of 'dark' respiratory processes in heterotrophic and photolithotropic plant growth. *Annals of Botany N.S.*, **40**, 587–602.

# 4
# Water Relations

Liquid water is absolutely necessary for life as we know it. Firstly it is the solvent and reaction medium of the cell; secondly it is a reactant in many metabolic processes and thirdly, as the hydration water associated with macromolecules, it forms part of the structure of protoplasm, existing as 'liquid ice' in an ordered but labile structure. The physico-chemical properties of water are unique; even heavy water (deuterium oxide, $D_2O$ or deuterium hydroxide, DHO) is toxic. Additionally in plants, the turgor pressure of water-filled vacuoles gives mechanical rigidity to thin-walled tissues and some plant movements occur as a result of turgor pressure changes.

Although certain plant tissues may be able to endure periods of almost complete desiccation (see Chapter 7), they are while 'dry' metabolically inactive. When in an active state, most plant tissues contain 70–95% water. Plant cells can grow only when they are turgid and even small decreases in water content below full saturation may result in decreased growth rates. In the fungus *Polystictus versicolor,* the growth rate of the hyphae is directly proportional to their degree of hydration. Not only quantitative but qualitative changes in metabolism can be brought about by changes in water content. For example hydrolysis of starch to sugar occurs in many plant tissues as their water content decreases (Chapter 7).

## Water relations of plant cells

The amount of water present in a plant cell (or tissue, or whole plant) at any one time will be determined by a balance between internal forces promoting water entry and environmental factors tending to withdraw water. According to fundamental thermodynamic principles, any system tends to assume the state of lowest possible free energy. If there are two water-containing systems in contact, and the free energy of water differs in the two systems, water will move *from the system of higher free energy to that of lower free energy,* till the free energy of water is equal in both systems; at equilibrium, there will be no net change in the water content of either system, though an exchange or flux of water, equal quantities moving in each direction per unit time, may still proceed. We can thus say when:

Free energy of cell water <free energy of external water then net flux will be *into* the cell, whereas when

Free energy of cell water >free energy of external water net flux will be *out of* the cell.

In order to predict the direction of movement of water into/out of plant cells, we thus need a measure of the free energy of the water.

## Concept of water potential

The amount of free energy of water contributed by every mole of water in a system is the chemical potential of water, $\mu_w$. This chemical potential cannot be measured directly, but the chemical potentials of water in different systems can be compared, using for comparison the chemical potential of pure water ($\mu^o_w$) at atmospheric pressure and at the same temperature as the system being studied. We then define the *water potential* $\psi$ (psi) or $\psi_w$ of a system as the difference between $\mu_w$, the chemical potential of water in the system, and $\mu^o_w$, the chemical potential of pure water:

$$\psi = \frac{\mu_w - \mu^o_w}{\bar{V}_w} \tag{4.1}$$

It can be seen from this equation that the water potential of pure water at atmospheric pressure is zero. The term $\bar{V}_w$ is the partial molal volume of water, i.e. the volume of 1 mole water (18 cm$^3$). By dividing through by this value, the water potential is obtained in convenient units:

$$\psi = \frac{\text{energy per mole water}}{\text{volume per mole water}} = \text{energy per unit volume}$$

The basic units of $\psi$ are thus energy per unit volume, e.g. J cm$^{-3}$. But energy per unit volume can easily be converted to an equivalent force per unit area, i.e. pressure. Hydrostatic pressures and tensions (negative pressures) play an important part in controlling water movements in plants, and for many years it has been customary to express the numerical values for forces influencing water movement in pressure units. The values of $\psi$ are also frequently expressed in pressure units, as pascals (Pa) and its multiples, kPa (kilopascals, $10^3$ Pa) and MPa (megapascals, $10^6$ Pa). (See Table, p. 267 for definitions, and for relationships with older units of pressure, atmospheres and bars.)

Since the vapour pressure of water depends on its chemical potential, the water potential can also be expressed in terms of vapour pressure:

$$\psi = \frac{RT \ln \dfrac{e}{e^o}}{\bar{V}_w} \tag{4.2}$$

where R is the gas constant, T the absolute temperature, e the vapour pressure of water in the system and e$^o$ the vapour pressure of pure water at the same temperature. Measurements of vapour pressure can be used to obtain the water potential. The ratio of vapour pressures expressed as a percentage, $\dfrac{e}{e^o} \times 100$, is the *relative humidity* (R.H.), a measure often employed in quantifying the water status of the atmosphere.

The values of $\psi$ of plant cells generally vary from 0 downwards, i.e. are negative. If a cell is completely water-saturated, the free energy of the cellular water equals that of pure water, i.e. $\mu_w = \mu^o{}_w$; then $\mu_w - \mu^o{}_w = 0$ and $\psi = 0$. This situation may occur naturally in a submerged aquatic plant in equilibrium with almost pure water. Any decrease in the chemical potential of cellular water below that of pure water makes $(\mu_w - \mu^o{}_w)$ negative and hence $\psi$ is usually negative. In a reasonably well-watered mesophytic tissue, $\psi$ would be $-0.1$ to $-1.0$ MPa. A positive value for $\psi$ of a cell is possible only under the unlikely circumstance that the cellular water is at a higher chemical potential than pure water.

The water potential of a cell is determined by three kinds of forces. *Hydrostatic pressure*, $\psi_p$, in excess of atmospheric *increases* the free energy and raises the water potential. *Osmotic forces* decrease the free energy and *osmotic potential* is denoted by $\psi_\pi$ (psi, pi); since $\psi_\pi$ decreases the free energy, it is assigned a negative value. (Osmotic pressure is numerically equal to osmotic potential but with a positive sign; a solution with $\psi_\pi$ of $-5$ MPa has an osmotic pressure of $+5$ MPa.) Thirdly there are *matric* (= colloidal, interfacial, imbibitional) forces which also *decrease* the free energy and are given the symbol $\psi_m$, again with a negative value. The relationship between these quantities and $\psi$ is conventionally expressed by the formula:

$$\psi = \psi_p + \psi_\pi + \psi_m \qquad (4.3)$$

In plant cells $\psi_p$ is represented by the turgor pressure of the cell wall (wall pressure) tending to squeeze water out, and $\psi_\pi$ by the osmotic effects of low molecular weight solutes, mainly vacuolar. The matric potential derives from surface tension forces in the wall microcapillaries and water attraction by charged groups on cell wall polysaccharides, and also from colloidal hydration of macromolecules in the cytoplasm and occasionally in the vacuole. Thus the water content of plant cells is maintained by osmosis and by colloidal hydration (imbibition). Osmosis, which depends upon the presence of solutes and the semipermeable properties of the plasmalemma and tonoplast, controls water uptake and loss in vacuolated cells of high water content; imbibitional forces are important only in controlling water uptake and retention by tissues of really low water content such as those of 'dry' seeds.

The meaningfulness of *formula 4.3* has been questioned on the grounds that a cell is a heterogeneous system. Its individual compartments – wall, cytoplasm, vacuole – will all have the same $\psi$, but the values of $\psi_p$, $\psi_\pi$ and $\psi_m$ will vary between the compartments. Hence the term 'matric potential *of the cell*' could be argued not to be a real quantity. The formula is, however, useful in indicating the relationships between the components that contribute to water potential. What is *not* justifiable is to take, say, the $\psi_\pi$ *of the vacuole* and the $\psi_m$ *of the cell wall* and to sum these values in the calculation of the $\psi$ *of the cell*. That would be equivalent to attempting to obtain the average sugar concentration of a cell by adding the sugar concentration in the vacuole to the sugar concentration of the cytoplasm! The basic units of $\psi$ are, as stated, energy per unit volume (just as those for concentration are mass per unit volume), and the values for individual compartments are *not* additive.

The meaningfulness of the concept of matric potential has also been queried. It has been argued that since colloids lower water potential by causing decreases in hydrostatic pressure and by exerting at their charged surfaces forces on water molecules essentially the same as the forces exerted by low molecular weight solutes, the effect of colloids and the cell wall matrix could be incorporated in the $\psi_p$ and $\psi_\pi$ terms. Since matric forces are of relatively little importance in highly vacuolated cells, their contribution to water potential is often ignored and the cell is considered simply as an osmotic system, where:

$$\psi = \psi_p + \psi_\pi \qquad (4.4)$$

Water movement can then be considered as resulting from a balance between osmotic forces tending to draw water in and the hydrostatic pressure of the wall opposing water entry.

Some workers have cited experimental observations apparently indicative of the development of hydrostatic pressures in vacuolated cells in excess of those to be expected from osmosis and have reported water movement into growing cells in opposition to the expected direction of net water movement from osmotic considerations. On the basis of such observations it has been proposed that an 'active' pumping of water into plant cells can occur. However the evidence for such an 'active' component in water movement is very inconclusive and is rejected by most workers.

A cell with a positive turgor pressure is said to be, to some degree, turgid. When the cell is at full turgor, its water uptake capacity is saturated; it no longer has a negative $\psi$ or tendency to take up water. The osmotic pressure of the cell contents is then being balanced by an equal and opposite turgor pressure:

$$\psi = 0 \quad \text{and} \quad -\psi_\pi = \psi_p \qquad (4.5)$$

On the other hand, as water is lost from the cell, its $\psi$ decreases, both as a consequence of concentration of solutes in the cell, and of the reduction of turgor pressure. When the stage is reached where the protoplast no longer presses against the cell wall, the cell is said to be flaccid; then:

$$\psi_p = 0 \quad \text{and} \quad \psi = \psi_\pi \qquad (4.6)$$

A plant part in which the cells are flaccid is wilted. If still more water is removed, the effect on the cell depends on the mode of removal. If this is accomplished by immersion in a solution of low osmotic potential plasmolysis results; the protoplast shrinks away from the wall, letting the external solution fill the space left between it and the wall. If water is removed by evaporation, the protoplast generally remains adpressed to the wall, and the walls cave in as the contents shrink. The degree of flexibility of the wall may then determine the amount of water that can be removed. Loss of water beyond a certain limit, which varies with the particular cell and with factors such as the rate of dehydration, is fatal (see Chapter 7).

*Note:* while the water potential concept has been universally accepted, minor variations in terminology abound, e.g. *formula 4.3* is sometimes given in the form:

$$\psi = P - \pi + \tau \qquad (4.7)$$

Here $P \equiv \psi_\pi$, and $\tau$ (tau) $\equiv \psi_m$, while $\pi$(pi) stands for osmotic *pressure*, not osmotic potential. It is common to see subscripts such as $\psi_x$ for the water potential of the xylem, $\psi_l$ for that of the leaf, etc. The units in which $\psi$ is expressed are also varied. Useful lists of terms and units will be found in: Sutcliffe (1979), *Plants and Water*, Studies in Biology no. 14, Edward Arnold.

### Measurement of water potential

The water potential of a tissue can be measured by exposing replicate samples of the tissue to a graded series of water potentials, either by immersing the samples in solutions of known water potential, or by enclosing them in atmospheres of known vapour pressure. In a solution there is no excess hydrostatic pressure and the water potential is determined solely by osmotic forces:

$$\psi_p = 0, \quad \psi = 0 + \psi_\pi = \psi_\pi \qquad (4.8)$$

The water potential of a solution equals its osmotic potential. Water will move out of a tissue into solutions with water potentials more negative than that of the tissue and will move into the tissue from solutions with water potentials less negative than that of the tissue; in a solution whose water potential equals that of the tissue, there is no net change in tissue water content. Changes in the water content of the tissue are estimated by weighing the samples before and after immersion. If immersion of tissue in liquid is thought to be undesirable (errors can arise by penetration of liquid into tissue air spaces, giving misleading weight increases) the tissue sample can be exposed to atmospheres of known vapour pressure by suspension above the solutions in small chambers. The water potential in the atmosphere of each chamber will equal that of the solution in it and the tissue will react just as to the solutions, although equilibration with a vapour phase takes longer than when immersed in a solution.

Another method for measuring water potential is by means of the *thermocouple psychrometer*. The tissue is arranged to line the walls of a small chamber in which a thermocouple junction is enclosed, and which is incubated at a constant temperature. Water evaporates from the tissue into the chamber until, at equilibrium, the water potential of water vapour in the chamber is equal to the water potential of the tissue. A small drop of pure water is then introduced on to the thermocouple junction. As the water evaporates it cools the thermocouple, causing a current to flow and the magnitude of this is recorded. The current is proportional to the rate of cooling; the rate of cooling depends on the rate of evaporation which in turn depends on the vapour pressure in the chamber. The instrument is calibrated by measuring the currents produced in a series of atmospheres of known water potential and the water potential of the tissue can be accordingly obtained from the current reading.

## Water relations of whole plants and organs

The water relations of a whole plant are more complex than those of the individual cell. The formulae given above are applicable only at the cell level, or at the level of an uniform tissue; there is no such thing as say, the wall pressure of a plant or even of a leaf. A plant usually also contains at any given moment tissues of different $\psi$ so that there is a flow of water within and through the plant according to $\psi$ gradients between the soil, various parts of the plant, and the atmosphere.

Flowering plants are essentially land plants; the aquatic species represent a secondary evolutionary return to water. The evolution of land plants has depended upon the evolution of systems for the absorption, conduction and conservation of water. Their water is usually obtained by a root system invading a soil. Some tropical epiphytes absorb rainwater through special aerial roots and through scales and hairs on their leaves, and a few flowering plants such as *Tillandsia usneoides* derive their water entirely from atmospheric rain, mist and dew.

The photosynthetic mode of life of flowering plants requires the development of a large surface area for light absorption and for gaseous exchange, and in consequence of a large surface from which there is continual water evaporation or *transpiration*. The water potential of the atmosphere is nearly always very much more negative than that of the plant – often by tens of megapascals – and hence there is a great tendency for water loss from the plant. This loss of water by transpiration must be made good by absorption from the soil; there is a flow of water, the transpiration stream, from the roots to the transpiring surfaces. Only a very small fraction of the water absorbed by the roots is retained in the plant. For maize, an annual plant, this fraction has been estimated at less than 1% of the water absorbed during its growing season. During one single bright, sunny day, leaves may transpire several times their own weight of water. For example, a leaf of *Senecio jacobaea* growing on a sand-dune can transpire its own weight of water in 45 minutes. The water content of the aerial organs of a plant is generally lower in the daytime, when transpiration is higher, than during the night when the water content rises, the deficit being made good. The amount of water in the roots fluctuates much less (Fig. 4.1). If water loss exceeds water uptake, wilting ensues and death may follow.

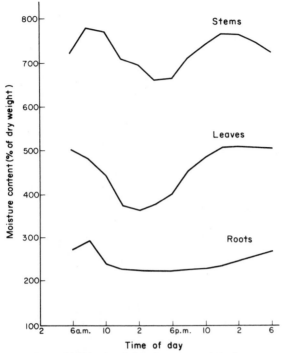

**Fig. 4.1** Diurnal fluctuations in the moisture content of the leaves, stems and roots of field-grown sunflower, *Helianthus annuus*. (From Wilson, Boggess and Kramer, 1953, *American Journal of Botany*, **40**, 97–100.)

### The absorption of water by roots

The root systems of plants are often very extensive. Roots may extend much further underground than the shoot spreads into the air (Fig. 4.2). The roots of an apple tree may go down to about 10 m; and even in herbaceous plants this kind of depth can be reached, e.g. in alfalfa *(Medicago sativa)*. Some measurements on the root system of a 4-month-old plant of rye *(Secale cereale)* grown singly in a box are reproduced in Table 4.1. The total external surface area of the rye shoot was only 4.61 m². If to this is added the internal surface area of the shoot, that is, the area of cells bordering the internal air spaces, then the total surface area of the shoot was 27.9 m². The root surface area was 22 times this value. Root hairs contribute considerably to the surface area of the root. Root hair area is included in the root area quoted in Table. 4.1; estimated separately, it was 1.7 times that of the remainder of the root surface. In other species, root hair areas amounting to 8–12 times that of the rest of the root have been recorded although many of these observations have been made on roots grown in moist air when the root hairs are more abundant than in the soil. Not only do the root hairs increase the absorbing surface, but they make very intimate contact with the soil, bending round soil particles and penetrating into crevices. The extent to

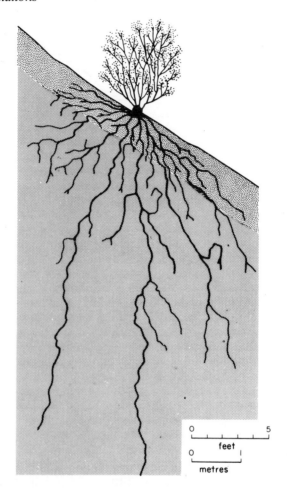

**Fig. 4.2**   Diagrammatic sketch of root system of a Californian woody shrub, the chamise (*Adenostoma fasciculatum*). (From Hellmers, Horton, Juhren and O'Keefe, 1955, *Ecology,* **36**, 667–78.)

**Table 4.1**   The extent and average daily growth rate of the root system of a 4-month-old plant of rye (*Secale cereale*) grown singly in a box. (After Dittmer, H. J., 1937, *American Journal of Botany,* **24**, 417–20.)

| | |
|---|---|
| Surface area of total root system (hairs included) | 639 m² |
| Area of root hairs | 402 m² |
| Length of total root system (hairs excluded) | 622 km |
| Length of root hairs | 10 620 km |
| Average daily growth rate of root (hairs excluded) | 4·99 km |
| Average daily growth rate of root hairs | 89 km |

which soil is filled by roots may be appreciated from the fact that the entire root system of the above-mentioned rye plant was contained in less than 0.06 m³ of soil.

Most of the absorption takes place near the growing root tips where the epidermis (piliferous layer) is thin-walled and root hairs are present. As the root tissues mature, the epidermis with its root hairs is replaced by a more impermeable suberized periderm. For most efficient water uptake, root growth must continuously regenerate the absorbing zone behind each growing root apex. Continuous growth is also necessary to enable roots to invade new soil. This can be very important under conditions of water stress for there is very little movement of water in soil beyond downward drainage directly after water addition. Water will not move to a substantial extent to the roots, and the only way in which the roots can reach a further supply of water is by growth. Positive hydrotropism may be important in directing root growth towards soil water (see Chapter 10). Roots can grow very rapidly; a rate of 10 mm per day is stated to be common for grasses; maize roots can extend 50–60 mm per day. The average daily increase in length of the total root system of the rye plant (Table 4.1) was calculated to be almost 5 km per day.

Though the root hair zones provide the chief absorbing surfaces water uptake through older regions is still appreciable, particularly during conditions of water shortage and at times when root growth is slow, such as winter. Points of emergence of lateral roots break the suberized layers and make regions of entry for water.

### The route of water movement

The main channel for upward movement of water in the plant is the *xylem*. When the tissues outside the xylem in a short length of woody stem are peeled off, the conduction of water beyond the stripped region continues unimpeded. The xylem vessels at least in young wood are filled with a watery sap and dyes and Indian ink can be seen to move in the xylem. During the later nineteenth century, several workers, notably Eduard Strasburger, demonstrated that the water passes through non-living tissues by showing that strong poisons moved up to the leaves of trees, and that a second aqueous solution could move up after the passage of the poisonous solution, which would have killed any living cells in its path. Water conduction continues through heat-killed stems. Chilling has no effect on the rate of water movement, as long as the temperature is kept above the freezing point of the xylem sap. When xylem vessels are experimentally blocked by mercury or cocoa butter, water movement is strongly inhibited. The chief conducting elements of the xylem are considered to be the vessels which can be regarded as well suited for mass flow of liquid. The diameters of vessel units range from 20 to 400 $\mu$m; even 500–600 $\mu$m is attained in some lianas. Vessel lengths range from a few centimetres to many metres; in some trees, some continuous vessels run right from the crown down to the roots. In woody perennials, the older parts of the xylem lose their conducting capacity; the vessels become air-filled or blocked by tyloses and deposits of gums and resins. The water moves then only through the young sapwood elements. In ring-porous species, conduction is usually confined to the current year's growth; in diffuse-porous wood, the few outermost annual rings are utilized.

The rate of water movement is very variable; Table 4.2 gives examples of maximum rates attained in a number of species. In any one plant the rate is also highly variable, depending on environmental conditions. Among the species quoted in Table 4.2, the order of increasing speed of movement is also the order of increasing anatomical development towards wider and more numerous vessels in the xylem. In conifers, which lack vessels and depend on tracheids only for transport, water movement is much slower than in angioperms and the conductivities for water of conifer woods are much lower than those of the woods of most angiosperms.

**Table 4.2**   The mid-day maximum speeds of the transpiration stream, measured by observation of dye movement or heat conduction. (After Huber, B., 1956, *Encyclopedia of Plant Physiology*, ed. Ruhland, W., **3**, 541–82, Springer–Verlag, Berlin.)

| Species | Speed $(m\ h^{-1})$ |
|---|---|
| Evergreen conifers | 1.2 |
| Mediterranean sclerophylls | 0.4–1.5 |
| Deciduous diffuse-porous trees | 1–6 |
| Deciduous ring-porous trees | 4–44 |
| Herbaceous plants | 10–60 |
| Lianas | 150 |

Since the xylem of the root is a central core of tissue, water must pass through the epidermis (piliferous layer), cortex, endodermis and pericycle to reach the xylem vessels (Fig. 4.3). It is still disputed whether this radial movement proceeds mainly through intercellular spaces, cell walls, cytoplasm or vacuoles. Movement through intercellular spaces does not appear very likely, since the spaces are generally air-filled. In the epidermis and the endodermis the cells are closely packed without air spaces; in these tissues at least, water must pass through some part of the cells. The radial walls of the endodermal cells at the level of most active water absorption of the root develop Casparian strips which resemble suberin in chemical reactions. It has therefore been frequently suggested that at the endodermis, if nowhere else, water is forced to pass through the protoplasts of living cells, and that this layer of cells thereby regulates water movement from soil to xylem. Direct measurement of the water permeability of endodermal cell walls has not yet been achieved, but there is evidence which supports the above suggestion. There is a view that the cytoplasmic contents of all the cells of the root constitute a continuous unit, the symplast, and that once water has entered the cytoplasm of a root cell, it can move from cell to cell via plasmodesmata without the necessity of crossing any more outer cell membranes (see also Chapter 5 and Fig. 5.2, p. 80).

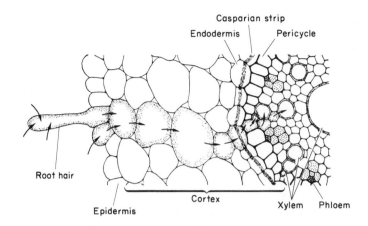

**Fig. 4.3** Diagram of root cross-section, showing the tissues through which water and minerals move when passing from the soil solution to the xylem. (From Esau, 1960, *Plant Anatomy,* Wiley, New York and London.)

In a herbaceous plant with separate vascular strands, any part of the root system normally supplies those parts of the stem and its appendages which are directly above it, these being the parts with which it is in direct vascular connection, so that lateral movement does not occur or is very restricted. If, however, part of the root system is deprived of water, the overlying aerial parts then receive their water supply from other root sectors indicating an activation of lateral movement. In trees with a continuous cylinder of wood, dye injection has shown that the path of water movement frequently winds in a spiral round the stem, but this follows the spiral arrangement of the conducting cells in the wood and involves longitudinal not lateral movement.

### The motive forces in water movement – root pressure and transpiration pull

The movement of water in the xylem vessels must involve either pumping under pressure or pulling (lifting) under tension. Capillary rise (a surface tension phenomenon) could not account for a rise of more than 1 metre in the finest vessels (many plants are much taller than this, some reaching 90–100 m!) and would still require the operation of an extracting force.

In certain circumstances, evidence has been obtained that the xylem sap is under positive hydrostatic pressure. Evidence in favour of this from inserting manometers into the wood has been questioned on the grounds that the pressure registered is an exudation pressure from injured living cells of the xylem not directly involved in water movement. However in a number of instances (when such pressures have been recorded) it has also been shown that decapitation of the plants just above ground level leads to continuing exudation from the exposed vessels in the root stump. A manometer fitted over such a bleeding root stump records the pressure known as the *root pressure*. In many cases root pressures of up to 0.1–0.2

MPa have been recorded. In excised tomato roots in culture root pressures apparently of up to 0.7–0.8 MPa occur. The development of this pressure is dependent on conditions favourable to the metabolic activity of the roots. No positive root pressure is found when the roots are subjected to treatments inhibiting metabolic activity such as lack of oxygen, addition of respiratory inhibitors, low temperatures, or starvation. Root pressure is regarded as the outcome of an osmotic mechanism: water from the soil moves through the root tissues along a gradient of osmotic potential ($\psi_\pi$) (p. 49), a negative $\psi$ being maintained in the xylem by continuous secretion of salts or other osmotically active substances into the xylem sap (see Chapter 5). In cases where root pressure can be demonstrated it has been shown that the $\psi_\pi$ of the xylem sap is more negative than that of the external solution (Table 4.3). When the $\psi_\pi$ of the solution external to the root is suddenly decreased, the rate of exudation falls and may become negative (liquid is drawn into the vessels at the cut surface); roots maintaining a root pressure have been described as behaving like sensitive quick-response osmometers.

**Table 4.3**   The effect of varying the osmotic potential ($\psi_\pi$) of the solution bathing the roots on the osmotic potential of the exudate from the stumps of decapitated plants of tomato (*Lycopersicon esculentum*). (After Kramer, P. J., 1959, *Plant Physiology*, ed. Steward, F. C., **2**, 607–703, Academic Press, New York and London.)

|  | **Normal salt plants** | | | **Low salt plants** |
|---|---|---|---|---|
|  | May 24 | May 25 | May 26 | June 5 |
| $\psi_\pi$ of culture solution, MPa | − 0.049 | − 0.054 | − 0.060 | − 0.082 |
| $\psi_\pi$ of stump exudate, MPa | − 0.240 | − 0.150 | − 0.218 | − 0.221 |

Drops of liquid sometimes appear at leaf tips and edges when plants are maintained in a humid atmosphere: this phenomenon, termed *guttation*, results from the forcing out of xylem sap under pressure through pores overlying vein endings. The salt concentration in the guttated fluid is nevertheless lower than in the xylem sap (suggesting more absorption of salts by the leaf cells) although markedly dependent upon the composition of the solution bathing the roots. Both bleeding from exposed root stumps and guttation exhibit a repeating rhythm during successive 24-hour periods, with maxima usually at mid-day.

Although the development of positive root pressure is well authenticated, there are many cases where it cannot be invoked to account for water movement. The observed pressures are as a rule too low to raise and move water against a frictional resistance to the required level. Some species apparently never develop root pressure. In deciduous trees of temperate habitats, root pressures are measurable in the spring before the leaves open; however, once the leaves expand and rapid transpiration starts, positive root pressures can no longer be detected. The *quantity* of water that can be moved by root pressure seems to be very limited. Wheat seedlings, which can transpire 2.6–3.0 ml water per hour, can exude by root pressure only

0.5 ml per hour. In many cases maximum bleeding rates are only 1–2% of the rate of water loss occurring by transpiration. Root pressure persists only whilst the water yielding capacity of the soil is high. In beans, root pressure uptake was stopped when the $\psi_\pi$ of the solution external to the isolated roots had a value of $-0.19$ MPa whereas intact plants could extract water from a solution of $-1.47$ MPa $\psi_\pi$. Moreover, during periods of rapid water movement, the evidence indicates that the xylem sap is not under positive pressure but under tension (negative pressure). If, when this is the case, a cut is made into the xylem under a dye solution, the dye moves instantaneously into the xylem and travels very rapidly both upwards and downwards from the cut. Lateral shinkage of tree trunks detected by sensitive calipers (the dendrograph), and shrinkage of individual xylem vessels observed microscopically, have been reported to occur as water movement begins or is speeded up by enhanced transpiration. These observations accord with the development of longitudinal tensions in the conducting vessels. There is a diurnal pattern of water movement in trees, water movement beginning in the morning (as transpiration begins) and starting first in the upper branches. In conformity with this there occurs a downward moving wave of contraction in diameter of the branches and trunk. It has been demonstrated that transpiring twigs can pull water against an artificial resistance more effectively than can a vacuum pump; for instance leafy twigs can raise a column of mercury to heights greater than supported by the atmosphere's pressure (76 cm). On the basis of such evidence it is generally accepted that water movement in plants, and particularly tall woody species, occurs primarily as a result of the water being pulled up to replace that lost in transpiration. We are therefore faced with the need to explain how water can be pulled up under tension to the height of the topmost leaves of a tall tree.

### The transpiration-cohesion theory

A theory of the ascent of sap based on the cohesive properties of water was advanced independently by Dixon and Joly (1894) and Askenasy (1895), although earlier writers had postulated on less convincing arguments that water movement under tension was involved. The transpiration-cohesion theory visualizes that the cells in contact with the external atmosphere or with the internal air space system of the plant are constantly losing water by evaporation. Their $\psi$ decreases and as a result water moves to them by osmosis from the more deeply-seated cells with which they are in contact. These in turn replenish their loss from still deeper-lying cells, till water is extracted from the xylem by cells in contact with the conducting vessels, especially at the veinlet endings in the leaf. This removal of water from the xylem can be made good only by the further uptake of water from the soil by the roots. There is thus considered to exist an uninterrupted column of water moving under tension from the soil to the leaves. The energy for the movement comes from solar heat, which causes water to evaporate continually by supplying the latent heat of evaporation.

In the classical account given above osmosis has been postulated to cause movement of water between the fine vein endings and the intercellular spaces of the mesophyll. However here, as in the case of the radial movement of water across the root from soil to xylem, it is disputable whether the water leaving the xylem passes mainly through cell walls, cytoplasm or vacuoles. Osmosis operates only where passage across cellular membranes is involved. According to more recent estimates, movement would be too slow if water had to diffuse from cell to cell; hence it is held that a large proportion of water movement passes through cell walls, which can deal with a movement of up to 50 times that which could flow under a similar tension through the protoplasts. Altogether, a view is emerging which regards the water in the plant as a continuous system but divided into a number of compartments: vacuolar, cytoplasmic, xylem sap, cell wall and extracellular water. Exchange of water between all these compartments and the environment is possible, but the rapidity of such movement is limited by protoplasmic membranes and cell walls. The extracellular water is the most accessible to environmental influences, the vacuolar water the least accessible. When transpiration induces water movement in the plant, water will move by preference along routes where it does not encounter the resistance of cell membranes, that is on the outside of cells if extracellular water is present, otherwise within cells walls, and in the xylem. Only if the transpiration rate is high, and there is a shortage of water in the plant, will water move from the less accessible compartments into the flow.

The maintenance of water columns under tension depends, according to the cohesion theory, on the cohesive attraction of water molecules. Theoretically, this cohesion is very great; pure water, under appropriate conditions, should withstand at least 130 MPa tension. Dixon and Joly claimed to have demonstrated the development of tensions of 20–30 MPa in sealed glass capillaries before the water columns in them cavitated (collapsed with vaporization). Later workers have queried the calculations of Dixon and Joly and concluded that the tensions developed in such capillaries are much lower. A more recent calculation taking into account the diameter of the xylem vessels, the nature of their walls, and the presence of solutes and dissolved gases in the xylem sap, puts the tensile strength of xylem sap *in situ* at only about 3 MPa. A tension of 0.1 MPa will raise water against an average resistance to flow in the xylem to a height of about 5 metres; thus a tension of 3 MPa would suffice to raise water to 150 metres and even the lower estimate given above for the tensile strength of xylem sap *in situ* is sufficient to allow water movement into the crowns of the tallest trees by a transpiration pull. The maintenance of water columns under tension also requires a high tensile strength of the xylem cell walls to avoid collapse.

The direct measurement of tensions in the xylem *in situ* in intact plants has so far proved physically impossible, since the water column under tension is immediately disrupted by the insertion of any instrument. Xylem tensions have, however, been estimated on cut branches. On cutting, the sap retracts from the cut as the tension is released; the branch is then enclosed in a pressure chamber with the cut end protruding, and gas pressure is applied until sap re-appears at the cut end. The pressure necessary to achieve this is taken to be numerically equal to the original xylem tension. With this technique, tensions of 0.5 to 8 MPa have been measured, the highest values occurring in halophytes and desert plants. The tension is greater in higher branches of trees than lower ones. Another indirect method is to measure the $\psi$s of leaf cells or of living stem cells in contact with the xylem. In wilted plants of tomato (*Lycopersicon esculentum*), privet (*Ligustrum lucidum*) and cotton (*Gossypium barbadense*), leaf $\psi$s have been found to reach $-4.1$, $-7.0$ and $-7.7$ MPa respectively, while values of up to $-14.3$ MPa have been obtained for wilting desert plants. The xylem is then assumed (without direct proof) to be under a tension of the same magnitude, since the tension must be balanced by the leaf cells if water is to move into the leaves. In the ivy (*Hedera helix*), leaf $\psi$s decrease progressively up the plant, in accordance with the expected tension gradient. Such numerical values for the *in situ* tension thus support the cohesion theory.

## Criticisms of the transpiration-cohesion theory

Reluctance in accepting the transpiration-cohesion theory has arisen because the great stability of the water-conducting system under natural conditions seems at variance with the metastable state of water columns under tension. Any break in the water column, or any introduction of an air bubble into a vessel, ought to stop the flow of sap through it; indeed, the breaking of a single twig might be expected to let air spread throughout the xylem of the whole plant. Yet clearly this does not happen in nature, nor do experimentally produced breaks in the column of xylem sap necessarily disrupt water movement. Twigs which are cut so as to allow air into the vessels will resume water uptake when the severed end is placed in water. Cuts can be made in woody stems, overlapping from opposite sides, one above the other, and impermeable barriers inserted in the cuts; sap movement in the xylem still continues, transpiration rate is unchanged, nor do the leaves above the cut show any sign of water stress – provided the cuts are a certain minimum distance apart (see below). In the winter xylem sap may freeze; the dissolved air is then trapped as bubbles in the ice and should remain as bubbles when the sap thaws – yet sap flow is resumed in the spring. Moreover water columns under tension in capillary tubes cavitate on the slightest mechanical shaking or tapping and the movement of plants by wind might therefore be expected to cause cavitation in the xylem. The hydrophilic nature of the vessel walls has been invoked as giving stability to the water columns but experiments have indicated that xylem sap in grapevine stem segments is much less stable towards tension than water in glass capillaries of comparable diameters. The xylem segments cavitated on centrifugation at tensions of 0.1–0.15 MPa, rarely enduring 0.3 MPa, whereas water in the capillaries withstood 1.0–1.7 MPa. A technique has been developed by which cavitation can be demonstrated in intact plants or organs. The minute and quite inaudible 'noise' produced by the cavitation of a single xylem vessel is amplified by means of suitable electronic equipment to an audible click. The clicks can then be counted and/or tape-recorded to give a record of the course of cavitation. With this method it has been shown that cavitation occurs in intact plants when water is withheld from the roots or when water uptake is inhibited by chilling the roots (Fig. 4.4). During rapid drying out one can count several hundred clicks per minute in a wilting leaf; the cavitation starts well in advance of any visible wilting. When the water supply is restored the cavitation ceases and the plants recover from wilting, showing that water uptake is resumed in spite of the preceding cavitation. Cavitation has been observed in the field under natural conditions. All these observations are difficult to reconcile with the cohesion theory. However, no more satisfactory theory has been forthcoming and workers in this field have therefore considered how far such observations might still be compatible with a cohesion theory. It is quite clear that air *can* be introduced into the xylem in nature (by freezing, by cavitation due to water stress, by accidental breaks), or experimentally, without permanently or even temporarily stopping the overall flow. How is this to be reconciled with the need of continuous columns for the movement?

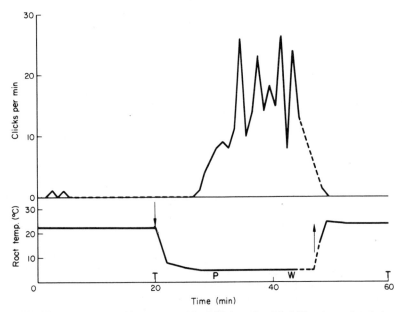

**Fig. 4.4** Cavitation caused in a water-cultured *Ricinus* plant by chilling the roots to induce a water shortage. After 20 min at 22°C, during which almost no cavitation occurred, the roots were cooled to 5°C and clicking, indicating cavitation, began after a short delay. On raising of the root temperature to 25°C, the clicking quickly ceased. The state of the plant is indicated on the time axis as T=turgid; P=partially wilted and W=wilted. (From Milburn, 1973, *Planta*, **112**, 333–42.)

To account for the resumption of sap flow in the spring after freezing the suggestion has been made that air bubbles introduced into the xylem by freezing and thawing might be gradually forced into solution again by root pressures, which are highest in early spring, though the root pressures actually measured in trees have in general been too low to make this possible. Another proposition is that the old vessels in which air bubbles have developed do not resume conduction in the spring, their function being totally taken over by the newly differentiating vessels each season. Some studies of dye transport support this view. Scholander (1958) has made the interesting suggestion that the xylem should be regarded as 'a flooded continuous micropore system scattered with elongated macrocavities (vessels)'. The 'micropore system' is represented by the xylem cell walls from which water cannot be centrifuged out at 1700 **g** but can be squeezed out under pressure. As long as the whole system is water-filled, water movement is regarded as proceeding largely through the vessels since these offer the paths of least resistance. If however an airlock develops in a vessel, the water flow can proceed round it in the wall micropores. In support of this concept Scholander cites experiments on lianas, where the introduction of air into the xylem did not diminish the *rate* of water transport, but did increase the *resistance* to flow; this was presumably compensated by the development of correspondingly lower $\psi$s in the leaves. Moreover, the airlocks in these lianas did not spread along the whole length of the vessels, suggesting the operation of valves, confining the air to segments about 60 cm to over 100 cm long. These valves can hardly be anything else but vessel cross-walls. These are perforated, but the pores, being very small, may stop the receding meniscus of a broken water column and prevent the next vessel unit from becoming air-filled. The mechanism is clearly of great importance as demonstrated by the experiments on overlapping cuts mentioned above. To allow water flow in the presence of such cuts, these must be separated by a minimum distance which depends on the vessel length of the species: the distance must be long enough to leave some intact vessels between the cuts. Apparently only the vessels actually cut through become air-filled, the flow of sap passes sideways around these and continues in the intact vessels at the side of each cut (Fig. 4.5). The cohesive system as a whole would not be seriously reduced in cross-sectional area by the presence of air in occasional segments. While in trees some vessels extend to over 10 m, most are much shorter; e.g. in *Ilex verticillata*, with a maximum vessel length of 1.3 m, 99.5% of the vessels were found to be under 5 cm long. Further the total amount of xylem in a plant is generally much in excess over that needed to satisfy the water needs of the plant, as shown when large amounts of wood are removed. Cuts in stems severing up to 90% of the cross-section can be made without any detectable change in leaf water potential or transpiration; only the velocity of sap flow near the cut is increased.

When water movement is controlled by transpiration pull, roots act as passive absorbing surfaces. The roots offer a resistance to water movement as evidenced by the observation that the transpiration stream moves faster when the roots are cut off under water, or killed by heat. An active role was at one time postulated for the roots, on the grounds that transpiration is slowed down by chilling the roots, by depriving them of oxygen and by

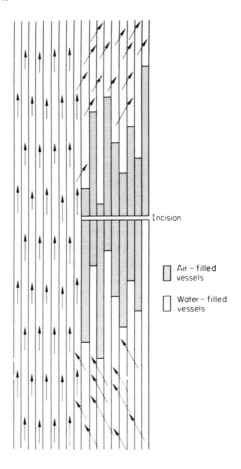

**Fig. 4.5** Suggested pattern of sap- and air-filled vessels and the water flow pathways (arrows) after a stem incision, as seen in longitudinal section. Cross walls of uncut vessels are not shown. To allow sap flow in the presence of two overlapping cuts, the cuts must be spaced widely enough to leave some intact vessels between the cuts. (From Mackay and Weatherley, 1973, *Journal of Experimental Botany, 24*, 15–28. By permission of the Clarendon Press, Oxford.)

treating them with carbon dioxide or respiratory inhibitors. However these findings can be explained on the basis that all these treatments decrease the permeability of roots to water and a lowering of temperature additionally increases the viscosity of water. As the suction tension applied to the roots increases, their resistance decreases, so that regions further back from the tip are utilized under strong tensions. It has also been suggested that if rapid withdrawal of water from the roots causes flaccidity of root cells then spaces between the cells are opened up, facilitating water entry.

It therefore seems that under conditions when transpiration is rapid, the transpiration pull developed is responsible for water movement. Positive root pressures if developed will be of importance only when transpiration is slow, at night, and in spring before leaf expansion. A movement of sap in trees and shrubs in spring can also occur by osmotically motivated flow within the stems. In a number of tree species water moves from the inner regions of the xylem into the bark and into buds before any root pressure develops. Such movement occurs also in rootless shoots: the inner wood then becomes quite dry; normally it would have received water by a 'slow movement, from the roots. Associated with this outward flow of sap there is a conversion of starch to sugar in the bark. A similar conversion in the xylem parenchyma and medullary rays of woody plants in spring may promote water uptake.

Since submerged aquatic plants are able to absorb water over the entire plant surface, there would seem to be no need in such plants for water conduction. Notwithstanding this there is a slow movement of water in their xylem probably mediated by a root pressure. The movement could function to transport ions, or it may be an evolutionary relic with no physiological significance.

## Factors controlling the rate of water uptake and movement

The rates of water absorption, water loss, and consequently of water movement through the plant, are determined by a combination of plant and environmental factors. The environmental factors can be classified as soil and atmospheric: the amount and availability of soil water, soil temperature and soil aeration on the one hand; atmospheric temperature, humidity, wind and light, on the other. The plant factors are the number and degree of opening of stomata; the areas of the evaporating and absorbing surfaces and their ratio, and the water permeability of the absorbing and evaporating surface.

### Soil conditions

The soil is a complex system of particles and pores of heterogeneous size. When a soil is saturated with water, all the pores are filled, but a well-drained soil does not remain water-saturated for long. Water drains away quickly under gravity from the larger spaces, but some is retained by capillary and surface tension forces within the smaller pores, and as adsorbed surface films around the soil particles. When a soil contains as much water as it can hold against gravity, it is said to be at *field capacity*. The amount of water present at this stage is a characteristic of the type of soil. Soils with fine particles have many small pores, and can hold more water than coarse soils. A clay soil at field capacity may hold 55% of water on a dry weight basis (i.e. 55 g water per 100 g dry soil), while a coarse sand may hold only 17% (Fig. 4.6). Once the water content has fallen to field capacity there is practically no more movement of liquid water in a soil, though water evaporates and escapes to the atmosphere. Since soil 'water' is a solution of ions, the plant must therefore absorb water from the soil against surface and osmotic forces.

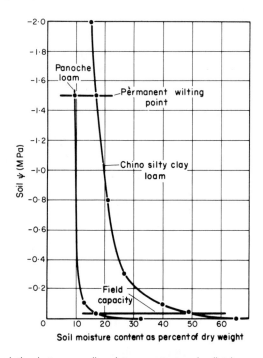

**Fig. 4.6**   The relation between soil moisture content and soil $\psi$ in a sandy soil (Panoche loam) and a clay soil (Chino silty clay loam). (Adapted from Kramer, 1949, *Plant and Soil Water Relationships*. Used by permission of McGraw-Hill Book Co., New York and Maidenhead.)

The $\psi$ of a soil at field capacity is very high, almost zero, and uptake by plants can proceed freely. As water is removed from the soil, the soil $\psi$ decreases progressively (Fig. 4.6). At first, correspondingly lower $\psi$s develop in the absorbing cells of the root. However as the water content of the soil falls, a stage is reached when the $\psi$ of the soil becomes so low (i.e. soil moisture stress so high) that the plant can no longer obtain enough water to compensate for transpiration losses, and it wilts. The soil moisture has reached the ***permanent wilting point*** (PWP), defined as the stage when the plant will not recover from wilting (even in a saturated atmosphere) unless water is added to the soil. Numerically, the PWP is expressed as the percentage of water left in the soil. Temporary wilting may occur before that, the plant wilting by day when transpiration is high, but recovering at night, when water uptake catches up with water loss. Once the PWP is reached, removals of very small amounts of water cause very large decreases in the soil $\psi$.

When the concept of the PWP was initially introduced, it was considered to be a characteristic of the soil rather than of the plant. Experiments seemed to indicate that different species growing in the same soil reached permanent wilting simultaneously; the soil $\psi$ at the PWP in different soils appeared the same, about –1.5 MPa. The water contents at which this is reached of course vary with the soil type, just as do the water contents at field capacity (see Fig. 4.6). More recent work indicates that this view is an oversimplification and the results in Table 4.4 show that each of the three species investigated reduces soil moisture to a different level at the PWP. This suggests that, for each plant, the PWP reached is when the turgor pressure of its leaf cells falls to zero. The lower leaves may reach permanent wilting before the upper. Table 4.4 shows moreover that, contrary to earlier views, there is still a measurable amount of water uptake by a plant after the PWP is passed. A progressive decrease in the $\psi$ of the plants can be seen in the above results; the lower value of leaf $\psi$ as compared with leaf sap osmotic potential is believed to be the result of wall tensions developed by cohesion between walls and protoplasts.

Most of the water uptake of a plant occurs when the soil moisture content is between field capacity and permanent wilting point. As the soil dries out, the forces opposing plant water uptake increase progressively and growth rate can be impaired by high water stress even before wilting is reached. This reduction of growth rate can aggravate the effect of water shortage, for the

**Table 4.4** Osmotic potential ($\psi_\pi$) of expressed leaf sap, leaf $\psi$, soil $\psi$ (all expressed in MPa) and soil water content as percentage of soil dry weight as three plant species were allowed to dry out gradually. The 'first PWP' is the stage when the first leaves wilted permanently; at 'ultimate wilting' all leaves had wilted. The experiment was concluded when the plants were dying. (Adapted from Slatyer, R. O., 1957, *Australian Journal of Biological Sciences*, **10**, 320–36.)

| Species | At first PWP | | | | At ultimate wilting | | At conclusion of experiment | | | |
|---|---|---|---|---|---|---|---|---|---|---|
| | Leaf | | Soil | | Leaf | Soil | Leaf | | Soil | |
| | $\psi_\pi$ Sap | $\psi$ | $\psi$ | Water content % | $\psi$ | $\psi$ | $\psi_\pi$ Sap | $\psi$ | $\psi$ | Water content % |
| *Lycopersicon esculentum* | − 1.8 | − 1.9 | − 2.0 | 11.8 | − 3.2 | − 3.2 | − 3.5 | − 4.1 | − 4.5 | 9.8 |
| *Ligustrum lucidum* | − 4.7 | − 4.5 | − 4.8 | 9.7 | − 5.0 | − 6.0 | − 6.6 | − 7.0 | − 11.0 | 6.9 |
| *Gossypium barbadense* | − 3.8 | − 4.3 | − 3.8 | 10.2 | − 4.8 | − 5.0 | − 6.7 | − 7.7 | − 10.7 | 7.0 |

slowing down of root growth decreases the rate at which new areas of soil are tapped by the roots.

Soil aeration affects water uptake. An adequate oxygen supply is necessary for root growth and a lack of oxygen and a high concentration of carbon dioxide are reported to decrease the permeability of roots to water. Temperature similarly affects root growth and root permeability, both being decreased at low temperatures. Since the viscosity of water increases as the temperature is lowered, low soil temperature may, for this reason, under certain conditions, considerably reduce the water uptake via the root resulting from transpiration pull. Active water uptake leading to positive root pressures develops only in well-aerated soils of favourable moisture content and temperature.

### Atmospheric conditions

The daily course of water absorption closely follows, with a time lag, the course of transpiration (Fig. 4.7). Thus the atmospheric factors which determine the rate of transpiration also determine the rate of water uptake. The rate of transpiration, i.e. the outward diffusion of water vapour, is subject to the same laws as the inward diffusion of $CO_2$. Transpiration is directly proportional to the water potential gradient between the leaf and the air, $\Delta\psi$, and inversely proportional to $R_a$ (boundary layer resistance) and

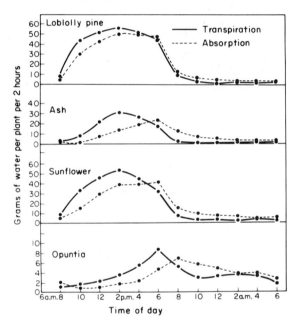

**Fig. 4.7**  The relation between the rate of transpiration and the rate of absorption of water by plants. (From Kramer, 1937, *American Journal of Botany,* **24**, 10–15.)

$R_s$ (stomatal resistance). Denoting the rate of transpiration by T, we have:

$$T = \frac{\Delta\psi}{R_a + R_s} \qquad (4.9)$$

Or, since atmospheric water status is often expressed in terms of vapour pressure, e, which is proportional to $\psi$ (*equation 4.2*),

$$T = \frac{\Delta e}{R_a + R_s} \qquad (4.9')$$

There is no term in this equation to correspond with the mesophyll resistance which is involved in $CO_2$ diffusion, for water vapour does not pass through cells on its way to the outside, nor is its movement limited by chemical reactions.

Transpiration rate increases with increasing temperature, a rise in temperature resulting in a steeper concentration gradient of water vapour out of the leaf, i.e. $\Delta e$ increases. The air spaces within the leaf are normally at near saturation vapour pressure, $c.$ 100% R.H. The absolute concentration of water vapour at a given R.H. increases with increasing temperature, i.e. air at 100% R.H. at 20°C will contain more water vapour than air at 100% R.H. at 10°C. Thus a rise in leaf temperature increases the vapour pressure in the leaf without a corresponding rise in the external air. The gradient is approximately doubled for a 10°C rise in temperature. Wind stimulates transpiration, decreasing $R_a$ as it sweeps away the water vapour accumulating at the leaf surface. By causing leaf bending it may cause mass flow of air into and out of the leaf thereby enhancing water loss. Light has no direct effect on water loss, but through the control it exerts over stomatal opening (Chapter 3) it has a profound effect on the water relations of plants. Atmospheric conditions thus tend to promote higher transpiration rates by day than by night, and in summer than in winter.

*Plant factors*

The plant factors which control the rate of passage of water through the plant are to a great extent those that control the rate of transpiration. However, the water uptake capacity of the roots can become limiting, as shown in Fig. 4.8; when half the leaves are removed from a plant, the

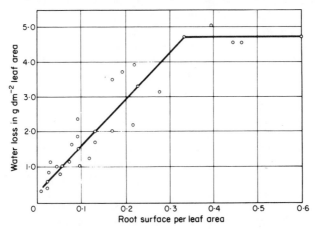

**Fig. 4.8** Effect of variation in ratio of root surface to leaf area on rate of transpiration of rooted lemon (*Citrus limonia*) cuttings. Transpiration was reduced in proportion to reduction in root surface below a root surface to leaf area ratio of about 0·35, but independent of root surface above this ratio. (From Kramer, 1956, *Encyclopedia of Plant Physiology*, ed. Ruhland, **3**, 188–214. Springer–Verlag, Berlin.)

remainder transpire more rapidly, being now able to draw on the whole root system for water. Plants also differ in the $\psi$s they can develop (Table 4.4).

If comparison is made between plants of different species, or different individuals of the same species, the following shoot characters are found to favour rapid transpiration: an outer surface with a thin cuticle and devoid of hairs; a high stomatal frequency (number of stomata per unit area), and a large surface area. The area of cells exposed to internal air spaces may be determinative rather than the external surface area. Table 4.5 shows that the *Citrus* species with the higher ratio of internal to external leaf surface area has the higher rate of transpiration as calculated per unit of external surface, although per unit of internal surface the transpiration rates of the two species are almost equal.

**Table 4.5** The effect of internal leaf surface area on the rate of transpiration, measured over a period of two months. (After Turrell, F. M., 1936, *American Journal of Botany*, **23**, 255–64.)

| Species | Transpiration (g dm$^{-2}$ of external surface) | Ratio of internal to external surface | Transpiration (g dm$^{-2}$ of internal surface) |
| --- | --- | --- | --- |
| *Citrus limonia* | 45.57 | 22.2 | 2.05 |
| *Citrus grandis* | 37.91 | 17.2 | 2.15 |

If on the other hand the water relations of an individual plant under different environmental conditions are considered, the degree of opening of the stomata is often the most important single factor controlling the rate of transpiration. By far the greater proportion of the water lost comes from the leaf air spaces via the stomata, although the stomatal area may be only 1–2% of the total leaf surface. This can be seen by comparing the rates of transpiration with the stomata open and closed (Fig. 4.9). Where there is a mid-day closure of stomata this is always accompanied by a reduction of transpiration rate. The transpiration of leaves with stomata confined to the lower surface shows a much greater reduction of rate when the lower surface is vaselined than results from vaselining the upper surface.In plants with a thick cuticle, like *Laurus nobilis*, water loss through the epidermis itself, the 'cuticular transpiration', may be as low as 2% of total transpiration. When the cuticle is thinner, cuticular transpiration can constitute 5–45% of the total and the 'average' cuticular transpiration for mesophytes is 10–25% of the total. Closing of the stomata will therefore, according to species, reduce transpiration to 2–45% of that occurring when the stomata are open.

Transpiration is most rapid when stomata are fully open and slowest when they are fully closed. At intermediate apertures, other factors determine whether the degree of stomatal opening affects transpiration rate or not. If the atmospheric conditions do not favour rapid transpiration, the full transpiration rate may be reached when the stomata are only partly open. But when external factors favour rapid transpiration, the rate increases with increasing stomatal aperture right up to maximum opening (Fig. 4.10).

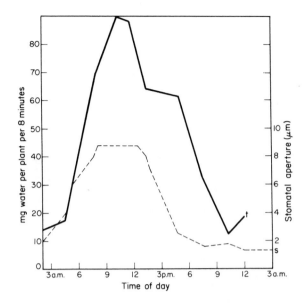

**Fig. 4.9** The relationship between the daily course of transpiration and stomatal aperture in *Verbena ciliata*. The rate of transpiration, t, follows closely change in stomatal width, s. (From Lloyd, 1908, *The Physiology of Stomata*, Carnegie Institution, Wash.)

This is very similar to the way in which stomatal aperture affects $CO_2$ uptake: compare Fig. 4.10 with Fig. 3.6. Being situated at the point in the soil $\rightarrow$ plant $\rightarrow$ air pathway where the drop in $\psi$ is greatest, the stomata can exert very effective control over water movement when, in moving air, boundary layer resistance is low. In still air, however, the boundary layer resistance is frequently more important than stomatal resistance. Further it may be noted that partial stomatal closure cuts down the rate of transpiration more than the rate of $CO_2$ diffusion: there being no mesophyll resistance controlling water vapour movement, the stomatal resistance assumes relatively more importance.

As already mentioned in Chapter 3, stomata are highly sensitive to water stress and react by (partial) closure at quite low levels of water deficit, protecting the plant from further water loss, and in wilted plants the stomata are usually shut. The early stages of wilting are, however, often accompanied by a widening of the stomatal aperture, the guard cells being pulled apart by the shrinking of surrounding epidermal cells, which lose water more rapidly. In extreme wilting, too, the protective mechanism may break down as shrinking epidermal cells again pull the pores open.

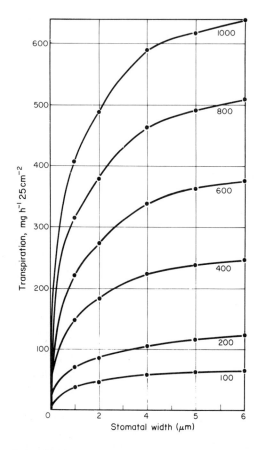

**Fig. 4.10** The relationship between transpiration rate and stomatal aperture in birch (*Betula pubescens*), under different conditions of evaporation. The number on each curve represents the rate of evaporation of water, in mg, from 25 cm$^2$ blotting paper surface under the same conditions. At low rates of evaporation, full transpiration rate is reached at a stomatal aperture of only 2 $\mu$m; at higher rates of evaporation full transpiration rate is scarcely reached with a stomatal aperture of 8 $\mu$m. (From Stålfelt, 1932, *Planta*, **17**, 22–85.)

FURTHER READING

DIXON, H. H. (1914). *Transpiration and the Ascent of Sap in Plants*. Macmillan, London.

FOGG, G. E. (ed.) (1965). The State and Movement of Water in Living Organisms. *Symposium of the Society for Experimental Biology*, **19**.

KRAMER, P. J. (1969). *Plant and Soil Water Relationships: a Modern Synthesis*. McGraw-Hill, New York.

MEIDNER, H. and MANSFIELD, T. A. (1968). *Physiology of Stomata*. McGraw-Hill, New York and Maidenhead.

MEIDNER, H. and MANSFIELD, T. A. (1976). *Water and Plants*. Blackie, Glasgow.
MEYLAN, B. A. and BUTTERFIELD, B. G. (1972). *Three-dimensional Structure of Wood*. Chapman and Hall, London.
MILBURN, J. A. (1979). *Water Flow in Plants*. Longman, London.
STOCKING, C. R. (1953). Histology and development of the root. *Encyclopedia of Plant Physiology*, ed. Ruhland, W., **3**, 173–87. Springer–Verlag, Berlin.
STREET, H. E., ELLIOTT, M. C.and FOWLER, M. W. (1975). The Physiology of Roots. In *Interactions between Soil Microorganisms and Plants*, ed. Dommergues, Y. and Krupa, S. Elsevier Sci. Pub. Co., Amsterdam.
SUTCLIFFE, J. (1979). *Plants and Water*, 2nd edition. Studies in Biology No. 14, Edward Arnold, London.


SELECTED REFERENCES

GREENIDGE, K. N. H. (1958). Rates and patterns of moisture movement in trees. In *The Physiology of Forest Trees*, ed. Thimann, K. V., 19–41. Ronald Press, New York.
MACKAY, J. F. G. and WEATHERLEY, P. E. (1973). The effects of transverse cuts through the stems of transpiring woody plants on water transport and stress in the leaves. *Journal of Experimental Botany*, **24**, 15–28.
MILBURN, J. A. and McLAUGHLIN, M. E. (1974). Studies of cavitation in isolated vascular bundles and whole leaves of *Plantago major* L. *New Phytologist*, **73**, 861–71.
PASSIOURA, J. B. (1980). The meaning of matric potential. *Journal of Experimental Botany*, **31**, 1161–9.
RICHTER, H. (1978). A diagram for the description of water relations in plant cells and organs. *Journal of Experimental Botany*, **29**, 1197–1203.
SCHOLANDER, P. F. (1958). The rise of sap in lianas. In *The Physiology of Forest Trees*, ed. Thimann, K. V., 3–17. Ronald Press, New York.
THUT, H. F. (1932). Demonstrating the lifting power of transpiration. *American Journal of Botany*, **19**, 358–64.
ZIMMERMANN, M. H. (1964). Effect of low temperature on ascent of sap in trees. *Plant Physiology, Lancaster*, **39**, 568–72.
ZIMMERMANN, M. H. and JEJE, A. A. (1981) . Vessel-length distribution in stems of some American woody plants. *Canadian Journal of Botany*, **59**, 1882–92.

# 5

# Mineral Nutrition

## Essential mineral elements

Of the naturally occurring 92 elements of the periodic table, only a limited number is essential to plants. An element is classed as essential to a plant if the plant is unable to complete its normal life cycle without it, and no other element can substitute for it. The effect of the element on the plant should also be direct, i.e. it should not act by promoting the uptake of another essential element, or by retarding the absorption of a toxic one. To test for the essentiality of an element, plants must be grown in an environment totally free from that element.

The list of elements essential for plants has been built up over many years and is perhaps even now not complete. By the mid-eighteenth century it had been established that plants needed minerals derived from the soil and natural waters. In the nineteenth century, the development of sand culture and water culture techniques led to the recognition of the quantitatively most important elements; e.g. Sachs (1860), by means of water culture experiments, showed the essentiality of nitrogen, sulphur, phosphorus, calcium, potassium, magnesium and iron, for a number of plant species. As purer chemicals have become available, more elements have been added to the list, these elements having been inadvertently supplied as impurities in earlier experiments. Some elements are required in such minute amounts that they are very difficult to eliminate from culture media to below the level required by plants. Even distilled water, glass of containers, gaseous aerial pollution or atmospheric dust, may provide enough of certain elements to support plant life. Such problems must be overcome by using specially purified water, spectroscopically pure chemicals, and filtering of the air supply. The essentiality of chlorine was not established until 1954, chlorine tending to be ubiquitous in air and water. There is the further problem of the store of minerals already present in the propagule used to initiate the culture; the internal supply may be fully sufficient for the early stages of growth. The current list of elements essential for flowering plants stands at 17–20 (Table 5.1). The precise number depends on the species, some elements apparently being essential for certain plants only; it also depends on how strictly the criteria for essentiality are applied (see functions especially of silicon, p. 83, cobalt, p. 84 and molybdenum, p. 84). Some

**Table 5.1** The elements required by flowering plants. Elements shown in parenthesis have been found to be essential for a limited number of plant species only. Except for C, H, O, the elements are obtained as mineral ions.

| Essential elements | | Beneficial elements |
|---|---|---|
| Macronutrients | Micronutrients | |
| C | Fe | Na |
| H | Mg | Cl |
| O | Mn | Si |
| N | Cu | Se |
| S | Zn | Al |
| P | Mo | Rb |
| Ca | B | Sr |
| K | Cl | |
| (Si) | Co ? | |
| | (Na) | |
| | (Se) ? | |

elements, hitherto known to be required by only a few plants, may perhaps eventually be discovered to be more generally needed when other species are rigorously tested.

As indicated in Table 5.1, the essential elements are classified as *macronutrients* or *micronutrients*. The macronutrients are required in large amounts relative to the micronutrients; in culture solutions, macronutrients are supplied at $10^{-3} - 10^{-2}$ M, while micronutrient concentrations may be as low as $10^{-7}$ M. Most of the micronutrients become toxic at quite moderate concentrations, above, say, 0.1–1 mM; only sodium and chloride can be tolerated at fairly high levels. In addition to essential elements *beneficial* elements can also be recognized (Table 5.1). While not absolutely necessary for survival, beneficial elements do promote plant growth. Sodium and chlorine benefit many species; aluminium is beneficial to tea (*Camellia sinensis*), *Miscanthus sinensis*, *Polygonum sachalinense*, and a number of grasses. Selenium is required for species of *Astragalus* growing in selenium-rich soils, for *Lupinus albus* and for *Phleum pratense*; but it should perhaps be classed as beneficial, its essentiality not being absolutely proven. Rubidium and strontium probably owe their beneficial effect to an ability to replace some (but not all) of a plant's requirements for potassium and calcium respectively: rubidium enhances growth particularly in potassium-deficient media. Similarly, sodium can substitute to some extent for potassium and calcium.

Non-essential elements are also taken up by plants; any element present in the environment will be found in at least small amounts in plants. For plants grown in soil, large amounts of aluminium, silicon and sodium are frequently present as these are common elements in soils. Though inessential, such elements are far from being inert in the plant. They may fulfil some of the metabolic functions of essential elements and, being taken up as soluble ions, they influence the ionic balance and osmotic potential of the cells. Many non-essential elements are toxic in quite low amounts and their uptake is detrimental to plants and the animals which feed on them.

## Mineral nutrients in the soil

With the exception of carbon, hydrogen and oxygen which are derived from water and $CO_2$ and are not further considered here, plants acquire their elements as inorganic ions; for flowering plants, except free-floating aquatic species, the chief source of mineral ions is the soil. Soils are of many kinds and are always complex physical, chemical and biological systems. Pure silica sand can be regarded as biologically inert and is used in sand culture to give physical support to the growing plant and, by allowing free drainage, to ensure very simply an adequate aeration of the root system. Once a sand is enriched by natural additions of inorganic and organic matter it begins to take on the properties of a soil. Natural soils contain organic matter and clays and it is these which confer upon the soil its capacity to hold ions by adsorption, to show *ion exchange capacity*. These soil constituents usually carry a net negative charge. Positively charged cations ($Ca^{2+}$, $Mg^{2+}$, $H^+$, $K^+$, $Na^+$ and $NH_4^+$) are adsorbed at negatively charged groups in the clay and organic particles and adsorbed cations account for most of the exchangeable ions of the soil. These exchangeable ions are not held rigidly at the surface of the soil particles but loosely in an electric field of force arising at the surfaces by the ionization of reactive groups and are surrounded by water molecules. They are freely exchangeable for other ions and can be extracted by washing the soil in a strong solution of a salt such as $NH_4NO_3$ when the $NH_4^+$ ions will displace the other cations from the surface of the soil particles. By contrast the adsorption of anions is relatively unimportant and effectively the anions $Cl^-$, $NO_3^-$ and $SO_4^{2-}$ are entirely free in the soil solution. Hence these ions are very readily leached from soil and availability of $NO_3^-$ and $SO_4^{2-}$ depends particularly upon the rate of decomposition of soil organic matter. The situation with regard to phosphate ions is more complex. They are to some extent adsorbed at soil particle surfaces (note the high charge density of the trivalent phosphate ion), they can replace hydroxyl and silicate ions in clays and can react with iron ($Fe^{3+}$) and aluminium ($Al^{3+}$) ions. Exchangeable phosphate is however only of significance in acid soils of low $Ca^{2+}$ content, where the very low soluble phosphate level is maintained by continuing release from almost insoluble, more complex, phosphates.

The mineral elements which pass from the soil into the plant can be regarded as derived directly from the displaceable liquid phase of the soil, the *soil solution*. The concentration of this soil solution rises as the water content of the soil falls but except under dry conditions is always very dilute. In several instances it has been shown that a solution corresponding in ionic composition to the soil solution will support good growth of crop plants provided it is frequently renewed or applied as a flowing solution so that it does not become depleted. Thus for instance the content of phosphate in the soil solution is always low: 1 ppm or less. Nevertheless plants can meet their phosphate requirement in water culture solutions containing as little as 0.1 ppm. However, with such low soluble phosphate levels it is essential that phosphate ions are continuously released from the solid phase of the soil as they are removed by plant roots from the soil solution. It has been calculated that the phosphate of the soil solution may have to be renewed 10 or more times a day to meet the phosphorus demand of a growing crop. Because nitrate is almost entirely present in the soil solution and is rapidly absorbed by plants its rate of renewal is often quite critical. This renewal involves release of the cation $NH_4^+$ from the organic fraction of the soil and its rapid conversion to nitrate by nitrifying bacteria.

Soils as media for plant nutrition vary according to: (1) the composition of the soil solution; (2) the capacity of the solid phase to renew the solution by ion exchange (release of essential cations in exchange for $H^+$ or of essential anions for $OH^-$ or $HCO_3^-$), by chemical decomposition or by microbial action; (3) the degree of aeration of the soil (availability of $O_2$ and rate of release of respiratory $CO_2$); (4) the physical resistance of the soil to root penetration. In consequence within any climatic region each distinctive soil type supports a characteristic flora. These soil (edaphic) factors are thus of importance in plant distribution and in the development of plant communities. The problems raised here are therefore usually studied in more detail in ecology.

## The uptake of minerals by the root

### Cation exchange

Uptake of ions by plant roots has been shown to consist of two phases, adsorption and absorption (accumulation). The adsorption phase is a physico-chemical phenomenon, is non-metabolic and is predominantly concerned with cations. It is an exchange process, $H^+$ generated by respiration being released into the soil solution in exchange for metal cations. The cation exchange properties of roots appear to depend upon cell wall constituents, particularly the uronic acids of the cell wall. In general, cation exchange values are higher for the roots of dicotyledons than for those of monocotyledons. The movement of cations from soil colloids to the root will depend upon the relative binding capacity of the soil colloids and the cell wall colloids and the rate at which ions are moved from the outer cell walls into the interior of the rest. The extent to which a given cation (say $Ca^{2+}$) can be exchanged depends upon the other cations adsorbed. If these other cations (complementary ions) are strongly bound, the $Ca^{2+}$ will be more completely released than if they are weakly bound.

The most generally accepted concept of the path of ions from soil to root postulates an intervening liquid phase represented by the soil solution. In 1939, two workers, Jenny and Overstreet, put forward the theory that when a root surface makes contact with a soil particle interchange of ions take place by contact-exchange directly between the two solid surfaces. However, the significance of such contact-exchange remains very doubtful and many results can be quoted in support of the view that the ionic environment of the root is best represented by the composition of the soil solution.

## Ion accumulation

The uptake of ions is normally a process of salt accumulation, the movement taking place from a lower concentration in the external medium to a higher salt concentration within the tissues. Such accumulation can proceed from extremely low external concentrations as evidenced, e.g. by the uptake of phosphate as discussed above. In contrast to water uptake, which proceeds along a gradient of decreasing water potential, along a free energy gradient, ion uptake can proceed *against* the free energy gradient. Since the free energy of ions in solution depends not only on ion concentrations but on electric charge, ions may be said to move into plant cells against the *electrochemical potential gradient*. Salt accumulation is dependent on respiratory energy and is described as an active, or metabolic, process, as opposed to passive diffusion along an electrochemical potential gradient. Further evidence for metabolic control is provided by the selectivity of the process: the relative concentrations of different ions inside the cells differ, often strikingly, from their relative concentrations in the external medium (Fig. 5.1). The uptake of a particular ion is often markedly affected by the

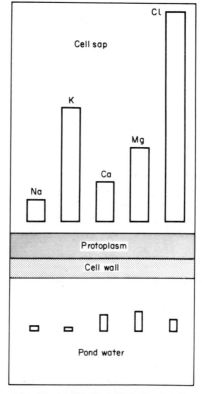

**Fig. 5.1** Relative concentrations of different ions in pond water and in the vacuolar sap of the cells of the alga *Nitella* growing in the pond. (From Hoagland, 1944, *Inorganic Plant Nutrition.* Chronica Botanica, Waltham, Mass.).

presence of other ions in the soil or culture solution; various groups of ions can be distinguished, the members of which compete with one another in the uptake process (e.g. $Ca^{2+}$ and $Mg^{2+}$), but not with members of other groups. Cellular membranes are very effective diffusion barriers to ions and active transfer of ions across cellular membranes is mediated by membrane-located chemical 'ion pumps', but their possible mechanism of action is not considered further here; it is an important aspect of cellular physiology and biochemistry.

### The movement of ions within and across the root

Ions move from the root to the shoot in the xylem (Chapter 6). The overall path of ions across the root is the same as taken by water (Fig. 4.3), i.e.:

epidermis → cortex → endodermis → stele→ xylem vessels

Some ions are of course retained by the root cells, and used in metabolism or stored in vacuoles.

The ions present in the root are divided among a number of compartments (*cf.* water, p. 60). A proportion of ions is washed out by water within minutes; some cations not easily removable by water are rapidly exchanged for other cations; the remainder is retained and only very slowly released. The ions quickly released into water are described as occupying the 'water free space' (WFS), which seems to correspond in volume to the cell walls of the tissues. The cations exchangeable for externally applied cations appear to be associated with fixed negative charges on cell wall polysaccharides and at the protoplast surfaces; they are said to occupy the 'Donnan free space' (DFS) – named after Donnan who first described the ionic equilibrium established between two compartments containing soluble inorganic ions, when one of the compartments also holds immobile organic anions. The ions most strongly retained by the cells will have been transported in through cellular membranes.

The plasmalemma, the membrane at the surface of the protoplast, is generally regarded as the chief primary barrier to ion movement. There is, however, no universal agreement about this and some workers consider the vacuolar membrane, the tonoplast, (and organelle membranes) to act as the primary barriers, leaving the cytosol to be included in free space. Ion pumping has in fact been observed by plasmalemma, tonoplast, mito-chondrial and plastid membranes. Probably the plasmalemma is the primary barrier for ion movement, with tonoplast and organelle membranes forming further barriers, but for some ions the plasmalemma may be a more effective barrier than for others. It has been proposed, too, that WFS and DFS should not be equated with spatial tissue compartments, but that the ions should merely be divided according to their behaviour into water-leachable, exchangeable, and retained, components. The complexity of ion compart-mentation complicates investigations of the pathways of ion movement to the xylem.

The protoplasts of the living root cells are interconnected by fine protoplasmic threads, the plasmodesmata, of submicroscopic dimensions (diameter $c$. 50 nm), but very numerous: one meristematic cell may have $10^5$ plasmodesmata connecting it to its neighbours and the number per mm² of mature endodermal inner tangential wall has been estimated at $6 \times 10^5$. A protoplasmic continuum, the *symplast*, can be envisaged, stretching through the root from the epidermis through cortex, endodermis and pericycle to the stelar parenchyma (Fig. 5.2). The continuum of cell walls, intercellular spaces and lumina of xylem vessels (and other dead cells) constitutes the *apoplast*.

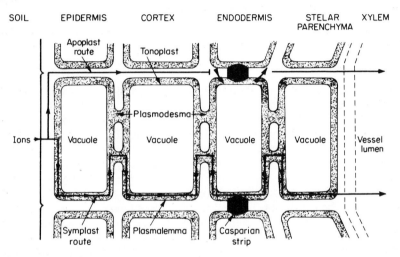

**Fig. 5.2** Possible pathways for ion movement across the root (highly diagrammatic). The cytoplasm, forming the symplastic continuum, is stippled. The *symplast route* is shown at the bottom; the ions are finally discharged into the apoplast, i.e. xylem vessel walls and lumen, through the plasmalemma of the xylem parenchyma cells. The *apoplast route* (top) is interrupted by the Casparian strips (black) which can be bypassed only by entry into the symplast (curved arrows). The passage of ions into vacuoles is not indicated; most of the vacuolar ions remain in the root cells.

If the plasmalemma is assumed to be the effective barrier to ion diffusion, a symplastic ion movement pathway can be visualized, with ions being actively pumped across the plasmalemma into the outer cells of the root and then passing from cell to cell via plasmodesmata, with protoplasmic streaming aiding movement within the boundaries of single cells. This would take ions as far as the xylem parenchyma. The concentration of ions in the living root cells is much higher than in the xylem fluid; this concentration gradient is maintained by active pumping into the outer root cells, the soil solution being of a much lower ion concentration than even the xylem sap. The final step in this salt transport pathway, the release of salts into the xylem vessels (i.e. from the symplast into the apoplast), could be an active secretory process; alternatively it may result from leakage, perhaps because the xylem parenchyma cells, deep within the root, suffer some shortage of oxygen and lack energy to maintain membrane semipermeability. The concept of an active secretion of salts into the xylem is supported by the observation that inhibitors of ATP synthesis such as DNP (2,4-dinitrophenol) inhibit the movement of salt into the xylem vessels. The view that salts leak into the conducting cells receives support from the fact that freshly isolated root stele separated from the cortex by shearing along the line of the endodermis will, in contrast to the cortex, rapidly leak ions into a bathing solution.

The cell walls of the root cells offer an alternative apoplastic pathway for ion movement; the ions would move along this partly by diffusion, but largely along with the flow of water moving into the transpiration stream. At the endodermis, however, this pathway is interrupted by the suberized Casparian strips (Fig. 5.2), which block the apoplast across the radial walls, and in older endodermal cells, suberin lamellae are laid over tangential walls as well. The endodermal cells are, however, joined by plasmodesmata to cortical cells on one side and cells of the pericycle on the other. For many years it has been postulated, from knowledge of the structure of endodermal cells and the impermeable nature of suberin, that the Casparian strips (and suberin lamellae of older cells) are ion barriers and that to cross the endodermis ions must enter the endodermal cytoplasm. Some direct evidence for this is now available from electron microscopy. Solutions of salts of lanthanum, lead and uranium have been used as tracers of ion movement, these elements casting electron-dense images because of their high atomic weights (La = 139, Pb = 207, U = 238). The ions also form insoluble deposits during fixation procedures and are not easily lost during the preparation of specimens. These three elements can all be seen to move through the cortex in cell walls, and into the endodermal cell walls *as far as the Casparian strips* but not beyond the strips. Lanthanum and uranium are apparently unable to cross the plasmalemma and are excluded from the stele; lead, however, has been reported to enter endodermal cytoplasm and also the stele. The above observations strongly support the belief that (*i*) the Casparian strip is an effective barrier to ion movement; and (*ii*) the endodermis can be crossed by ions only if they can enter the symplast. The weakness of the argument lies in extrapolating from data obtained with these ions of high atomic weight, to the behaviour of the 'physiological' ions of much lower mass. There is, however, indirect evidence for similar behaviour by the nutrient ions. In barley roots, for example, $PO_4^{3-}$ and $K^+$, which enter the symplast with ease, are translocated to shoots even from the basal parts of the roots where the endodermis is fully developed. $Ca^{2+}$ penetrates the symplast only with difficulty and is translocated to the shoots mainly from apical regions where the endodermal cell walls are still permeable.

The endodermis can thus regulate the entry of ions into the stele. When the cortex was stripped from a segment of attached barley roots so as to break the endodermal cells across at the Casparian strips, the concentrations of $^{32}P$ and $^{85}Sr$ in the transpiration stream became equal to those in the external medium. The endodermis also acts as a barrier against the outward leakage of ions from the stele. Experiments with $^{45}Ca$ in barley roots have shown that over 60% of the radioactive ions taken up were exchanged from the cortex for unlabelled Ca in 10 minutes; but from the pericycle, the cell layer just inside the endodermis, only 19% was lost at the same time.

In summary: the mineral ions which are absorbed by roots can travel to the xylem both in the symplast and in the apoplast. In the extreme apices of the roots, the apoplastic pathway is continuous, but once the endodermal Casparian strips have developed (this may occur within a few mm of the tip), the endodermis can be crossed only via the symplast. In regions of the root where secondary rootlets emerge, the endodermis may be interrupted, giving freer access to the stele again. Control of what passes into the xylem is possible at various points along the pathway: the transfer of ions to the symplast anywhere between the epidermis and the endodermis; the retention by living root cells; perhaps a 'filtering' action by the endodermis. The sap composition is further modified by the absorption of ions from the flowing stream by the xylem parenchyma cells lining the route.

When the rate of water transport from the soil to the xylem is increased, particularly by enhanced transpiration, salt uptake and transport are also increased. A generally accepted explanation of this is still lacking. It has been suggested that the additional ions entering the xylem represent a mass flow of ions in the water stream – a passive flow of ions in a water current. Alternatively it has been suggested that at low transpiration rates the concentrations of ions in the xylem vessels are high enough to limit ion movement into them, whereas at a higher transpiration rate the xylem fluid is more dilute and hence salt release takes place at a higher rate. On this explanation the increased water flow is considered to increase the 'active' movement of ions from the symplast. A still further possibility is that the tension generated by a high transpiration rate lowers the resistances to water and salt movement thereby increasing salt release whether this is a metabolically controlled secretion or a leakage due to impaired 'active' salt retention. Certainly there is evidence that tensions developed in the xylem vessels are transmitted across the root diameter.

## Physiological roles of the mineral elements

For most of the essential elements their role in plants is at least partly known, but for some our understanding is very incomplete. One difficulty often encountered, and well illustrated by potassium, is the identification of the aspects of metabolism for which the element is truly essential. Sodium, for example, appears to be able to substitute for some of the activities of potassium, but cannot completely replace the need for this element. Many enzymes can, with different effectiveness, be activated by either $Mg^{2+}$ or $Mn^{2+}$ ions although both elements have other activities where one cannot substitute for the other. Again the demonstration of one essential role does not preclude the element from having other vital functions. It is beyond the scope of this book to consider in detail the roles of each essential element and in the discussion below, emphasis is placed on some of the less well known functions of the elements, and on those elements for which the roles are still disputed, and form subjects of current research.

The macronutrient mineral elements are constituents of major structural components of protoplasm. *Nitrogen* is found in proteins and nucleic acids, the key molecules of life, as well as in pigments, coenzymes and very varied secondary products. *Sulphur* is also present in proteins, where disulphide bridges (–S–S–) stabilize tertiary structure while sulphydryl groups (SH) participate in the active sites of many enzymes. Membrane sulpholipids contain sulphate groups. *Phosphorus* is contained in nucleic acids and membrane phospholipids. As a component of the adenosine phosphates (ATP, ADP, AMP) and related nucleotides, the phosphate group is involved in energy transduction, and intermediary metabolism involves many phosphorylated intermediates. There is in active cells a continuous turnover of phosphate from organic combination to the inorganic form and back again. *Calcium* as calcium pectate participates in cell wall structure, but it also associates as $Ca^{2+}$ ions with cell membrane phospholipids and is necessary for maintenance of normal permeability of the plasmalemma. It further acts as activator of some enzymes – amylases, ATPases and phospholipases. Another element which plays a structural role is *silicon*, which in the form of silica gel gives the cell walls of grasses, including cereals, their characteristic rigidity. Silicon is not known to participate in any biochemical reactions within the cells and silicon-requiring plants can be nursed to maturity in culture in the absence of the element. Lack of silicon, however, results in wilting, withering of leaves and necrosis and under natural conditions such silicon-deficient plants would have little chance of survival; hence it seems reasonable to consider silicon an essential element for the species which incorporate it in their cell walls.

Many of the micronutrients have been identified as enzyme activators or as parts of prosthetic groups of enzymes. Iron and copper are present in respiratory and photosynthetic oxidoreductases; iron is also necessary for chlorophyll synthesis. Magnesium and manganese activate many dehydrogenases and phosphate transfer enzymes, and magnesium is of course a part of the chlorophyll molecule. In higher plants zinc has been identified as essential for the three enzymes carbonic anhydrase, alcohol dehydrogenase and superoxide dismutase (an enzyme important in degrading highly dangerous superoxide radicals formed during certain oxidative reactions). But some seventy zinc-requiring enzymes are known from animals and microorganisms and the true number of higher plant enzymes needing zinc may well be much greater than three.

*Potassium* is something of a mystery element. It is the chief cation of protoplasm and as such balances the charges on cytoplasmic anions, organic cations being few. Plant cells show a great preference with respect to accumulation towards potassium as opposed to the chemically very similar sodium. The only known metabolic role for potassium is as activator of some enzymes; but other elements which act as enzyme activators are required in micronutrient quantities only. The affinity of proteins for potassium is, however, low, so that fairly high potassium concentrations may be needed to make potassium-enzyme complexes. In plant cells which function in movements involving turgor changes, $K^+$ ions are concerned in turgor control – stomatal guard cells, and pulvinar cells (hinge cells) of leaves and petioles. In such cells increases and decreases in turgor are achieved by $K^+$ uptake and expulsion respectively, water then following the ions according to the water potential gradient.

*Sodium* is chemically very similar to potassium and to some extent it can interchange with potassium; e.g. in *Commelina benghalensis* sodium can replace potassium in the control of turgor of stomatal guard cells. For succulent halophytes sodium acts as an osmoregulatory ion and cell turgor is maintained in the face of a low water potential in the saline growth medium by vacuolar accumulation of high concentrations of $Na^+$ (and $Cl^-$) ions. Sodium is essential in species which carry out the $C_4$ type of photosynthesis (p. 43), but it is not known in what reaction(s) sodium is involved in the $C_4$ plants.

*Chlorine* occurs in cells as the chloride anion. It is necessary for the $O_2$-evolving system of photosynthesis (Photosystem II). It also is the main inorganic anion of cells and as such is associated with $Na^+$ and $K^+$ in turgor control.

*Molybdenum* is sometimes omitted from the catalogue of essential elements. It is vital for the nitrate reduction system, the complex of enzymes which reduce nitrate to the metabolically usable ammonia. The process of nitrogen fixation also requires molybdenum. If nitrogen is supplied as ammonia, as ammonium salts or in organic form, molybdenum becomes redundant. Under natural conditions, however, nitrate is the chief nitrogen source for higher plants and the nitrate reductase system is of universal occurrence.

The need for *cobalt* also is arguable. This element is involved in nitrogen fixation, so that it is certainly beneficial for plants with nitrogen-fixing nodules, but these plants can survive without the nitrogen-fixing symbionts. Vitamin $B_{12}$, an essential cobalt-containing metabolite in animals, has been identified in plants but its functional role is not unequivocally established. If cobalt is essential, it is required in only most minute traces.

The element whose role in plants has aroused the greatest amount of controversy is *boron*. Its position among the essential elements is not disputed. The essentiality of boron was demonstrated for broad bean (*Vicia faba*) by Warington as long ago as 1923, and subsequent studies have confirmed the essentiality of boron in many instances. In the absence of boron, all types of plant meristems blacken and become necrotic; the cells derived from such meristems fail to differentiate normally, but may swell to an abnormal extent. Anthers seem to be especially sensitive to boron deficiency, breakdown of pollen grain formation occurring; pollen germination, too, is often dependent on a supply of boron. Pollen tubes arising in the absence of boron may swell and burst. Boron-deficient roots lose their gravitropic sensitivity, and their ion uptake capacity deteriorates. Yet no metabolic function has been identified for the element.

Boron is present in soils and plants as boric acid, $H_3BO_3$, a weak acid which readily forms complexes with *cis*-polyhydroxy molecular configurations such as occur in certain sugars, polyhydric alcohols and diphenols. Boron deficiency is usually associated with increased concentrations of carbohydrate in leaves and there is evidence that this accumulation takes place before the phloem becomes non-functional by failure of sieve tube differentiation and necrosis. These observations led Gauch and Duggar (1953) to advance the hypothesis that the main role of boron is to facilitate the translocation of sugars across cellular membranes by formation of sugar–borate complexes. The necroses of terminal apices would be a consequence of sugar starvation following impaired sugar translocation. However, when plants are transferred to a boron-free medium other symptoms of deficiency occur more rapidly than a decrease in the rate of sugar translocation. Moreover sucrose, the most commonly translocated sugar, forms only very weak complexes with borate.

Boron-deficient cells have been reported to exhibit enhanced activities of oxidative enzymes – polyphenol oxidase, aldehyde oxidase, catalase and peroxidase, which can mediate browning of cells, and it has been suggested that boron is either an inhibitor of these enzymes, or acts by reducing the levels of their natural substrates by forming borate–diphenol complexes. However, enhanced activity of the above enzymes is observed in a number of other mineral deficiencies and may be a secondary consequence of the onset of necrosis.

D. H. Lewis (1980) has attempted to explain the role of boron starting from the well-known fact that endogenous boron is an essential element only in vascular plants. (In some marine algae, exogenous boron is believed to stabilize cell wall polysaccharides.) Hence, Lewis argued, the primary role of boron should be in a process unique to vascular plants – a process which he postulated to be promotion of lignin biosynthesis. The suggested mechanism involves an inhibition of catechol oxidase by boron, and thus accounts for the activation of this enzyme and consequent browning of tissues in the absence of boron. The phenolic compounds which would accumulate have toxic effects. The weaknesses of this hypothesis are that (*i*) the involvement of boron in the biochemical pathway for lignin synthesis has not been demonstrated experimentally and (*ii*) boron may exert effects on plants rather more rapidly than would be expected from an inhibition of lignification and an accumulation of phenols.

Another proposal for the role of boron is that it acts on the hormone system of vascular plants, a system unique to this plant group. Failure of meristems in the absence of boron could result from a disturbance of hormonal balance; boron deficiency has been shown to stimulate increases in auxin levels. But equally well the hormonal disturbances could be secondary effects following interference with growth.

Some effects of boron are rapid; e.g. enhancement of ion uptake by addition of boron to boron-deficient roots is considerable by 20 min and recovery is complete within 1 h. Such fast action suggests a primary role of boron on cellular membranes – changes in permeability, or action on membrane-bound transferases. In favour of an effect of boron on membranes are also reports that boron maintains a positive charge on cell surfaces.

Clearly hypotheses abound when established facts are lacking. The problems of boron have been considered at some length to illustrate the difficulties involved in the study of the essential roles of elements. As stated, knowledge of the functions of potassium, sodium, chlorine and cobalt is also scanty. Even where there is some appreciation of the biochemical role(s) of an element, there may well be other functions as yet undiscovered, and complex interactions take place between mineral ions and metabolism over and above the primary roles of the minerals. Several examples of interactions between mineral ions and growth hormone metabolism have been reported. In sunflower plants, deficiency of nitrogen, phosphorus or potassium in the rooting medium has been found to decrease the flow of cytokinins from the roots to the shoots. Macronutrient deficiency can thus act on plants not only through a shortage of elements for bulk biosyntheses, but via the hormone supply. The biosynthesis of the gaseous hormone ethylene is promoted by $Ca^{2+}$ ions, but another step in the synthesis is blocked by $Co^{2+}$ ions. Ethylene acts antagonistically to auxin in a number of physiological effects. Thus $Ca^{2+}$ which promotes the synthesis of ethylene antagonizes auxin effects, while cobalt enhances the effects of auxin. Many more such interactions remain undoubtedly to be worked out.

### Deficiency diseases

A number of economically important and symptom-characteristic plant diseases are now recognized as due to mineral element deficiencies. Examples of such diseases are 'tea-yellows' (sulphur deficiency); 'grey-speck' of oats, 'marsh spot' of peas and 'speckled yellows' of beet (all due to manganese deficiency); 'exanthema' of fruit trees, 'mottle leaf' of *Citrus*, 'white bud' of maize and 'sickle-leaf' of cocoa (all due to zinc deficiency); 'heart-rot' of beet, 'stem-crack' of celery, 'brown heart' of swede, 'corky core' of apple, 'top-sickness' of tobacco (all due to boron deficiency); and 'scald' disease of beans and 'whiptail' of *Brassica* crops (due to molybdenum deficiency). Even where the symptoms of a particular element deficiency have not been given a disease name, the symptoms are usually characteristic for the species and a reliable means of identifying the deficiency. However such identification may only be possible by the time the crop is virtually ruined and earlier diagnosis may often be achieved by rapid techniques of leaf analysis or by studying the curative effects of foliar applications of salt solutions.

The symptoms of mineral deficiencies in a wide range of crop plants have been accurately described and recorded by colour photography to assist rapid diagnosis (*The Diagnosis of Mineral Deficiencies in Plants – by Visual Symptoms*, T. Wallace, H.M.S.O., London, 1951). Certain plant species which rapidly develop characteristic deficiency symptoms for particular elements are now used as *indicator plants* for testing soils suspected of being mineral deficient.

FURTHER READING

BOWLING, D. J. F. (1981). Release of ions to the xylem in roots. *Physiologia Plantarum*, **53**, 392–7.

CLARKSON, D. T. and HANSON, J. B. (1980). The mineral nutrition of plants. *Annual Review of Plant Physiology*, **31**, 239–98.

HEWITT, E. J. (1966). *Sand and Water Culture Methods used in the Study of Plant Nutrition*, 2nd edition. Commonwealth Agricultural Bureaux, Farmham Royal.

HEWITT, E. J. and SMITH, T. A. (1975). *Plant Mineral Nutrition*. English Universities Press, London.

STREET, H. E. (1966). The physiology of root growth. *Annual Review of Plant Physiology*, **17**, 315–44.

SUTCLIFFE, J. F. and BAKER, D. A. (1978). *Plants and Mineral Salts*, 2nd edition. Studies in Biology, no. 48, Edward Arnold, London.

SELECTED REFERENCES

BROWNELL, P. F. (1979). Sodium as an essential micronutrient element in plants and its possible role in metabolism. *Advances in Botanical Research*, **7**, 117–224.

DUPONT, F. M. and LEONARD, R. T. (1977). The use of lanthanum to study the functional development of the Casparian strip in corn roots. *Protoplasma*, **91**, 315–23.

LEWIN, J. and REIMANN, B. E. F. (1969). Silicon and plant growth. *Annual Review of Plant Physiology*, **20**, 289–304.

LEWIS, D. H. (1980). Boron, lignification and the origin of vascular plants: a unified hypothesis. *New Phytologist*, **84**, 209–29.

ROBARDS, A. W. and ROBB, M. E. (1974). The entry of ions and molecules into roots: an investigation using electron-opaque tracers. *Planta*, **120**, 1–12.

TANADA, T. (1978). Boron – key element in the action of phytochrome and gravity? *Planta*, **143**, 109–111.

# 6

# Transport of Metabolites

In unicellular organisms all aspects of vital activity proceed within the confines of a single cell. In simple multicellular plants, such as filamentous algae, all the vegetative cells are functionally equivalent. By contrast a flowering plant is a complex organism with cells and organs specialized for diverse functions; its separate parts are not self-sufficient but inter-dependent. Nutrients must be supplied by one part to another, and communication between distant organs via growth-regulating substances is necessary to ensure harmonious development.

The compounds transported in a flowering plant are water, organic nutrients and co-factors, mineral nutrients and hormones. Water transport has been discussed in Chapter 4 and some aspects of xylem transport of mineral ions in Chapter 5. In rooted plants xylem transport is essentially a unidirectional stream from root to shoot. Organic and mineral nutrients however undergo multidirectional movement. Organic compounds, synthe-sized primarily in photosynthetic organs (leaves and green stems), are exported to all non-photosynthetic parts; some of these parts may later re-export the products of their own biosynthetic activity. The mineral elements which are taken up from the soil move at first upwards into growing regions, but minerals, too, may later be withdrawn from such regions, particularly as the organs senesce, for transport to other sites. Hormones and other as yet unidentified chemical stimuli are transported, apparently in very small amounts commensurate with their high physiological activity, and in some instances in a strictly unidirectional (polarized) manner. In a flowering plant there is accordingly a complex traffic of chemicals in many directions, each organ simultaneously receiving some metabolites and exporting others. Over small distances, chemicals can move within and between cells by diffusion and by cytoplasmic movement, supplemented by active transfer across cell membranes. Such movements are classed as *short distance transport* (within cells) and *medium distance transport* (between cells). *Long distance transport* proceeds between organs, through the vascular tissues and this forms the subject of the present chapter. The term *translocation* has been used by some authors to include all long distance movement of compounds within xylem and phloem. Here, in common with the more general usage, the term is used to describe the multidirectional transport which occurs in the *phloem*.

# The transporting system

Various observations establish that xylem and phloem are the channels of transport. Their anatomy clearly suggests such a function. They contain elongate cells in longitudinal files and penetrate intimately into all parts of the plant as an uninterrupted network. Leaves are particularly well supplied with these vascular tissues; in leaves of many dicotyledons, the maximum distance between any mesophyll cell and a vascular bundle is only 60–70 $\mu$m in the plane of the leaf (Fig. 6.1). The highly developed vascular supply of leaves correlates with high activity of leaves in the exchange of metabolites and in transportation. During the vegetative growth of the plant its leaves are the chief recipients of water and minerals. Carbohydrates are manufactured there by photosynthesis and from the leaves flow sucrose, amino acids and other metabolites to growth centres and sites of food storage. In root tips, the apical meristem is separated from functional phloem cells by only 250–750 $\mu$m and from functional xylem elements by 400–10 500 $\mu$m. Acropetal differentiation of the vascular tissues similarly follows the upward growth of the stem apex. As axes thicken, primary vascular tissues become supplemented by differentiation of profuse amounts of secondary vascular tissues derived from the vascular cambium.

Visual evidence of transport in the vascular tissues comes from observations on dye movement. By appropriate choice of dye and method of

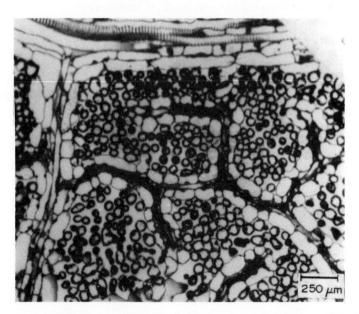

**Fig. 6.1** The fine venation in a small portion of a leaf of a dicotyledon (*Phaseolus aureus*), as seen in a section parallel to the leaf surface. Xylem is visible in the larger vein at the top. Chloroplasts (darkly stained) can be seen in mesophyll cell sections; bundle sheath cells have sparse contents ($\times$ 415).

application transport can be shown by this technique in both xylem and phloem. Fluorescent dyes (e.g. fluorescein), which are detectable in low concentration by virtue of their fluorescence, have been used to determine the speed of solute movement. Further information comes from experiments in which the continuity of vascular tissues is interrupted by their excision ('ringing' or 'girdling' experiments). Thus in woody plants it is easily possible to cut down to the vascular cambium and to remove the outer tissues ('bark') which include the phloem, leaving only the central xylem cylinder and pith intact. It is then found that regions of defoliated stem separated from all leaves by bark rings become deficient in carbohydrate, but water and minerals can pass the 'bark rings', indicating that carbohydrate transport occurs in the phloem, whereas movement of water and minerals is in the xylem. The more difficult operation of removing a section of the xylem, and creating a water-filled cavity inside the phloem, has also been accomplished. The growth of the shoot above such a 'bark ring' was almost equal to that of the control shoots indicating that the phloem is able to transport organic and possibly also significant amounts of mineral nutrients. These ringing experiments thus indicate that organic materials move in the phloem and that minerals are transported in both the xylem and the phloem.

## The composition of xylem and phloem sap

Xylem sap can be obtained as an exudate from bore holes in the trunk or main stem or from decapitated root stumps at times when it is under pressure; at other times it can be sucked out with a vacuum pump. Alternatively xylem sap can be centrifuged or pressed out from lengths of plant axis. It is more difficult to obtain adequate and uncontaminated samples of phloem sap. The quantity of tissue available is smaller, and there is risk of contamination with protoplasm not only from the sieve tube elements which contain the sap, but from companion cells and phloem parenchyma. Sap does sometimes exude from cuts made into the phloem of trees. A more elegant, and at present the most reliable method, of obtaining sieve tube sap, is to use aphids to tap individual sieve tubes. These insects feed on phloem sap by inserting their stylets into sieve tubes from the outside; when the insertion has been accomplished, the insect is cut away under anaesthesia. Sap will then continue to drop from the stylets for up to several days, at a rate of 1–2 $\mu$l per hour. This liquid has been assumed to represent more or less unadulterated sieve tube sap. There is evidence that the saliva initially released by the aphid exerts no digestive function on the phloem. However even with this technique there remains the possibility of seepage of liquid from surrounding tissues into the tapped sieve tube units.

Sap analyses vary quite markedly from species to species; there are also seasonal and daily variations in sap constituents, as well as irregular fluctuations resulting from fluctuating environmental conditions. Nevertheless the composition of xylem sap always differs very significantly from that of phloem sap.

*Xylem sap* is a clear liquid with a viscosity close to 1.0 and with a pH of about 5. The solute concentration is low; the total dry matter does not normally exceed 1%; values of 0.1 to 0.5% are common. Its $\psi$ is seldom below $-$ 0.2 MPa. During periods of rapid transpiration and water uptake, the solute concentration of xylem sap falls to a very low level; since transpiration is faster by day than by night, the xylem sap is more dilute by day. Of the solids present about two-thirds to three-quarters are organic, including amino acids, amides and organic acids. The carbohydrate content is usually only 0.02–0.05%, but may be almost undetectable. The remaining solids are mineral salts. Some of the metal ions are in the form of organic chelates. In perennials, the solute concentration in the xylem sap is at a maximum in spring (Fig. 6.2). In a few trees, e.g. sugar maple (*Acer saccharum*), and birch (*Betula* sp.), the xylem sap in early spring may contain up to 8% sugar. This high sugar content is however maintained only for a limited period, being derived from reserve carbohydrates stored in the stem during winter. The organic nitrogen compounds of the xylem sap are indicative of the assimilation of nitrate and ammonium ions into organic form in the root system. When nitrate occurs in quantity there is usually little organic nitrogen in the sap indicating that the shoot is synthesizing its amino acids from nitrate.

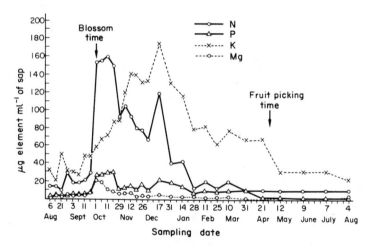

**Fig. 6.2** Seasonal variation in the contents of nitrogen, phosphorus, potassium and magnesium in apple (*Pyrus malus*) xylem sap. The nitrogen is mainly present in organic combination. The data were obtained in New Zealand, so that September is spring. (From Bollard, 1958, in *The Physiology of Forest Trees,* ed. Thimann, 83–93. Copyright 1958. The Ronald Press Co., New York.)

*Phloem sap* is more viscous and, in contrast to xylem sap and most vacuolar plant saps, is usually alkaline (pH 7.5 to 8.6), although in perennials it may be faintly acid early in the spring. The solute concentration is much higher than in xylem sap; sugars generally make up 90% of the solids, being present at concentrations of 2 to 25% w/v. Its $\psi$ ranges from $-0.6$ to $-3.4$ MPa. In the majority of species, sucrose is the main sugar present, with traces of the oligosaccharides, raffinose, stachyose or verbascose, but in a few species one of the last sugars predominates. Amino acids are regularly present, amounting to 0.2 to 12% (exceptionally up to 50%) of the transported solutes. Mineral elements are also found in phloem sap, generally at higher concentrations than in xylem fluid, although it is not always known whether all the elements are present as inorganic salts, or in organic combination. Potassium is the predominant cation and is often present at 0.03–0.06 M, but may reach 0.5 M. ATP is a regular constituent. Protein is usually detectable and numerous enzyme activities have been detected in sieve tube exudates, but it is not clear how many of these are true sap components: leakage from companion cells is very probable. In small quantities, many other compounds have been detected including organic acids, alkaloids, vitamins, hormones, viruses, and artificially applied chemicals.

The sugar concentration of phloem sap exhibits regular daily variation. In the cotton plant, the highest concentration is recorded in the latter part of the day. In a number of trees, the highest concentration occurs at night and close to the leaves, a concentration wave moving down the tree. Such changes in concentration may reflect turgor changes in the axis as well as changes in the amount of sugars supplied by the leaves. The greatest volume of phloem exudate from cuts on trees is generally obtained on sunny afternoons. In several tree species, the greatest amount of exudate from cuts is obtained in late summer, no flow occurring till about mid-June. Marked seasonal changes are found with respect to amino acid content; this is high in spring, falls in the summer, and rises to a second peak in the autumn when leaf proteins break down prior to abscission; the nitrogen translocated out of leaves at this time is deposited in organic form in the stems where it persists during the winter.

Analyses of xylem and phloem sap thus support the evidence from the ringing experiments in pointing to the phloem as the main channel for the transport of organic materials, and to the xylem as the main path for the movement of minerals.

### The movement of radioactive compounds

Radioactive labelling makes possible the detection of the labelled compound in quantities far below the limit of chemical detection, and the movement of the radioactive element is not masked by movement in the plant of the same element in the normal unlabelled form. For instance one can follow the passage of radioactive phosphorus ($^{32}P$) *into* a leaf rich in phosphate and from which there may be a large net export of phosphorus. The radioactivity can be measured after an extract of the plant has been prepared, or it can be detected *in situ* by scanning with a Geiger counter, or by preparing an autoradiograph (the plant material is placed in close contact with a sensitive photographic emulsion, and the site of the radiation is revealed when the photographic plate is subsequently developed).

Supplying radioactive $^{14}CO_2$ to leaves results in the appearance of radioactivity in the phloem, showing that the products of photosynthesis are exported out of leaves in the phloem; chemical analysis shows that it is the radioactivity of the carbon atoms of sucrose which accounts for most of the radioactivity of the phloem sap. A similar result is obtained if radioactive sugars are applied to the leaves. When radioactive ions are applied to the roots the radioactivity appears in both the xylem and the phloem. This might mean concurrent upward transport in both tissues, or result from lateral transfer from one to the other, for xylem and phloem are in close contact. To decide between these two possibilities, Stout and Hoagland in 1939 performed stripping experiments on young stems of house geranium (*Pelargonium zonale*) and willow (*Salix lasiandra*). Two parallel longitudinal slits were made along 22 cm lengths of stem, and the xylem and phloem separated by inserting a strip of paraffined paper; the whole treated length of stem was then wrapped in paraffined paper to protect the tissues against desiccation, and the radioactive ions – $^{42}K^+$, $^{32}PO_4^{2-}$, $^{82}Br^-$ and $^{24}Na^+$ were applied to the roots. It was then found that where the xylem and phloem were left in contact, radioactivity was present in xylem and phloem alike; but in the stripped areas, only the xylem was appreciably radioactive. From this, the conclusion was drawn that the upward transport of ions occurs in the first place in the xylem, but that continual lateral transfer occurs into the phloem.

Some of the ions transported to the shoot become immobile and remain in the tissues in which they are first deposited; others move out again, passing from older to younger leaves, back to the roots, even out into the soil once more. Movement in the xylem is a strictly one-way upward traffic. Any transport of ions in the opposite direction can pass only through the phloem; when radioactive salts are fed to leaves and exported from them it is the phloem of the stem which becomes radioactive.

We can conclude that organic compounds move in both directions in the phloem and upwards in the xylem. Minerals move upwards mainly in the xylem; their downward movement is confined to the phloem. Upward transport of minerals to bud primordia may be in the phloem, due to more precocious differentiation of phloem in the bud traces.

### Patterns and control of the direction of metabolite transport

The longitudinal course and interconnections between the vascular bundles in the stem and root determine between which organs transport occurs. Lateral transport from one vascular strand to another hardly occurs in either the root or the stem. Leaves export photosynthetic products to the young leaves directly above them, and to parts of the root system directly below them. This can be most readily demonstrated by applying radioactive $CO_2$ to a leaf and tracing the movement of radioactivity to other leaves. If labelled $CO_2$ is confined to one longitudinal half of an exporting leaf, only the corresponding half-leaves above this may receive radioactivity. This applies also to xylem transport: mineral deficiencies have been induced in individual halves of tobacco leaves by withholding nutrients from appropriate parts of the root system. Some adjustment from the normal pattern of transport may be induced. *Coleus* plants have been grown with the root system separated into two halves, in separate pots. When phosphorus was withheld from one pot, leaves above that root half became phosphorus-deficient. However if this pot was left unwatered, no wilting of the leaves occurred, and phosphorus deficiency symptoms did not appear, showing that water and minerals were being transported laterally, presumably through cell walls, from the half receiving both water and the phosphate. The pattern of phloem transport too can undergo some adjustment in response to injury. In apple trees, leaves and fruits can be separated longitudinally by *c.* 3 m of defoliated stem without affecting the size and sugar content of the fruit. Lateral diversion, however, is possible only if the plant has a suitably anastomosing vascular system, as occurs for example in the beetroot. Generally the removal of leaves from one side of a plant results in asymmetric growth as a consequence of one-sided transport.

The direction of movement of compounds in the xylem is always upwards from the roots; the flow is never reversed. But the direction of phloem translocation is variable, and is under close metabolic control. From the mature leaves, organic materials must be supplied to all parts of the plant that are not self-supporting. The direction is towards *growing* regions. In a young vegetative plant, the direction of this flow is mainly to the rapidly growing root system although some movement also occurs to the shoot apices and developing leaves. As the plant becomes larger, lower leaves transport predominantly to roots, upper leaves to the growing apices of the shoots; in such older plants translocation proceeds more upwards by day and downwards at night. When the reproductive stage is reached, almost all the transport is directed to the flowers and fruits; this includes not only organic nutrients, but mineral elements moving out from the leaves. At the start of the growing season, perennials translocate nutrients from storage regions – tubers, roots, bulbs, stems – to growing tips. Similarly in a germinating seed, translocation is from the storage tissues to the growing regions.

Phloem translocation is often described as directed 'from source to sink', that is from the regions of primary production to the regions of growth or storage. This is, however, an oversimplification. No nutrients are translocated into a mature leaf which is shaded so that it cannot photosynthesize at a sufficient rate to meet its need for energy-rich organic molecules; such a leaf would be expected to act as a 'sink', but in fact it starves to death. Sugar transport in the phloem is sometimes faster than its utilization in the storage or reproductive organs to which it is moving, and temporary accumulation of sugar occurs in petioles or fruit stalks without preventing or reversing the flow.

Translocates can be attracted to a cut stem stump by applying auxin to it, and sugar movement towards growing wheat ears can be reversed by placing auxin on the stem at a point remote from the ears. It has therefore been suggested that actively-growing meristematic regions and fruits attract translocates by virtue of their high auxin content, whilst the failure of old leaves to obtain translocates correlates with their low auxin level. It is not clear, however, whether the hormone has a direct effect on translocation, or whether it acts by promoting metabolic activity in the sink, and thereby stimulating the demand for assimilates.

The main directions of xylem and phloem transport are summarized in Fig. 6.3; the movement of hormones is discussed at the end of this chapter.

## The mechanism of phloem translocation

### The hypothesis of mass flow driven by an osmotic gradient

For the mode of movement of any substance in solution, two possibilities exist. Either it is carried along by *mass flow* (bulk flow) of the solution, or else only the solute moves, e.g. by diffusion, while the solvent remains stationary. Movement in the xylem is undisputedly a mass flow of the whole solution, and the motive force is either the tension pull of transpiration, or root pressure. The mechanism of phloem translocation, however, still remains one of the great unsolved mysteries of plant physiology. It is not certain whether mass flow of phloem sap occurs, or whether solutes are transported independently, and consequently there is equal uncertainty as to the motive force involved.

Speeds of phloem transport have been obtained by calculation from the rates of increase in weight of growing tubers and fruits, or from direct measurement of the rate of movement of fluorescent dyes or radioactive tracers. The velocities reported have commonly ranged from 20–100 cm h$^{-1}$, but up to 450 cm h$^{-1}$ has been claimed. These values immediately rule out diffusion as the mechanism of transport, being $10^4$–$10^5$ times too high for diffusion. The total quantities translocated per unit time are also high. Sugar moves into a growing fruit of pumpkin (*Cucurbita pepo*) at a rate of at least 1.7 g h$^{-1}$. The rate of sugar movement across 1 cm$^2$ of sieve tube area per hour (the rate of mass transfer) has been calculated to be about 2 g for petioles and 20 g for fruit stalks. Phloem translocation is a highly efficient process, with both a rapid velocity and a high rate of mass transfer.

One of the earliest theories for the mechanism of phloem translocation was Münch's hypothesis (1930), which postulated a mass flow along a turgor pressure gradient, induced by a physiologically maintained gradient of $\psi$. A model for the system is shown in Fig. 6.4 and is easily constructed in the laboratory from dialysis sacs and pieces of glass tubing. Two semipermeable reservoirs A and C are joined by a connecting tube B. If a concentrated sugar solution is placed in A and a more dilute solution in C, and both are immersed in water, the water uptake into A will be faster than into C. There will be a mass flow of liquid along tube B from A to C, and water will be forced out of C; a manometer, M, inserted on tube B will register a hydrostatic pressure. In the model, flow will stop when the sugar concentrations in A and C have equalized but, in a living plant, a continuous flow could be maintained by a continued secretion of sugar into the phloem

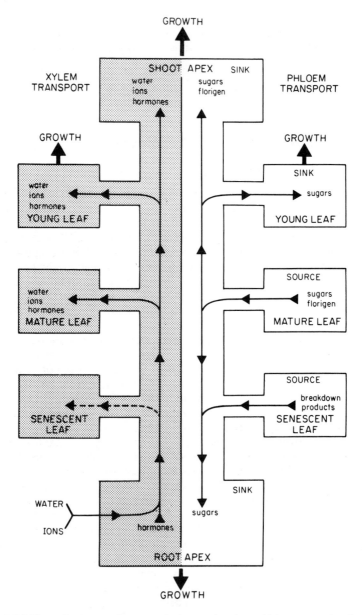

**Fig. 6.3** The mainstreams of transport in a vegetative plant: phloem transport on the right, xylem transport on the left. In the reproductive stage, most of the phloem transport is directed towards the developing fruits.

at the source end (A) and removal at the sink to which it is transported (C), the water being released into the xylem at C, and drawn from the xylem into the system at A.

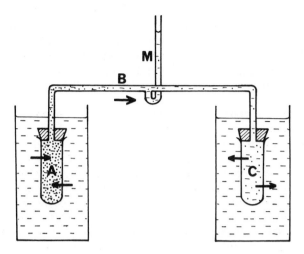

**Fig. 6.4** Diagram of an osmotic system in which mass flow of liquid can occur, as postulated by Münch's hypothesis. See text for explanation.

Many observations can be quoted in favour of a mass flow in the phloem. When one single sieve tube unit is pierced by an aphid stylet, the volume of sap exuded in one hour may be 1 $\mu$l, equivalent to the volume of some 2500 individual cells, and the flow may continue for days. On a more macroscopic scale, in tropical countries the phloem sap of palms and agaves is tapped for sugar production or fermentation. One tree can exude for several months, yielding several thousand litres of sap. This is manifestly a mass flow of liquid! Moreover the sugar concentration remains steady during the period of exudation in both aphid stylet exudate and palm sap, so that the flow of liquid cannot be attributed to a leakage of water from the nearby xylem into the punctured sieve tube cell(s). If mass flow is rejected, all the above-mentioned exudations of sap must be ascribed to an injury reaction unrelated to normal flow in the phloem. The phloem has been shown to conduct a locally applied heat pulse longitudinally in the direction along which solutes are translocated – and this applies to completely undamaged phloem.

The Münch hypothesis requires that sieve tube contents should be under positive turgor pressure, and that there should be a gradient of turgor (and of solutes), decreasing from source to sink. The sieve tubes are certainly under positive turgor pressure; this is demonstrated even by the simple fact that sap does exude from cuts. Aphid stylets contain only a very narrow channel, offering considerable resistance to the flow of the viscous sap, and it has been calculated that a pressure of 1–3 MPa is required to force the sap through the stylets. Attempts to measure the turgor of sieve tubes directly, either by inserting sensitive micromanometers, or by pressure cuffs (analogous to the apparatus used for measuring human blood pressure), have yielded values of about 0.4–2 MPa. Such measurements are in practice fraught with great difficulties; nevertheless not only have positive pressures been measured, but some workers have succeeded in demonstrating gradients of turgor passing down trees. Gradients in sugar concentration have been observed by sampling the phloem at intervals. In soybean plants, for example, the sugar concentration in the sieve tubes of the leaflet stalk was found to be 10.5–12.5% when the sieve tubes of the root contained only 4.4–6.3%.

All this is good evidence for the Münch mass flow hypothesis. But the hypothesis cannot be accepted until it has been shown to be feasible quantitatively as well as qualitatively: until it is shown that mass flow *at the experimentally measured velocities* can be driven *by the actually existing pressure gradients*, through channels *of the dimensions provided by the sieve tubes*. Difficulties are encountered when the quantitative aspects are considered. Many calculations have been made, but none has yet succeeded in unequivocally proving or disproving the feasibility of turgor-driven mass flow, for all calculations have had to make assumptions about some of the parameters involved. Quite reliable values are available for velocities, and reasonable estimates for solute and turgor pressure gradients; but the dimensions of the channels available are still a matter of controversy.

### The structure of phloem

Mass flow requires continuous open channels, the wider the better, for the narrower the diameter, the greater is the frictional resistance, and the higher is the force required to move liquid along. The phloem of flowering plants consists of sieve tubes, built of longitudinal files of individual cells, the sieve tube elements; associated with the sieve tube elements there are often companion cells, one or several per sieve tube cell; and the phloem also contains phloem parenchyma and fibres. The *sieve tubes* have been shown to be the conducting channels by means of tracers (p. 90); they are the only cells in the phloem with a structure suggestive of a conducting function, and certainly the only ones which might offer a possible route for mass flow. In the mature state they usually lack nuclei and the only cytoplasmic organelles identified are plastids, mitochondria and endoplasmic reticulum elements; however, these organelles do not occupy much of the cellular volume. Most of the lumen is filled with the sap containing P-protein (P for phloem), sometimes visible as longitudinal strands and clumping to make 'slime plugs' on injury. The cells are, however, still bounded by a functional semi-permeable plasmalemma. The sieve tube diameter is 10–50 $\mu$m; this would readily allow mass flow, but at intervals of 150–1000 $\mu$m the tubes are interrupted by the transverse or oblique cell walls of the individual cells, the *sieve plates*, 0.5–2 $\mu$m thick. The plates are pierced by the *sieve plate pores* and the crucial question for mass flow is: are these pores open channels, of sufficient cross-sectional area to permit transport at the observed rates?

In mature sieve tubes the pores are usually 2–6 $\mu$m in diameter but may be finer down to < 1 $\mu$m; exceptionally, as in the Cucurbitaceae, their diameter reaches 10 $\mu$m. If it is assumed that the entire sieve plate pore is a hole open to flow, a hydrostatic pressure gradient of 0.06–0.10 MPa m$^{-1}$ has often been regarded as adequate to drive mass flow at the measured rates of transport, though the value has been put as high as $\geqslant 0.5\,\text{MPa}\,\text{m}^{-1}$. If, however, there is a cytoplasmic lining to the pores, or any other obstruction, much higher pressure gradients become necessary, and if the postulated conducting channels are of submicroscopic dimensions, the required forces assume impossible magnitudes; one estimate is $28\,\text{MPa}\,\text{m}^{-1}$. Current measurements of pressure gradients range from 0.02 to 0.55 $\text{MPa}\,\text{m}^{-1}$: it is clear that turgor-driven mass flow is possible only if the pores are more or less fully open.

Sieve plates and their pores are usually seen to be lined with a special gluco-polysaccharide, callose, which may narrow the pore diameter to $0.1\,\mu\text{m}$. With the consideration of callose we reach the first disputable point: it is arguable how much of the callose seen in any particular preparation was there in the living transporting cells, for it has been shown that injury, e.g. resulting from excision, can cause callose deposition on the sieve plates within seconds. The amount of callose present can affect pore diameter very strongly.

If the mass flow theory is accepted the sieve plates are regarded as safety devices to prevent loss of sap on wounding. From most species there is little exudation of sap from cut phloem in spite of the fact that the sieve tube contents are under considerable hydrostatic pressure, for as soon as the phloem is damaged, callose deposits and slime plugs block the plates near the cut. By repeated rubbing or beating, the system can be desensitized so that cutting no longer induces blockage – an effect that is utilized when sugar palms are tapped.

The contents of the pores cannot be accurately examined in the living state by light microscopy. The pore diameters are too near the limit of resolution of the light microscope, and moreover it is physically impossible to get a clear unobstructed view of a sieve plate (or pore) in living phloem: the plates lie within cells, and to avoid damage, a whole vascular strand must be viewed – sieve tubes cannot be dissected out undamaged. Electron microscopy overcomes the problems of resolution and obstruction, yet the results remain inconclusive. The sieve tubes are extremely delicate – highly hydrated, under high pressure, and very susceptible to fixation damage. It is generally agreed that the sieve tubes are lined by a plasmalemma that also lines the pores; that the few organelles present tend to be associated with the lateral walls; that there is no tonoplast; and that the P-protein is fibrillar, with the fibrils tending to lie parallel to the long axis of the cells. But according to the method of preparation and the species studied, the appearance of the pores varies from quite empty, or containing parallel but spaced-out P-protein filaments, to being completely packed with plugs of P-protein fibrils, and sometimes also endoplasmic reticulum. (For an illustrated account of phloem structure the reader is referred to another book in this series by E. G. Cutter: see *Further Reading*). Consequently there are current several models for sieve tube fine structure (Fig. 6.5). Proponents of the mass flow theory regard the more or less empty pores as representing the natural state, and dismiss the densely-filled pores as artifacts, containing material swept in as turgor was released on cutting, or

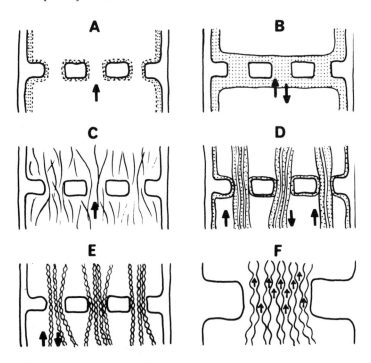

**Fig. 6.5**   Models of sieve tube structure showing the region of a sieve plate in longitudinal section (diagrammatic). Cytoplasm is shown stippled; arrows denote possible directions of flow. Modes of translocation possible with each model are noted.
A—empty lumen, open pores; mass flow driven by osmotic gradient, electro-osmosis.
B—membrane-covered, cytoplasm-filled pores: active transport, protoplasmic streaming.
C—filamentous meshwork in lumen, open pores: mass flow, spreading along interfaces, electro-osmosis.
D—membrane-bound strands with finer contractile tubules: pumping within tubules by contractile waves; protoplasmic streaming (Thaine's model).
E (low power) and F (enlarged view of one pore)— fine contractile lipoprotein tubules: pulse flow inside tubules, mass flow outside; spreading along interfaces (Fensom's model).
Bidirectional transport in the same sieve tube is possible in B, D and E.

coagulated on fixation, and squeezed into dense plugs by wound callose. Opponents of the mass flow theory argue that the empty-looking pores have their natural contents destroyed by fixation! Freeze-fracture studies have failed to resolve the conflict. Here the tissue is frozen instantaneously by being plunged into a liquid nitrogen cooled bath at $c. -196°C$, and the frozen tissue is fractured to expose the cells. All chemical fixatives are avoided and freezing takes only a fraction of a second compared with many minutes for chemical fixation; the possibility of any changes occurring is therefore minimized. R.P.C. Johnson (1978) has carried out some painstaking freeze-fracture studies of the phloem of the water lily *Nymphoides peltata*, freezing the petiole bundles while the bundles were still attached to the plants, and were actively translocating, as shown by applying $^{14}CO_2$ to the leaves and monitoring the bundles for radioactivity. The results were variable: even in the same sieve tube plate, some pores were empty, others had sparse filaments, some had close-packed longitudinal filaments; some had fibrils lying across them. Even the 'instantaneous' freezing may not be fast enough to prevent all movement of contents. Freezing takes at least 0.01 s; this time may seem very short, but R.P.C. Johnson has calculated that, should there be mass flow at 40 cm h$^{-1}$ (a moderate value), this is equivalent to a movement of 1.1 $\mu$m in 0.01 s, and several $\mu$m through the pores (to allow for their smaller cross-sectional area) – appreciable distances compared with the dimensions of the sieve plate pores. Between the start of freezing and the final solidification, there is still time for structural disturbances to take place.

Studies of fine structure have failed to resolve whether mass flow is feasible or not. Can physiological studies of the properties of the translocation process, of its reaction under various treatments, cast light on the nature of the mechanism?

### Physiological arguments for and against the mass flow hypothesis

According to the mass flow hypothesis, the sieve tubes act as passive conducting channels, with no necessity for the expenditure of metabolic energy on the way; the energy input takes place at the source and the sink, where the compounds translocated are respectively loaded into, and removed from, the phloem. Phloem tissues are, however, reported to have high rates of respiration; ATP is a regular constituent of sieve tube exudate, at an average concentration of 0.4 mM. In contrast to the concentration of sucrose which falls along the direction of translocation, the concentration of ATP remains constant, and the turnover rate is high. This suggests an expenditure of ATP all along the pathway. To check whether respiratory energy is in fact utilized throughout the phloem, limited lengths of petioles or stems on intact plants have been treated with low temperature, anaerobic conditions, or respiratory inhibitors such as cyanide, and the effect of these treatments on translocation through the treated area has been noted. In chilling-resistant plants, translocation through a chilled sector slows at first, but after a few hours the normal rate is resumed, or almost so; this has been the case even when the chilled length has been extended to 65–70 cm. This observation favours the Münch hypothesis, though the initial drop in rate is puzzling. In chilling-sensitive plants, low temperatures do very severely inhibit translocation, but this can be attributed to general cellular damage. Similarly, anaerobic conditions applied to 15 cm lengths of petioles of squash (*Cucurbita*) have failed (after a transient blockage) to inhibit translocation. In other instances, however, localized anaerobiosis has proved inhibitory. Inhibition of translocation by cyanide and DNP (2,4-dinitrophenol), both resulting in an inhibition of respiratory ATP production, has been observed, and could be interpreted as demonstrating the need for energy during translocation. But again, the application of these inhibitors could be causing drastic injury to all living phloem cells rather than specifically inhibiting the apparatus of translocation.

One observation which would conclusively disprove the Münch mass flow hypothesis would be the demonstration of simultaneous bidirectional transport in the same sieve tube, for mass flow can proceed in only one direction at a time. Bidirectional transport in the same vascular bundle does occur – this can be demonstrated by applying tracers above and below a sampling point; but there is no proven case of bidirectional transport in one *single, intact,* sieve tube. Fensom and his co-workers have introduced radioactive sucrose, radioactive potassium ions, and tritiated water, into individual sieve tube cells of *Heracleum sphondylium* by injection from very fine capillary pipettes, and have found the radioactive labels to spread *both* ways from the point of application at speeds of up to 450 cm h$^{-1}$; yet it is questionable whether the punctured phloem still functioned normally.

Still another line of investigation has involved the comparison of speeds of movement of different compounds in the same direction. In mass flow, it is argued, solvent and all solutes should move at the same rate. The speed of movement of different compounds, as judged by the time taken for radioactivity from the labelled compounds to move a certain distance, does actually vary, and importance has been attached to the observation that tritiated water (the solvent) seems to move more slowly than labelled solutes. This however, has later been shown to be because of a greater leakage rate of the labelled water out of the phloem: the more a compound leaks away sideways, the longer it will take to transport enough longitudinally for the radioactivity to be detectable. In xylem, where mass flow is not doubted, tritiated water seems to move more slowly than solutes. Even if truly different rates of movement in the same sieve tube were proved for two compounds, this would not disprove mass flow: substances could be separated out during flow by a process analogous to chromatography, e.g. by adsorption to P-protein.

As indicated above, neither structural nor physiological studies have disproved the feasibility of mass flow; but enough doubt has been cast on the hypothesis to invite serious consideration of alternative mechanisms.

**Alternative hypotheses for phloem translocation**

*(i)   Spreading along interfaces*

This concept visualizes a process in the phloem similar to the rapid spreading of oil at a water–air interface, to give a monomolecular film of translocated solute(s). This film would then remain intact, with molecules continually added at the source end and removed at the sink. Movement of molecules along an interface is much faster than diffusion. The P-protein fibrils would provide the surfaces for adsorption; any energy requirement along the pathway would be for maintenance of protoplasmic structure; no specific function in the transport process is assigned to the sieve plates. The spreading of certain dyes in the phloem could be quoted as evidence for such a mechanism, and perhaps also the two-way spreading in the microinjection experiments of Fensom (above). But the observed rates of mass transfer of sucrose are high; there does not appear to be enough P-protein to carry so much sucrose. Moreover, compounds of very different nature are carried in the phloem and a surface suitable for adsorption of, say, sucrose would not be suitable to carry amino acids. This hypothesis also fails to account for the movement of larger particles – viruses, and carbon black, which do move in the phloem.

### (ii)  Protoplasmic streaming and transcellular strands

In its original form, the protoplasmic streaming hypothesis considered compounds to be carried by streaming from one end of a sieve tube element to the other, and then to be transferred across the sieve plate by 'activated diffusion' or 'active transport'. Energy would be required for both the streaming and the transfer process at the plates. A second hypothesis involving protoplasmic streaming was proposed by R. Thaine, who claimed to observe, in living phloem, cytoplasmic strands 1–7 $\mu$m thick passing through sieve plates from one cell to the next (Fig. 6.5) and containing actively streaming cytoplasm. Subsequently he succeeded in demonstrating by electron microscopy of phloem of *Cucurbita*, structures which can be interpreted as membrane-bound strands passing through sieve plate pores. Thaine postulates the strands to be continuous throughout the length of a sieve tube, and to contain hollow tubules. Movement of compounds can occur both in the streaming cytoplasm, and within the tubules where peristaltic waves of contraction would speed the flow along. Thus different compounds could move inside and outside the tubules at different speeds. In the same cell, strands can – Thaine claims – be seen streaming in opposite directions, so that bidirectional flow could be accommodated. (Indeed the protoplasmic flow *must* reverse when it reaches the end of a sieve tube, or streaming cannot be maintained.) The sieve plates would act as supports to the long, thin strands.

The main criticism of this hypothesis is that all other workers have failed to see such strands. Thaine believes this to be due to the extreme fragility of the structures. Another criticism, that the streaming rates are too low (3.5 cm h$^{-1}$) compared with the velocity of translocation (20–450 cm h$^{-1}$), has been countered by postulating the peristaltic tubules, pumping rapidly; but these are quite hypothetical (see also (iii), below).

### (iii)  Pumping tubules and lashing filaments

These concepts have been advocated mainly by the school of D.S. Fensom. It is proposed that the sieve tubes contain contractile protein tubules, up to 60 nm in diameter when expanded. These would pump fluid along by pulses of contraction and many tubules all pulsing in unison could further propagate a mass flow outside the tubules. In addition, contractile protein filaments, attached to the outsides of the tubules and/or sieve tube walls, could beat like miniature cilia to waft the flow along. Energy would be required for the protein tubule contractions and filament lashings; bidirectional flow would be feasible if different tubules transported in different directions.

Fensom considers a 'shimmering' appearance of translocating sieve tubes, and the apparent vibration of plastids in them, to be suggestive of such activity. But there is no direct evidence, and the universal contractile protein actin is absent from sieve tubes, whilst P-protein has not been shown to have the properties of contractile proteins.

### (iv) Electro-osmosis

The electro-osmotic theory of phloem transport can be credited mainly to D.C. Spanner, who over a period of years has developed the idea into an elaborate hypothesis, of which only an extremely simplified account can be given here. (See Spanner, 1979, *Further Reading*). Electro-osmosis means the movement of ions along an electric potential difference (PD) gradient through charged pores, carrying with them a flow of water – and other solutes if present. In the phloem, electro-osmosis thus involves a mass flow of solution, initiated by a turgor gradient, but the sieve plates are now regarded as 'pumping stations'. An electric PD is postulated to be built up across the sieve plate; the pores are negatively charged, and the positively charged potassium ions (which occur in high concentrations in sieve tube sap) sweep through the pores, carrying the water and sucrose and other solutes, with them. In the original proposition, the PD gradient was thought to result from a continuous circulation of $K^+$ around the sieve plate: $K^+$ would be secreted into the sieve tube (from companion cells) above the plate; below, the $K^+$ would be taken up into the companion cells again. The hypothesis was, however, severely criticized on the grounds that the amount of potassium circulation involved required an impossibly high energy expenditure; and that the circulated ions amounted to an electric current sufficient to vapourize the cell contents! Spanner has now modified the hypothesis and suggests that the ions recirculated are hydrogen ions, for which the energy expenditure should be more moderate. He also visualizes the plates as 'firing' in pulses, building up a PD and discharging it. Potassium ions are still assumed to play an important part in moving through the charged-up plates and setting up the electro-osmotic flow. The attractiveness of this hypothesis is that it offers a very specific role to the sieve plates; the P-protein, too, is given a role in maintaining negative charges in the pores, and the high potassium content of sieve tube sap becomes explicable. At the same time, there is a mass flow, so that observations in favour of mass flow are compatible with the theory. Experimental proof, however, is lacking.

The possibility that several mechanisms operate in the phloem should not be discounted. There is great variability in the diameters of sieve plate pores. Possibly in plants such as *Cucurbita*, where the pores are very large, osmotically driven mass flow is the main mechanism; while in species with finer pores other mechanisms gain importance. This might explain some of the conflicting reports about, for example, the effects of anaerobiosis applied to a length of axis, which sometimes has seemed to inhibit translocation, sometimes not.

## The companion cells

The companion cells retain their nuclei and have a normal cytoplasm, particularly rich in mitochondria and ribosomes. Their precise function is uncertain, but it is generally assumed that they exercise some control over the activities of the enucleate sieve tube cells, to which they are joined by numerous plasmodesmata. The 'sieve element-companion cell complex' is sometimes considered as one functional unit. The high respiration of phloem is believed to reflect the activity of companion cells, since there are few mitochondria in mature sieve tubes. The companion cells may also be involved in the transfer of sucrose and other metabolites into and out of sieve tubes. The sucrose concentration in companion cells has been found to be as high as in the sieve tubes and in the minor veins of leaves, where much phloem loading occurs, the companion cells are particularly large. Contrary to the concept of the companion cells fulfilling a vital role in solute transfer into the sieve tubes are observations that radioactive phosphate and sulphate can move directly into exposed sieve tubes, and that phloem may lack companion cells altogether.

## Phloem loading, unloading, and transfer cells

The process of passage of solutes into the phloem at the source is generally referred to as 'loading' of the phloem. Loading occurs in the minor veins of leaves, which penetrate the mesophyll as a dense network (Fig. 6.1). The loading is a selective, and active process, disrupted by inhibitors of energy metabolism. It proceeds against a concentration gradient; when radioactive $CO_2$ is applied to leaves, the minor veins accumulate higher levels of radioactivity than the photosynthesizing mesophyll. The sugar concentration in the sieve tubes of the leaf is about twice that of the mesophyll. It is believed that the sugar moves towards the veins first in the cytoplasm and via plasmodesmata, but is then discharged into the apoplast (cell walls) of the fine veins. From the apoplast, the sugar is actively transported across cell membranes into the phloem. At the sink, it is thought that sucrose entering the apoplast is hydrolysed by invertase in the cell walls, and the monosaccharides cannot pass back into the phloem (which is highly selective in its sugar uptake), and are absorbed by the sink cells.

The possible role of the companion cells in mediating the passage of compounds into the sieve tubes has already been mentioned. A similar role may be played by the *transfer* cells. Transfer cells are characterized by wall ingrowths which greatly increase their cell wall and the plasmalemma surface area. They are common in minor veins, especially in some members of the Leguminosae. In the leaves of *Pisum*, the transfer cells differentiate at about the same time as export of photosynthate from the leaves begins, and fail to develop in the dark. Transfer cells differentiate in cotyledons of several leguminous species early during germination, and in root nodules – regions where there is much metabolite transport activity. Cells with similar wall protuberances are found in glands, where active secretion across cellular membranes is known to occur. Hence the transfer cells may also be active in loading and unloading of phloem. (They are associated with the xylem, too, and may secrete compounds into xylem sap, and absorb solutes from it.) But if transfer cells are involved in phloem loading, they are not indispensable, for they are not present in all species; no transfer cells have been reported from grass leaves, for example.

### Developmental changes in the phloem

In most perennials with secondary growth the sieve tubes and companion cells die after one growing season, though phloem parenchyma cells and fibres survive longer. In perennial monocotyledons lacking secondary growth, such as the palms, sieve tubes persist for many years, retaining structural integrity. It is very difficult to check whether any particular sieve tube in a full-sized tree is still functional, but by means of radioactive labelling, translocation has been demonstrated in palm sieve tubes 6–7 years old. In a few dicotyledons also, such as the lime (*Tilia*) and the grapevine (*Vitis*), the conducting cells survive for several seasons. In the lime, some sieve tube pores remain open throughout the year, though the pore diameters are restricted in the winter by callose; in the grapevine, a deposit of 'dormancy' callose closes the pores completely in the winter, but decreases when transport becomes active. Temporary depositions of callose, disappearing within days or even hours, take place, and may be a normal means of controlling the direction of transport during the season of active translocation. Some observations on diurnal changes in the direction of translocation would become explicable if, at a particular node of the stem, the phloem becomes blocked by such callose formation for several hours during the day.

## The transport of hormones and stimuli

All the known plant hormones have been identified in both xylem and phloem sap (with the exception of the gaseous hormone ethylene.) Externally applied hormones, too, find their way into the conducting tissues. When the synthetic auxin, 2,4-D (2,4-dichlorophenoxyacetic acid, p. 160) is applied to leaves, it moves out with the sugars in the phloem; when it is applied to the roots, it moves upwards with the transpiration stream in the xylem. The physiological significance of the long-distance transport of hormones in vascular tissues is not clear. The concentrations of the hormones in the saps may be appreciable and there is some evidence that the cytokinins and gibberellins moving up in the xylem sap and presumably synthesized in the roots, are of importance in growth control in the shoot, and in the maintenance of healthy leaves. But shoot organs are also themselves capable of synthesizing cytokinins and gibberellins.

The movement of hormones is, however, not confined to the vascular tissues but can proceed through parenchymatous cells, and this medium-distance transport is thought to be of physiological importance, as will be discussed in Chapters 9 and 10. It has been known for a long time that the movement of auxins involves a specific mechanism. The most striking feature of this natural auxin transport is its *polar* character. In shoots auxin moves 'down' a plant, or *basipetally*, away from the apical region in which it is synthesized. The polarity is inherent in the organism and in many cases is unaffected by the orientation of the plant: when the shoot is inverted the transport still proceeds away from the apex (see also Chapter 9). It should be noted that while net transport is basipetal, some acropetal movement does occur. In roots the situation is less clear-cut. *Acropetal* movement (towards the tip) has been identified, and it is believed that roots receive auxin from the shoot. But radioactive auxin applied to the root cap moves basipetally. It has been suggested that auxin in roots may move acropetally in the stele and basipetally in the outer tissues. Not all tissues of an organ are necessarily involved in the polar transport; acropetal movement is strongly localized in the stele of maize roots and *Nicotiana* stems, the cortex and pith lacking polarity and transporting at a lower rate. On the other hand, in coleoptiles polar movement of auxin occurs in the cortical parenchyma.

The rate of polar hormone transport is much lower than the rate of transport in the vascular tissues, not reaching more than 2.0 cm h$^{-1}$. The transport can occur against a concentration gradient and is generally regarded as dependent on the presence of living cells, though it has been reported that auxin moved faster in boiled *Coleus* petioles than in living; polarity was, however, completely abolished by boiling. This has led to the idea that the polarity does not involve an active transport in, say, the basipetal direction in shoots, but rather an active inhibition of movement in the opposite direction, in living tissues. This view has not gained much credence. Polar transport is also dependent upon aerobic metabolism of the cells. Polar auxin transport is strongly and specifically inhibited by the 'antiauxin' TIBA (2,3,5-tri-iodobenzoic acid) and also by prolonged exposure to ethylene. Some tissues, such as the hypocotyl hook of the French bean (*Phaseolus vulgaris*) at a certain developmental stage, are unable to carry out the transport of auxin. The capacity for polar transport seems to diminish in tissues as they age, and its maintenance also depends on a continued auxin supply; when the auxin-producing tip is removed from e.g. a sunflower stem, it fairly rapidly loses its capacity to transport auxin.

For a considerable time it was believed that polar transport occurred only with natural auxins and a few artificial compounds with auxin activity. More recently polar transport of gibberellins also has been demonstrated, occurring basipetally in both roots and shoots and cytokinins have been found to move basipetally in hypocotyl and petiole tissue, acropetally in roots. Very much less is known, however, of the transport of these other hormones compared with that of auxins.

The mechanism of polar transport is not understood. Electric gradients have been proposed as determining the polarity; experiments, however, have indicated that these gradients may be the result of polar auxin transport, but are certainly not its cause. As the auxin moves along a longitudinal file of cells, the apical and basal ends of each cell must somehow be differentiated so that, at each cell junction auxin moves preferentially *out of* the basal end of a cell and *into* the apical end of the next cell (in the case of basipetal transport). There is some evidence that the active secretion is localized at the basal ends of the cells, uptake by the apical ends being passive. If the hormone remains in solution in the cytoplasm, it could be carried to the basal end of the cell by cytoplasmic streaming, ready for exit into the next cell though polar transport still occurs even if cytoplasmic streaming is inhibited. The polarity is established during cellular growth and differentiation and is probably a special aspect of the general polarity acquired by cells during their differentiation. Once established, the polarity is tenaciously maintained. In cuttings inverted and forced to root at the shoot end by hormone treatment the original polarity of IAA transport has persisted for several months, in spite of the reversal of the positions of the root and shoot apices.

The photoperiodic stimulus has not been isolated as a chemical. It is believed to be transported in phloem since it moves strictly along the path of carbohydrate transport, and the movement is enhanced by the application of sugar to a photoperiodically induced leaf (see Chapter 12), presumably because this speeds up phloem translocation from the leaf. The rate of movement is reported as $0.2–0.4$ cm h$^{-1}$. This is about a hundredth of the rate of carbohydrate transport (see p. 95). However since the estimation of rate of movement of the photoperiodic stimulus is based on a subsequent appearance of the flowering response, one is really measuring the time needed for accumulation of the stimulus to the level where it will elicit a response in the subsequent absence of the induced leaf or leaves. The time required for the stimulus to reach the apex is presumably much shorter, and the true rate of transport is much faster.

FURTHER READING

CUTTER, E. G. (1978). *Plant Anatomy: Experiment and Interpretation. Part I. Cells and Tissues,* 2nd edition. Contemporary Biology Series. Edward Arnold, London.

FENSOM, D. S. (1981). Problems arising from a Münch-type pressure flow mechanism of sugar transport in the phloem. *Canadian Journal of Botany,* **59**, 425–32.

LÜTTGE, U. and HIGINBOTHAM, N. (1979). *Transport in Plants.* Springer-Verlag, Berlin.

MOORBY, J. (1981). *Transport Systems in Plants.* Longmans, London.

PIRSON, A. and ZIMMERMANN, M. H. (eds) (1975). *Transport in Plants I. Phloem Transport. Encyclopedia of Plant Physiology, New Series, Vol. I.* Springer-Verlag, Berlin.

RICHARDSON, M. (1975). *Translocation in Plants,* 2nd edition. Studies in Biology no. 10, Edward Arnold, London.

SPANNER, D. C. (1979). The electroosmotic theory of phloem transport: a final restatement. *Plant, Cell and Environment,* **2**, 107–21.

SELECTED REFERENCES

DE MARIA, M. E., THAINE, R. and SARISALO, H. I. M. (1975). Fine structure of sieve tubes prepared mainly for observation in the electron microscope by a cryogenic method. *Journal of Experimental Botany*, **26**, 145–60.

DEMPSEY, G. P., BULLIVANT, S. and BIELESKI, R. L. (1975). The distribution of P-protein in mature sieve elements of celery. *Planta*, **126**, 45–59.

DRAKE, G. and CARR, D. J. (1978). Plasmodesmata, tropisms and auxin transport. *Journal of Experimental Botany*, **29**, 1309–18.

FISHER, D. B. (1978). The estimation of sugar concentration in individual sieve-tube elements by negative staining. *Planta*, **139**, 19–24.

GOLDSMITH, M. H. M., GOLDSMITH, T. H. and MARTIN, M. H. (1981). Mathematical analysis of the chemiosmotic polar diffusion of auxin through plant tissues. *Proceedings of the National Academy of Science, U.S.A.*, **78**, 976–80.

GUNNING, B. E. S. and PATE, J. S. (1974). Transfer cells. In *Dynamic Aspects of Plant Ultrastructure*, ed. Robards, A. W., 441–80. McGraw-Hill, London.

JOHNSON, R. P. C. (1978). The microscopy of P-protein filaments in freeze-etched sieve pores. Brownian motion limits resolution of their positions. *Planta*, **143**, 191–205.

MILBURN, J. A. (1970). Phloem exudation from castor bean: induction by massage. *Planta*, **95**, 272–6.

SHAW, S. and WILKINS, M. B. (1974). Auxin transport in roots. X. Relative movement of radioactivity from IAA in the stele and cortex of *Zea* roots. *Journal of Experimental Botany*, **25**, 199–207.

STOUT, P. R. and HOAGLAND, D. R. (1939). Upward and lateral movement of salt in certain plants as indicated by radioactive isotopes of potassium, sodium and phosphorus absorbed by roots. *American Journal of Botany*, **26**, 320–4.

WARK, M. C. (1965). Fine structure of the phloem of *Pisum sativum*. II. The companion cell and phloem parenchyma. *Australian Journal of Botany*, **13**, 185–93.

WARK, M. C. and CHAMBERS, T. C. (1965). Fine structure of the phloem of *Pisum sativum*. I. The sieve element ontogeny. *Australian Journal of Botany*, **13**, 171–83.

WEILER, E. W. and ZIEGLER, H. (1981). Determination of phytohormones in phloem exudate from tree species by radioimmunoassay. *Planta*, **152**, 168–70.

# 7

# Resistance to Stress

Plants growing in the field habitually encounter a number of environmental stresses, e.g. drought and frost. The ability to withstand such stresses frequently becomes the limiting factor for plant growth, survival, and geographical distribution. The study of the reactions of plants under stress is of great scientific interest. What confers upon one plant organ the ability to withstand freezing in liquid nitrogen at $-196°C$, when another cannot endure freezing at $-2°C$, and yet a third succumbs to chilling at $+5°C$? Failure to give definitive answers to such questions exposes major gaps in our understanding of living plant cells. Moreover, study of the behaviour of plants under stress is of practical importance, since agricultural yield is only too often drastically reduced by stressful external factors.

In this chapter, the resistance of plants to water (deficit) stress, low temperature, high temperature, salinity, and heavy metals is discussed. Additional stresses, which are precluded from treatment here by lack of space, include those arising from flooding, radiation, and various toxic chemicals such as aerial pollutants. A comprehensive account is given in the monograph by Levitt (1980); see *Further Reading*, p. 134.

The literature on stress physiology contains numerous variations in terminology. The following general terms are used here, and more specific terms are defined in appropriate sections; some common alternatives are given in parentheses.

*Stress resistance:* the ability to endure an *externally* applied stress.

*Stress avoidance:* the ability to prevent an externally applied stress from producing an equivalent *internal* stress in a plant.

*Stress tolerance:* the ability to endure an *internal* stress induced by the externally applied stress.

The relationship between these qualities may be represented diagrammatically (Fig. 7.1).

*Hardening (acclimation):* the development of stress resistance, stimulated by the application of mild and/or gradually increasing stress. Some species are capable of hardening towards a particular stress, others are not. The term 'hardiness' has been applied both in the sense of 'resistance' or 'tolerance' as defined above.

**Fig. 7.1**

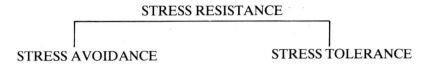

STRESS RESISTANCE

STRESS AVOIDANCE              STRESS TOLERANCE

## Resistance to water stress (drought resistance, drought hardiness)

### Measurement of stress: definition of resistance

Except in the most humid environments, plants are likely to suffer some degree of water stress many times during their life cycles. For human populations, drought has been a scourge down the ages and hardly a year passes without reports of drought-induced famine in some part of the world. Growth and photosynthesis, the two physiological activities most critical to crop yield, are among the first plant activities to be inhibited by water deficit stress. According to a current estimate, the global losses in potential agricultural yield caused by water deficit stress exceed losses from all other causes combined.

### *Measurement of stress*

Quantification of the external water stress is straightforward; the water status of the environment is expressed as the external water potential, ($\psi$); as the relative humidity (R.H.) or as vapour pressure (e). Tissue dehydration may also be measured in terms of the change in tissue $\psi$, or the tissue water content (percentage of tissue weight). For comparisons of water deficits in different tissues, however, the $\psi$ and the water percentage both suffer from the disadvantage that at full turgidity, individual tissues differ widely in water contents and in the value of $\psi$. A fully turgid leaf may contain 85% water whereas a fully imbibed seed storage cotyledon holds only 60%; dehydration to, say, 55% water means a relatively much greater water loss for the leaf than for the cotyledon. In comparisons of water stress resistance, the *relative water content*, RWC, is sometimes used to indicate the degree of water deficit:

$$RWC = \frac{\text{tissue water content}}{\text{tissue water content at full turgidity}} \times 100$$

$$= \frac{\text{(fresh weight)} - \text{(dry weight)}}{\text{(fully turgid fresh weight)} - \text{(dry weight)}} \times 100$$

The RWC thus indicates the proportion of the saturation water content still *retained* by a tissue. Some workers prefer to employ the *water saturation deficit*, WSD, a measure of the proportion of the saturation water content *lost* from the tissue: WSD = 100 − RWC.

When tissue water contents become very low (below *c.* 10%), as in air-dry material, the $\psi$ values, too, become extremely low and it becomes difficult to obtain accurate measurements of $\psi$. In such circumstances, the water status of a plant tissue, and its dehydration resistance, may be expressed as the equivalent R.H. with which it is in equilibrium. R.H. and $\psi$ values are interconvertible (see p. 48) but, at very low $\psi$ levels, the conversion becomes inaccurate.

It should also be noted that water stress resistance is sometimes discussed in terms of the level of external stress that can be endured, but at other times as the internal water deficit that the tissue can withstand.

## Criteria of resistance

The water stress resistance of a tissue is defined as the maximum level of stress that it can survive. This level is, however, by no means straightforward to measure. Long-term survival can be tested only with intact plants, but many experiments are conducted on isolated organs. Survival of such material must be judged by its ability to exhibit some function(s) of living cells, e.g. the accumulation of a 'vital' dye such as neutral red; retention of membrane semipermeability; or the enzymatic reduction of tetrazolium salts. But, as stated by Alexandrov in 1964, 'between the cell which is alive and that which is dead there are numerous intermediate transitions.' Some vital functions may persist after others have been irreversibly impaired. Injury symptoms may take hours, days or even weeks to become manifest; plants may apparently recover only to die a few days later. The ability to withstand a specified degree of dehydration is profoundly influenced by the speeds of desiccation and rehydration and by the time spent in the dehydrated state; it may vary with such conditions as irradiance or temperature. A detached organ may show a different degree of resistance than the same organ on the intact plant. Any value of resistance that is quoted holds strictly only under the particular conditions of observation. Since replicate plants, and individual cells in a tissue, vary in their resistance, the degree of resistance is frequently expressed as the $LD_{50}$, the level of stress that just kills 50% of the individuals, or cells. Deleterious effects are produced by water stress long before the lethal limit is reached. In spite of these difficulties in measuring resistance, great differences have been demonstrated in the resistances of different tissues.

## Water stress avoidance

The generalized diagram showing the different aspects of stress resistance (Fig. 7.1) can be made specific for water stress, and expanded, as shown in Fig. 7.2. Even under conditions of acute water shortage, some plants can reduce their rate of water loss to a very low level by means of morphological features and physiological mechanisms, so that a relatively high $\psi$ is maintained within their tissues. The adaptive xeromorphic features include

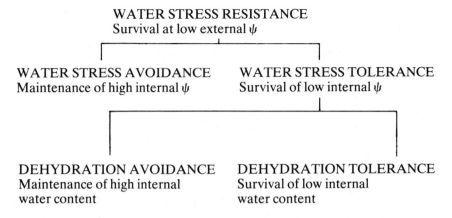

WATER STRESS RESISTANCE
Survival at low external $\psi$

WATER STRESS AVOIDANCE
Maintenance of high internal $\psi$

WATER STRESS TOLERANCE
Survival of low internal $\psi$

DEHYDRATION AVOIDANCE
Maintenance of high internal
water content

DEHYDRATION TOLERANCE
Survival of low internal
water content

**Fig. 7.2** The categories of water stress resistance in plants, based on the definitions by Levitt (1980), defining water stress as equivalent to low $\psi$. If water stress is defined as low water content, the dehydration-avoiding plants would be included in the category of stress avoiders.

thick cuticles, low surface/volume ratios, possession of succulent water storage tissues, sunken stomata, rolling of leaves, daytime closure of stomata, hairiness, and high reflectance. Such plants are the water stress avoiders and this category includes many xerophytes, including desert species, e.g. cacti. The cells of many such plants cannot withstand dehydration to any marked degree. Water stress avoidance mechanisms are clearly of supreme importance in arid habitats. In this chapter, however, emphasis is placed on the physiology of water stress tolerance. The different types of resistance mechanisms are not mutually exclusive; a plant growing in a xeric habitat may have a water-conserving mechanism as a first line of defence, and be able to tolerate desiccation when conditions become more extreme. Many resurrection plants for instance (see below) are also typically xeromorphic drought avoiders.

**The range of water stress tolerance in flowering plants; developmental changes**

Most angiosperms pass through a stage of very low water content as mature seeds, which can dry out till their water content is in equilibrium with atmospheric humidity (R.H. 20–50%). This brings their water contents to 5–20%, corresponding to $\psi$ values of $-100$ to $-10$ MPa. The water content of some seeds has been experimentally reduced by storage in a vacuum or by brief heating above 100°C to below 1% without loss of ability to germinate. Seeds of birch (*Betula* spp.) have remained viable for 1½ years with water contents of 0.01–0.4%. Not all seeds are so highly resistant towards desiccation, e.g. seeds of silver maple (*Acer saccharinum*) are shed from the parent plant with a 58% water content and die if this falls below 30–34%, and many species from the humid tropics have dehydration-sensitive seeds. Nevertheless the possession of 'dry' seeds is very widespread among the flowering plants.

During germination, the desiccation tolerance of seeds is lost. The early stages of imbibition are still reversible; seeds in the imbibition stage of germination (p. 9) can be dried back to the air-dry condition again without damage. Wheat seedlings can be dried to a RWC of 2% up to the stage when their coleoptile is 3–4 mm long, without lethal effect, although the emerged radicles are already unable to endure this and are replaced by initials in the embryo. Thereafter the wheat seedlings become progressively more sensitive to desiccation. Similar observations have been made with seedlings of other species.

After the seed and very early seedling stages, most flowering plants can never again endure desiccation to really low water contents. An exception is provided by the so-called **resurrection plants**, which tolerate extreme dehydration and revive rapidly when rewetted. Over 80 species of resurrection plants are known among the angiosperms, mainly from the hot dry regions of S. Africa (*Chamaegigas intrepidus, Myrothamnus flabellifolia, Talbotia elegans, Xerophyta* spp.); and Australia (*Borya nitida,* several grasses). During periods of drought, the leaves of these plants naturally reach air-dryness, (R.H. 50% to 30%), corresponding to $\psi$ of c. −95 to −160 MPa; the most tolerant can survive drying in air of 0% R.H. In comparison, the dehydration tolerance of leaves of many mesophytic crop plants falls into the range of R.H. 98–85%, or $\psi$ of c. −3 to −22 MPa. A small percentage of water still remains bound in the tissues even when the external humidity is at zero. While numerically the resurrection plants make up an insignificant fraction of flowering plant species, their striking water stress tolerance makes them important subjects for research. Among bryophytes, lichens and algae an ability to endure desiccation to the air-dry condition is quite common.

In perennials of temperate and arctic climates, water stress tolerance in stems and evergreen leaves shows a regular seasonal fluctuation, being greatest in the winter months. During the winter, plants are subjected to water stress owing to diminished uptake by the roots, even when the soil $\psi$ is high. Long-term growth with a low water supply can induce xeromorphism and many species can be hardened by short periods (a few days) of moderate water stress, to endure subsequently a more severe water deficit. The resurrection plants may need hardening before they exhibit maximal tolerance.

### The effects of water stress on plant cells

Almost every process in plant cells has been found to be affected by water stress. It becomes therefore very difficult to pinpoint the primary lesion(s) directly due to water deficit and to distinguish between primary events and indirect effects consequent upon the primary disturbance(s).

## Metabolic disturbances

As successively greater water stress is applied, successive metabolic disturbances become apparent (Fig. 7.3). Growth is extremely sensitive to water stress and growth rate may begin to decrease as soon as water content falls below full saturation. The inhibition of growth and associated biosynthesis may induce further deleterious physiological effects through the accumulation of metabolic intermediates that cannot be utilized.

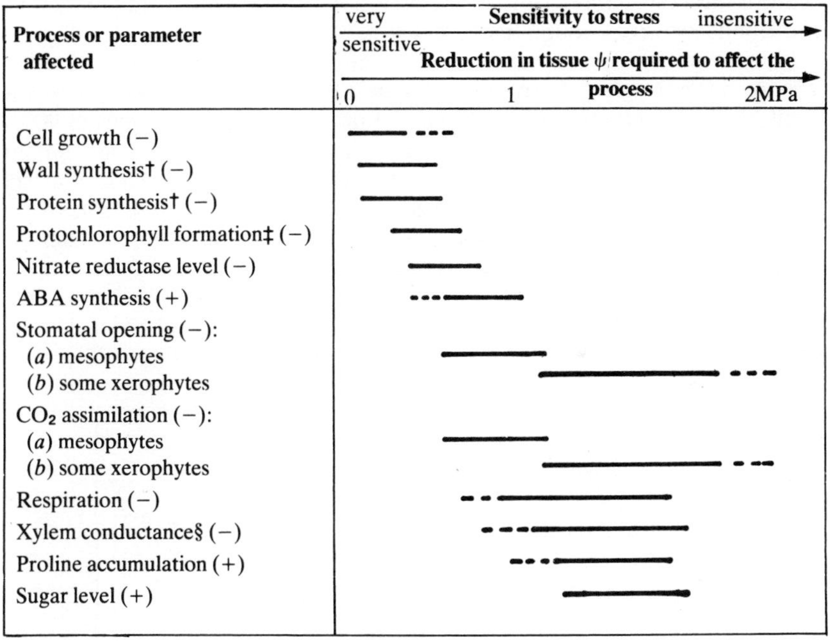

| Process or parameter affected | very sensitive | Sensitivity to stress | insensitive |
|---|---|---|---|
| | Reduction in tissue $\psi$ required to affect the process | | |
| | 0 | 1 | 2MPa |
| Cell growth (−) | ▬▬ ▬▬ | | |
| Wall synthesis† (−) | ▬▬▬ | | |
| Protein synthesis† (−) | ▬▬▬ | | |
| Protochlorophyll formation‡ (−) | ▬▬▬ | | |
| Nitrate reductase level (−) | ▬▬▬ | | |
| ABA synthesis (+) | ▬▬▬▬▬ | | |
| Stomatal opening (−): | | | |
| (a) mesophytes | ▬▬▬▬ | | |
| (b) some xerophytes | ▬▬▬▬▬ ▬▬ | | |
| CO₂ assimilation (−): | | | |
| (a) mesophytes | ▬▬▬▬ | | |
| (b) some xerophytes | ▬▬▬▬▬ ▬▬ | | |
| Respiration (−) | ▬ ▬▬▬▬▬▬ | | |
| Xylem conductance§ (−) | ▬▬▬▬▬▬▬ | | |
| Proline accumulation (+) | ▬▬ ▬▬▬▬▬ | | |
| Sugar level (+) | ▬▬▬▬▬ | | |

† fast growing tissue
‡ etiolated leaves
§ should depend on xylem dimension

**Fig. 7.3** Generalized sensitivity to water stress of plant processes or characters. Length of the horizontal lines represents the range of stress levels within which a process becomes first affected. Dashed lines signify deductions based on more tenuous data. The reduction in tissue $\psi$ is in comparison to $\psi$ of well-watered plants under mild evaporative demand. In the left column, (+) indicates that water stress causes an increase in the process or character and (−) signifies a decrease. (Adapted from Hsiao, Acevedo, Fereres and Henderson, 1976, *Philosophical Transactions of the Royal Society B*, **273**, 479–500.)

Photosynthesis ('CO₂ assimilation' in Fig. 7.3) is affected at quite moderate levels of water deficit in mesophytes, and rapidly declines to zero. This, if at all prolonged, will lead to a shortage of nutrients, especially as respiration is less sensitive and continues at lower water levels; in some tissues, respiration rate actually rises at moderate water deficits. Once the respiration rate does decline ATP shortages could occur and uncoupling in response to water stress has been reported. Overall protein content tends to fall; soluble protein levels may decrease by 40–60% and the levels of some individual enzymes, e.g. nitrate reductase, decrease; the activities of some hydrolytic enzymes on the other hand rise. The increased hydrolase activity combined with inhibited synthesis probably accounts for the overall decline in protein content. The fall in xylem conductance (Fig. 7.3), reflecting cavitation under increased tension (p. 60) aggravates water shortage. Over a period of time, a combination of such physiological effects could lead to an overall metabolic disturbance severe enough to cause death.

Marked effects of water stress become apparent when the extent of water loss from the tissues is so small that a direct effect of lowered water content would not be predicted. Growth may be inhibited at RWC values of 90%. Such a decrease in the amount of water in the cell, and the concomitant increase in cell solute concentration, would scarcely be expected to affect activities of cellular enzymes. *Turgor pressure*, however, diminishes very rapidly in the initial stages of water loss (Fig. 7.4), unless there is osmotic adjustment (see below, p. 118). It is currently believed that this decrease in turgor is the primary event that inhibits growth. Plant cells will grow only when the protoplast exerts a positive pressure on the cell wall and a yielding of the wall under turgor pressure is a prerequisite for growth (p. 172).

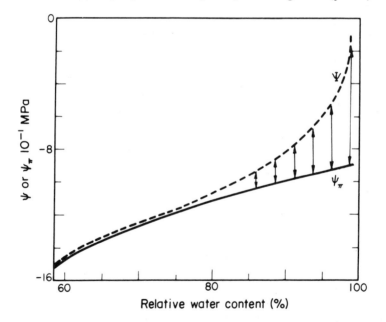

**Fig. 7.4**   Generalized relationships of water potential ($\psi$), osmotic potential ($\psi_\pi$) and relative water content for leaves of herbaceous crop plants. Length of arrows indicates the magnitude of positive pressure potential ($\psi_p$) which accounts for the difference between $\psi$ and $\psi_\pi$. (Source as Fig. 7.3.)

In mature tissues, water stress damage cannot be attributed to a turgor loss-mediated cessation of growth. The loss or decline of turgidity might affect the properties of cellular membranes: the permeability, and possibly enzymatic properties, of the plasmalemma and tonoplast could be different in the flaccid and turgid states. Stomatal opening also depends on turgidity of the guard cells, and the decline in rate of photosynthesis is to some extent secondary to stomatal closure: in both mesophytes and xerophytes, the commencement of photosynthetic decline coincides with the closure of stomata (Fig. 7.3). Yet there is also a more direct effect of water stress on photosynthesis. When submerged aquatics, which do not rely on stomatal entry of $CO_2$, are dehydrated in sugar solutions, their rate of photosynthesis falls. Chloroplasts, isolated from water-stressed plants and assayed *in vitro* with ample water supply, still show lowered rates of electron transport and lowered activities of $CO_2$-fixing enzymes. Mitochondria isolated from water-stressed plants also have impaired activities.

Once water contents have become appreciably lowered, the changes in concentration of cellular solutes could have significant effects on metabolic reactions – not perhaps so much through alterations in substrate concentrations as by alterations in the activities of cellular enzymes. Enzyme activity depends on the precise three-dimensional configuration of the enzyme protein, and electric charges play a role in maintaining the tertiary structure of the molecule. The concentration of charged solutes owing to water withdrawal could affect the charges on enzymes, and thereby alter their activities.

## Structural damage

Whilst at slight or moderate water stress the effects of dehydration can be described as physiological or metabolic, there can be no doubt that, with severe desiccation, death results from structural disorganization of the protoplasm. When cells dry out, the protoplast will be subjected to tension resultant on its contraction in volume and its adherence to the cell wall – stress which might tear the plasmalemma. In agreement with this idea are observations that gradual desiccation can be endured much better than sudden drying: this can be interpreted to indicate that cell structure can adjust to slowly increasing stress. Gradual remoistening is also beneficial; if dried tissue is abruptly flooded with water, stresses are again produced as the walls and outer layers of the protoplast hydrate before the inner parts of the cell. At a finer level, there is evidence for structural damage to macromolecules, i.e. denaturation, especially of proteins. In recent years, attention has been focussed on *membranes* as the key sites of damage, resulting from denaturation of membrane proteins. Levitt (1980) considers the reactions of protein SH (sulphydryl) groups to be vital. Many proteins contain a number of SH groups, and SS (disulphide) bridges derived from the oxidation of two adjacent sulphydryl groups often maintain the native configuration of a protein. If during desiccation proteins become unfolded, previously masked SH groups could come to the surface, while at the same time the withdrawal of water brings cellular proteins closer together. The

SH groups of adjacent molecules could then form SS links, binding the molecules into irreversibly denatured aggregates. The initial unfolding could result from mechanical stress, from disturbance of surface charges, and from loss of water of hydration. It has been proposed that, in very dry tissue, membrane lipids become oxidized by the formation of free superoxide radicals, $O_2^{\bullet-}$. In hydrated tissues, these radicals are removed by the enzyme superoxide dismutase.

Water deficit stress damage emerges as a multifaceted process, and the cause of injury varies according to the species, and the degree and duration of dehydration. If the water content remains high enough to permit the continuance of metabolic activity, the persistence of abnormal metabolism will result in a progressive increase in injury. Provided a lethal water content has not been reached, some leaves and epidermal strips have been found to endure a greater water loss better than a smaller one – presumably because, with greater water loss, metabolic reactions were slowed down more. Seeds, too, maintain their viability longer when stored at relatively low water contents. At very low water contents, however, structural damage to protoplasm becomes critical. Where tissues are very sensitive to structural injury moderate dehydration will be less injurious than extreme dehydration; when tissues are less sensitive towards structural injury a really low water content will have the beneficial effect of inhibiting the progress of metabolic disturbance. Even at very low water contents, injury may increase with time, i.e. adverse changes do still occur at water contents too low for detectable metabolism. Slow oxidations may be important in this context.

### The mechanisms of tolerance of water stress

In earlier investigations, emphasis was placed on observations that numerous desiccation-tolerant tissues have small cells, in which mechanical dehydration stresses would be less than in larger cells. Resistant tissues often are characterized by flexible walls which readily cave in or fold up during dehydration, thus minimizing the tensions on the protoplast. However, there are also many water stress tolerant tissues which do not possess particularly small cells, and hardening of tissues against water stress proceeds without changes in cell size or wall flexibility. Physiological and biochemical factors are therefore now thought to be of more universal importance than cell size and wall flexibility.

A decrease in the $\psi_\pi$ (osmotic potential) typically occurs as a response to water deficit; this decrease is mainly attributable to increased sugar concentration, but raised levels of amino acids (especially the imino acid proline), carboxylic acids, $K^+$ and $Cl^-$, may also contribute. The decrease in $\psi_\pi$ is not just a concentration effect: the total amount of solute per cell increases. Sugars are often derived from starch hydrolysis. Some species are able to maintain turgor in the face of water deficits of up to $-4$ MPa by lowering of their $\psi_\pi$ – a process termed 'osmotic adjustment' (Fig. 7.5). Artificial application of sugar solutions has sometimes alleviated desiccation injury. In addition to lowering the $\psi_\pi$, sugars may serve to stabilize

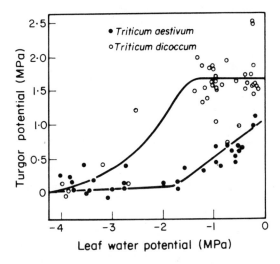

**Fig. 7.5** Relationship between turgor potential and leaf water potential for two wheat genotypes *Triticum aestivum* ssp. *vulgare* (AUS 3850) and *T. dicoccum* (AUS 3582). (Adapted from Turner and Jones, 1980, in *Adaptation of Plants to Water and High Temperature Stress*, ed. Turner and Kramer, 87–101. John Wiley & Sons Ltd, New York.)

protoplasmic macromolecules, preventing their denaturation by replacing water molecules in their hydration shells.

During hardening, increases occur in the amount of water strongly bound to colloids, and in the viscosity and elasticity of the protoplasm; the latter changes may help the protoplasm to survive mechanical tensions. In resurrection plants, protein synthesis occurs during hardening and it has been postulated that dehydration sensitive proteins may be (partly) replaced by more denaturation-resistant isozymes. Some of the metabolic 'disturbances' noted as responses to water deficit may actually be protective or repair activities, e.g. the increased respiration rate sometimes observed at moderate levels of dehydration could provide ATP for repair of desiccation damage. An increase in the level of ABA is a very widespread response to water stress. ABA is known to promote stomatal closure; additionally it has been proposed that ABA promotes the hardening process, but there is no direct evidence for this.

Since fine level structural damage is believed to be very important in water stress injury, the ultrastructure of resurrection plants has been extensively investigated for clues to their extreme resistance to desiccation. The degree of ultrastructural preservation, has, however, been found to vary between species of resurrection plants. In some instances fine structure is very well preserved in the air-dry state, in others drying results in extensive degradation of organelles. Some breakdown may be necessary for survival. When resurrection plants are rehydrated, the normal cell structure takes 1–2 days to be restored, although turgidity is regained in a few hours at most. Resistance may lie not only in an ability to prevent protoplasmic damage, but in a capacity for repair, only the very minimal level of structural preservation being maintained in the dry state.

## Resistance to low temperature stress (cold hardiness)

The geographical spread of many species is determined by their ability to survive low temperatures. The northernmost limit of a perennial plant in the northern hemisphere frequently coincides with a certain winter minimum isotherm. Although frost is not responsible for as much famine as drought, it still has a considerable impact on agriculture and results in large losses in yield. There is currently also a great deal of interest in freezing resistance in connection with long-term preservation of plant tissue cultures in the frozen state, to serve as gene banks.

Many species originating in warm climates cannot endure exposure to temperatures below 5°C or 10°C, sometimes even 15°C. This reaction, known as cold shock or chilling injury, and believed to result from the gelling of cellular membranes, is however not further discussed here. Discussion is confined to the effects of subzero temperatures, i.e. 'frost' or 'freezing' stress.

### Freezing resistance (frost hardiness)

The resistance of plants to subfreezing temperatures can be represented as in Fig. 7.6. Most plants do not possess any specific system of temperature regulation. In the centres of tree trunks, temperatures as much as 10°C above the ambient have occasionally been recorded, but usually even inside bulky organs, the temperature is close to air temperature, except in direct sunlight. On cold nights, leaf and bud temperature often falls below the ambient. Some adaptations for keeping the temperatures of plant organs above air temperature have been reported. At high altitudes in Africa and South America, where night temperatures throughout the year may fall to about − 10°C, several species produce giant rosettes up to 0.5 m in diameter. These rosettes diurnally form 'night buds', the adult leaves (which are tolerant to freezing) closing nightly over the central meristem. A large volume of mucilaginous fluid in the bud also acts as a thermal buffer, and the meristem temperature is maintained appreciably above air temperature during the night, usually not falling below zero. In one of these rosette species, *Lobelia telekii*, control of inflorescence temperature has also been claimed, although only limited data have been published. *L. telekii* bears a tall hollow inflorescence, partly filled with several litres of fluid. It is suggested that the nightly freezing of (part of) this fluid, occurring promptly at 0°C due to the presence of nucleating solutes, provides latent heat of fusion, which is circulated by air in the hollow to the upper part of the inflorescence, maintaining the flowers at near 0°C. These are, however, exceptional cases; while plants have many adaptations for avoiding water stress, low temperature resistance usually means tolerance to the low temperature. The positioning of perennating organs such as buds, rhizomes and bulbs below ground level, where the temperature does not fall as low as at the surface, has been regarded as a kind of freezing avoidance. These organs are, however, near temperature equilibrium with their immediate environment. The ability to survive internal *subzero* temperatures may be called *freezing resistance*, which in turn may be subdivided into freezing avoidance (no ice formed inside the plant) and freezing tolerance (tolerance of ice formation in the tissues) – see Fig. 7.6.

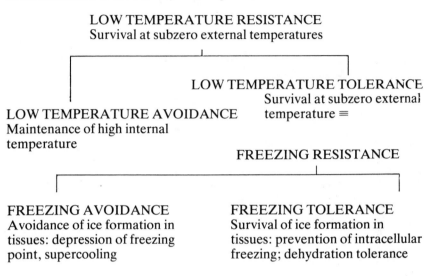

LOW TEMPERATURE RESISTANCE
Survival at subzero external temperatures

LOW TEMPERATURE TOLERANCE
Survival at subzero external
temperature ≡

LOW TEMPERATURE AVOIDANCE
Maintenance of high internal
temperature

FREEZING RESISTANCE

FREEZING AVOIDANCE
Avoidance of ice formation in
tissues: depression of freezing
point, supercooling

FREEZING TOLERANCE
Survival of ice formation in
tissues: prevention of intracellular
freezing; dehydration tolerance

**Fig. 7.6** The categories of low temperature resistance in plants.

### The measurement and limits of freezing resistance

The freezing resistance of a tissue can be defined as the lowest temperature that it can survive. Difficulties similar to those encountered in defining and assessing water stress resistance apply also to freezing resistance. The effects of cold depend not only on the lowest temperature reached but upon rates of cooling and warming, the time for which the minimum temperature persists and the subsequent treatment after warming.

Examples of minimum temperatures which can be endured by flowering plants are given in Table 7.1. Since the lowest temperature recorded in regions of the earth where plants grow is −60 to −70°C, it is clear from this table that certain plants and plant structures can endure at least for short periods temperatures below those which they will ever encounter in nature – indeed very close to absolute zero (−273°C). Plant species native to climates with a cold winter are more resistant than the species of warm climates. The cold resistance of perennials of temperate and arctic zones undergoes striking seasonal changes (Figs 7.7 and 7.8), being very much higher in the

**Table 7.1**   Some examples of low temperature resistance of flowering plants and organs. Where the time endured is given, it specifies the particular experimental conditions without implying that this is the longest time that can be endured. In the examples marked*, the temperature given shows the limit of resistance; in the other cases, temperatures lower than those shown were not tested. ? indicates that time endured or criterion of survival is not clearly indicated in the reference source.

| Plant material | Temperature (°C) | Time endured | Criterion of survival |
|---|---|---|---|
| Pollen grains | c. −273 | 2 hours | Germination |
| Seeds, vacuum dehydrated | c. −273 | 10 hours | Germination |
| | −253 | 77 hours | Germination |
| | −190 | 6 weeks | Germination |
| Winter leaves, *Pinus strobus* | −196 | 5 minutes | Tetrazolium test for dehydrogenases; appearance for 4 weeks |
| Bark cells, *Morus* twig | −196 | 160 days | Plasmolysis tests |
| Germinating seedlings, *Pisum sativum* | −183 | 1 minute | Subsequent growth |
| Bark cells, *Robinia pseudacacia* | < − 59 | ? | Vital staining; cell permeability tests |
| *Winter buds, temperate zone deciduous trees | − 21 to < − 40 | 2 hours | Absence of visible damage often up to 5 weeks |
| *January leaves, *Hedera helix* | − 25 | Varied; cooling rate kept constant | Appearance; tetrazolium test for dehydrogenases |
| *Winter leaves, *Quercus ilex* | − 13 | ? | ? |
| *Summer leaves, *Quercus ilex* | − 6 | ? | ? |

winter months than in summer; without the winter increase in resistance plants of these zones would be unable to survive the rigours of winter. The increase in hardiness occurs in the autumn as a consequence of the falling temperature and shortening day length and is associated with the onset of winter dormancy. The loss of hardiness follows from the relatively rapid rise in temperature in the following spring. Artifically, hardening can be induced at any season by chilling, and resistance destroyed in mid-winter by warming. In the field, cold injury is most frequent in spring when late frosts act on the non-hardy tissues. In severe winters, greater resistance is developed than in mild ones.

Resistance varies greatly between the organs of a plant. Underground structures are much less resistant than aerial parts, and their resistance is much more uniform over the year, because soil temperatures never fall as low as air temperatures and show less annual temperature variation. In a study of 43 species from a mid-European deciduous woodland, the minimum temperature endured by any underground organ was found to be −13.5°C, and the annual amplitude of variation, i.e. difference between the highest and lowest minimum survival temperature, varied according to

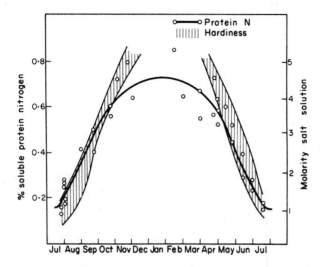

**Fig. 7.7** Changes during the year in hardiness and water-soluble protein nitrogen in the living bark cells of the black locust (*Robinia pseudacacia*). Hardiness is shown as the range of molarities of balanced salt solution whose dehydrating action the cells can survive. (From Siminovitch and Chater, 1958, in *The Physiology of Forest Trees,* ed. Thimann, 219–50. Copyright 1958. The Ronald Press Co., New York.)

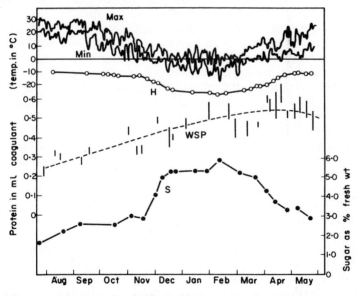

**Fig. 7.8** Annual changes in frost hardiness 'H' (measured as temperature endured), water-soluble protein content 'WSP' and soluble sugars 'S' in the leaves of ivy, *Hedera helix;* the actual temperature maxima and minima are given in the top curves. Viability in the hardiness test was judged by a tetrazolium test for activity of dehydrogenases. Protein is measured as ml coagulated by ethanol precipitation from leaf extract. (Adapted from Parker, 1962, *Plant Physiology, Lancaster,* **37**, 809–13.)

species from 4.5 to 8.5°C. Aerial parts from the same woodland plants could survive in temperatures below –40°C, and annual variation between highest and lowest minimum survival temperatures varied from 9° to 36°C. Buds increased in resistance with distance from the ground. These differences in resistance showed a clear correspondence with the temperatures that the organs might be called upon to endure in their environment.

**The cellular basis of freezing damage**

Freezing damage depends on ice formation within the tissues; injury could therefore be thought to result simply from mechanical disruption caused by ice crystals. As the ice crystals form, however, dehydration of the tissue occurs, as the water passes from the cells into the growing crystals, and freezing damage could therefore also be a form of desiccation damage involving effects on protoplasmic structure as discussed in the preceding section. Metabolic disturbances are less likely to be important in freezing injury because at subzero temperatures metabolic reactions become very slow. Some chemical reactions can nevertheless proceed in a frozen medium; indeed some enzymatic reactions are accelerated in ice, so that chemical (as opposed to structural) damage cannot be entirely dismissed in freezing damage.

Freezing can be intracellular, ice crystals forming inside the cells, or extracellular, with the ice confined to the extracellular (intercellular) spaces. Intracellular ice forms when cooling is sufficiently rapid, or when the tissue has supercooled, and is fatal unless the cooling rate is so fast that the cell contents 'vitrify', the ice crystals then being of submicroscopic dimensions. Vitrifying cooling rates can be achieved only under laboratory conditions. With slow cooling, ice formation always begins outside the cells. As the water vapour pressure in the intercellular spaces falls, a gradient is set up from the cells to the ice crystals, and water passes out of the cells to the growing crystals, leading to dehydration and shrinking of the cells. The lipid nature of the plasmalemma offers a high resistance to seeding crystals of ice entering the protoplast. Such slow cooling is therefore endured better than rapid cooling, and parallels what happens in plants frozen naturally in the field. Extracellular freezing is more common than intracellular freezing. Though extracellular ice can greatly compress and distort cells, and may, for example, separate the epidermis widely from the palisade tissue in some leaves, no cell rupture occurs. The presence of extracellular ice therefore does not in itself inflict any irreversible injury; the present view is that freezing damage during extracellular freezing is caused primarily by the withdrawal of water into ice crystals, and that frost injury is then a form of desiccation damage. In support of this it has proved possible to measure indirectly the frost resistance of tree bark by measuring its resistance towards dehydration by plasmolysing solutions (Fig. 7.7) or by drying out in air. As in desiccation at higher temperatures, the moment of injury may be during dehydration (freezing) or rehydration (thawing) or develop gradually in the dehydrated (frozen) state. Just as slow cooling is endured better than fast cooling, so, with slow thawing, tissues can survive freezing treatments that are fatal if followed by sudden thawing. The rapid flooding with water when the ice melts suddenly is apparently injurious to the dehydrated tissue (*cf.* dehydration stress, p. 117).

In spite of the fundamental equivalence of freezing and desiccation injury, the removal of a given quantity of water by freezing does not have precisely the same effect as the removal of an identical amount by drying at a higher temperature. One obvious reason for this difference is that at freezing temperatures metabolic disturbances are largely inhibited. Freezing resistance often parallels desiccation resistance closely, i.e. the more resistant a tissue is towards freezing, the more resistant is it also towards desiccation. Nevertheless the fatal degree of dehydration has a different value when achieved by freezing than when achieved by drying. The effect on the protoplasts is related not only to the total amount of water removed, but to the conditions under which it is removed.

Freezing has been shown to impair the activities of some enzymes, but many enzymes can withstand freezing and the most important effect of freezing is again thought to be damage to cellular membranes. Loss of cell semipermeability and outward leakage of cell contents are among the first signs of freezing injury, indicating damage to the plasmalemma. In isolated chloroplasts and mitochondria, freezing destroys the capacity for phosphorylation, an activity dependent on membrane integrity. Fine structural studies have shown disorganization of cellular structure on freezing. Protection against freezing damage can sometimes be afforded by external application of solutes that do not penetrate the plasmalemma, pointing to the plasmalemma as the potential site of injury. In addition to the strains set up in membranes owing to withdrawal of water into ice crystals, leading to protein denaturation and aggregation, membranes would suffer from the weakening of hydrophobic bonds, which become weak at low temperatures. Normally, hydrophobic forces of attraction keep the two layers of the bimolecular lipid leaflet together, and bind membrane proteins to the lipids.

**The physiological basis of freezing resistance**

From the above discussion it would be expected that the freezing resistance of a tissue will depend on the amount and state of water in its cells. The greatest cold resistance is encountered in dry seeds. Before subjection to the very low temperatures shown in Table 7.1, the seeds were dried as completely as possible in a vacuum, and the minute traces of water left were in a bound state inaccessible for freezing. During the period of frost hardening of perennials, there is a tendency for a decrease in the total water content of the tissues, and an increase in the proportion of bound water, due to an increased hydration capacity of cell colloids. Nevertheless many frost-hardy tissues contain a very appreciable amount of freezable water.

*Avoidance of freezing* may be achieved by lowering of the freezing point by the presence of solutes; hardening is associated with an increase in soluble carbohydrates (Fig. 7.8), most commonly sucrose, but in some cases other sugars or polyhydric sugar alcohols. The freezing point of cell contents is thereby lowered usually by just a few degrees; in halophytes, freezing points of −14°C have been recorded owing to the presence of high salt levels. Much lower temperatures can be endured without ice formation in tissues capable of *deep supercooling*. In the absence of nucleating substances around which ice crystals would start to grow, water can supercool down to a temperature at which ice forms spontaneously; for pure water this 'homogeneous nucleation point' is reached at −38°C, but in solutions the homogeneous nucleation point can be some degrees lower. In many temperate zone woody perennials, deep supercooling occurs in xylem ray parenchyma cells and in flower primordia in the buds; the apple is an example of such a species. (Other tissues in these species tolerate ice formation.) The northern growth limit of species relying on deep super-cooling is set by the −40°C winter minimum isotherm, for when the supercooled cells do freeze, they freeze very rapidly and intracellularly. The deep supercooling capacity is developed only during the winter months. It is not known how supercooling is achieved; ice formation occurs in the rest of the plant, but somehow ice is stopped from spreading into the tissues which supercool, e.g. in flower buds, ice spreading stops at a layer of undif-ferentiated cells between the flower primordia and their stalk. Possibly antinucleating compounds are present in the cells.

*Tolerance of freezing* enables tissues to endure lower temperatures than supercooling, right down to −196°C (though such tolerance is not shown by *all* tissues that can withstand freezing). Freezing tolerance depends on ice formation being confined to outside the cell membranes, and on the ability of the cells to survive desiccation and compression by the extracellular masses of ice. Hardening towards freezing may involve an increase in the permeability of cells to water, making it easier for water to leave cells and lessening the risk of intracellular freezing; but in some species, e.g. potato, hardening does not affect membrane permeability. Protoplasts of hardy cells have been reported to withdraw more easily from cell walls when water is lost, lessening mechanical strains as the protoplasts shrink. The formation of soluble sugars may help to stabilize colloids against dehydration damage as well as lowering the freezing point. Frost hardening of perennial tissues is associated with increases in soluble protein content, e.g. in the bark of the black locust tree (*Robinia pseudacacia*), and leaves of ivy (*Hedera helix*) and the protein level falls again in the spring, as the tissues lose their freezing tolerance. If the synthesis of protein in lengths of *Robinia* stem is inhibited by interrupting their nitrogen supply by ringing, hardening is diminished. It has been suggested that the chains of the proteins synthesized during hardening may fold into frost-stable configurations, with their hydrophilic groups to the inside; the internal hydrophilic bonds could then keep the molecules from denaturing when the hydrophobic bonds lose strength at low temperatures. Protein SH groups have been said to become less susceptible to oxidation to SS bridges, during hardening.

Since membrane damage is believed to be fundamental to freezing injury, much attention has been devoted in recent years to the behaviour of membrane lipids during the induction of freezing tolerance. One hypothesis is that an increase in the fluidity of cellular membranes, brought about by an increased proportion of unsaturated lipids, is an important component in the development of freezing tolerance. The advantages suggested for increased membrane fluidity are that more fluid lipids might flow together more easily to repair damage; proteins might retain their native configuration better in a more fluid membrane; and there would be a higher water permeability, allowing water to escape from the cells and freeze *extra*cellularly. Evidence for this hypothesis is, however, equivocal. In some cases, e.g. alfalfa and winter wheat, the degree of lipid unsaturation has been reported to increase during frost hardening. But in other instances, no increase in lipid unsaturation has been detectable, e.g. in black locust tree bark, and in *Brassica napus*. Increases in total amounts of cellular phospholipids have been reported to occur during cold hardening, and some authors consider this overall increase, possibly indicating a proliferation of cellular membranes, to be critical for hardening.

## Resistance to high temperature stress

Heat injury is less of a problem in agriculture than drought or frost damage; high temperatures more frequently cause problems indirectly by induction of water stress. Photosynthesis, however, has a rather low temperature optimum in plants of temperate climates, only 20–30°C, and therefore only moderately high temperatures can cause decreases in yield though no injury is apparent. In hot climates, high temperature does become a limiting factor for survival, but since the hottest regions of the earth are also the most arid, it may be difficult to dissociate the effects of heat stress from those of water stress in such habitats.

### The limits of heat resistance

The heat killing temperature, the *thermal death point*, of a plant tissue is highly dependent on the time of exposure: with shorter times, higher temperatures can be endured. For example, cells of *Tradescantia discolor* have been killed in 7 min at 60°C, 4 h at 50°C, and about 22 h at 40°C. Dry seeds have survived heating to 100°C from a few hours to a few days, and in some instances have even withstood exposure for some minutes to temperatures up to 138°C. For other organs of temperate plants, the thermal death point generally lies between 45–55°C, for exposures up to some hours. In studies of the survival potentials of plants in their natural habitats, it is reasonable to measure the temperature limit with exposures of some hours for, in the field, maximal heat would be experienced for that kind of time period, in mid-afternoon. Desert plants, such as the cacti, have thermal death points at over 60–65°C, whilst aquatic and shade plants may survive only up to 38–42°C.

From these values, it might be concluded that plants are safe from heat injury in temperate climates, where the air temperature is unlikely to exceed 40°C. The temperatures of plant tissues may, however, be well above the ambient, especially when the organs are bulky and unable to dissipate absorbed heat. In the sun, even thin leaves may reach temperatures 6–10°C above the air temperature in spite of transpirational cooling. Fleshy leaves have been found to have internal temperatures of 40–50°C with the air temperature at 20–30°C, and the cambium on the sunny side of a tree has been recorded to be at 55°C. Sunscald injury to the southern side of trees is a well-known phenomenon, and fleshy fruits, too, are liable to suffer heat injury on the side facing the sun. Soil surface temperatures can rise far above air temperature on sunny days and seedlings are sometimes killed by overheating at soil level.

**The causes of thermal injury and the basis of heat resistance**

Death that occurs within a few minutes or seconds at very high temperatures can be attributed mainly to protein denaturation, leading to a catastrophic collapse of protoplasmic organization. Coagulation of cell contents can be observed microscopically and many proteins are known to undergo denaturation at the temperatures that correspond to the thermal death points of plants. The $Q_{10}$ values of heat killing of plants are high, up to 2150(!); the only chemical process known to have such $Q_{10}$ values is protein denaturation.

The slower heat injury observed at less extreme temperatures (time of development measured in hours or days) is a more complex phenomenon involving numerous metabolic disturbances. Photosynthesis is very sensitive to high temperature, whereas respiration is relatively resistant and increases in rate with increasing temperature to near the lethal point; consequently there is a danger of starvation (*cf.* effects of water deficit, p. 115). There is net hydrolysis of proteins. Particular essential metabolites may run short, presumably because their synthesis is impaired more than their utilization; this is evidenced by the fact that protection against heat injury has in some cases been afforded by the application of e.g. certain vitamins.

An attempt has been made to put the rapid 'direct' type and the slower 'indirect' type of injury on a common basis by postulating that the slower injury follows the thermal denaturation of one or a few particularly heat-sensitive proteins. The loss of activity of this protein would cause a cumulative build-up of physiological imbalance. One possible candidate for the primary lesion is within the photosynthetic apparatus; electron transport via Photosystem II seems to be very heat-sensitive. Significance is attached to the fact that Photosystem II activity is obligately membrane-bound: the primary injury may be to a membrane. But in non-photosynthetic tissues, the primary lesion must be sought elsewhere.

*The basis of heat resistance*

Plants growing in hot climates show some adaptations for heat avoidance, i.e. for keeping their temperatures close to or below the air temperature. Desert plants commonly have external surfaces of high reflectance, which cuts down the amount of heat absorbed. Transpirational cooling is quoted much as a means of reducing shoot temperatures. In favourable circumstances some leaves may be cooled by transpiration to 10°C below air temperature. The extent of transpirational cooling is, however, much affected by environmental conditions and varies also according to species. Significant cooling through transpiration is feasible only when there is an ample water supply, while many heat-resistant desert plants are water conservers. Once leaf temperature has risen above that of the air, heat convection results in heat loss and for small leaves this effect may be important.

Heat resistance resides mainly in protoplasmic tolerance. For a number of enzymes, it has been demonstrated that the enzyme from a heat-tolerant plant is more thermostable than the same enzyme from a more sensitive species, and some improvements in the thermostability of an enzyme during the heat hardening of a species have been noted. It must, however, be recorded that, with other enzymes, no such correlations have been found in the thermostability of the enzyme and the heat tolerance of the plant. Maybe injury can be repaired as long as a minimal amount of metabolic apparatus is maintained intact. The properties of membrane lipids are also implicated in heat tolerance: increased tolerance is associated with more highly saturated lipids, i.e. lipids with high melting points, lipids which would confer more rigidity on a membrane. Because of the increase in thermal movements of the lipid molecules with rising temperature, membrane fluidity would increase at higher temperature. If a membrane becomes excessively fluid, membrane proteins may no longer be held in the specific spatial arrangements which are vital for the functioning of, e.g. electron transport in a chloroplast membrane.

**Relationships between different types of stress resistance**

The resistance to high temperature of a number of temperate zone evergreens shows two annual maxima. One is in the summer as would be expected, when the temperatures are highest. The second maximum, however, coincides with the midwinter maximum of cold and desiccation resistance. Hardening against cold and hardening against dehydration increase heat tolerance; hardening against freezing by low temperature treatment hardens also against water stress and vice versa. Further, during the time of greatest resistance towards freezing, water stress and high temperature, plants show high resistance also towards other injurious agents such as toxic chemicals and lack of $O_2$. This has led some authors to advocate the concept of a *general resistance* of protoplasm against all injurious influences. The concept is an oversimplification. Each type of stress carries with it specific problems. While cold hardening increases heat resistance, the reverse is not true; the midsummer maximum of heat resistance coincides with the minimum of freezing resistance. Some mechanisms of protection against one type of stress are incompatible with resistance towards another stress: a high proportion of unsaturated lipids in membranes is favourable at low temperatures, unfavourable at high temperatures. Yet a common denominator in tolerance of water stress, low temperature stress and high temperature stress appears to be the stabilization of cellular membranes.

## Resistance to salinity stress

Most of the water on the earth's surface is sea water, salt water containing 3% sodium chloride (*c.* 0.5 mol $l^{-1}$) – a level of salinity highly deleterious to most species of flowering plants. Around sea coasts saline habitats occupy large stretches in the form of the salt marshes of temperate regions and the mangrove swamps of the tropics. There are also salt-rich inland areas, the salt deserts and regions around salt lakes (e.g. the Dead Sea and the Great Salt Lake). Further areas have become saline as a result of irrigation by human agency. Salt stress is thus a problem experienced by plants in extensive areas and salinity limits the exploitation of considerable expanses of land.

The transition between saline and non-saline habitats is of course gradual. A convenient definition of a saline habitat is one where the concentration of NaCl equals or exceeds 0.5%, and plants able to grow normally at 0.5% or more of NaCl are classed as *halophytes*, as opposed to the salt-sensitive *glycophytes*. Nearly all agricultural crops are glycophytes. Many halophytes are facultative, but there are also obligate halophytes, needing at least 0.5% NaCl for growth, e.g. the temperate salt marsh plants *Suaeda maritima*, *Salicornia europaea*, and the mangroves *Rhizophora mangle*, *Avicennia germinans*.

There exist habitats where salts other than NaCl reach stressful levels, but NaCl salinity is by far the most widespread, and discussion is confined here to the effects of NaCl.

### The basis of salinity damage

There are two potential ways in which high salinity could exert a stress on plants. It could impose a water stress – the $\psi$ of sea water is about $-2.5$ MPa; or there could be direct toxic effects exerted by the ions, $Na^+$ and/or $Cl^-$. Attempts have been made to distinguish between these two possibilities by comparing the effects of salt solutions on plants with the effects of iso-osmotic solutions of non-penetrating organic compounds. The osmotic effect should be the same in the two types of medium; yet the deleterious effect is greater in the salt solutions. When the effects of different salts at equivalent concentrations are compared, the degree of injury is found to vary with the salt. Toxic effects may become apparent at salt concentrations too low to exert appreciable osmotic effects. The *in vitro* activity of isolated plant enzymes is inhibited by the presence of salts. Such observations have led to the conclusion that specific ion toxicity is involved in salt stress, though of course water stress may also be a contributory factor, especially at high salt concentrations. Nutrient imbalances can occur; in the presence of high NaCl concentrations, potassium uptake can be inhibited, leading to a lowering of the internal K/Na ratio, which is believed to be a physiologically significant feature.

**The physiology of salt resistance**

Some species achieve a certain degree of resistance to salinity by salt exclusion, i.e. by avoiding the uptake of excessive amounts into the tissues. Varieties of barley (*Hordeum vulgare*) and the grass *Agrostis stolonifera* differ in their resistance to salinity; when grown in saline media, the more resistant varieties take up less $Na^+$ and $Cl^-$ than the sensitive ones. The exclusion mechanism must depend on the properties of cellular membranes. Sometimes salt is absorbed into the root system, but very little is transported into the shoot. The extent of resistance that can be achieved by salt exclusion is however limited, permitting plants to endure only moderate levels of salinity. The halophytes able to grow in salinity levels of sea water (and even above that) accumulate high levels of $Na^+$ and $Cl^-$. That enables them to maintain a low internal $\psi$ and to avoid water stress deficits. For instance *Salicornia europaea,* when grown at a soil salinity giving a soil $\psi$ of $-1.6$ MPa, can maintain a shoot $\psi$ of $-6.5$ MPa.

The uptake of large amounts of NaCl solves the water stress problem for halophytes, but necessitates other physiological adjustments. When enzymes are isolated from halophytes, the enzymes are (with a few exceptions) just as susceptible to salt inhibition as the corresponding enzymes from glycophytes. It can be deduced that in the halophytes the ions must be segregated away from the cytoplasm, in vacuoles and cell walls. Direct proof of compartmentation has been obtained with X-ray analytical electron microscopy. It is possible to identify the presence of individual elements in specimens under the electron microscope by examining the spectrum of the X-rays emitted as the electron beam impinges on the specimen. Each element emits X-rays at specific wavelengths. By X-ray analysis, it has been found that for example most of the chloride is in vacuoles and cell walls in the leaves of the halophytes *Frankenia grandiflora, Suaeda maritima* and *Tamarix aphylla.*

The presence of high salt concentrations in the vacuoles and cell walls, while enabling the plant as a whole to absorb water, would nevertheless lead to a dehydration of the cytoplasm, if this did not contain osmotically active solutes to keep it at the same $\psi$ as the vacuole and wall. Osmotic balance is provided by the cytoplasmic synthesis of organic solutes, including amino acids, especially the imino acid proline, quaternary ammonium compounds, and sometimes sugars. Other common halophytic characters are succulence, thought to be a means of diluting the salt with large volumes of water; and the possession of *salt glands*. These glands, on the aerial parts of many halophytes, actively secrete $Na^+$ and $Cl^-$ ions to the exterior, enabling the plants to eliminate excess salt.

## Resistance to heavy metals

Metal-contaminated areas occur naturally (serpentine soils and surface mineral deposits) or are man-made (by mining and ore extraction, use of agriculture sprays, lead from 'anti-knock' additive of petrol contaminating roadside verges). The main elements involved are copper, zinc, nickel and lead. Study of the colonization of areas of high heavy metal content shows establishment of a characteristic flora (containing typical species which have been termed metallophytes), provided heavy metal levels are not completely toxic and the physical and nutritive conditions are not too severe. Some species are naturally confined to sites of abnormally high heavy metal content and are resistant to high levels of more than one heavy metal, e.g. *Viola calaminaria, Thlaspi alpestre, Minuartia verna*. Such species are at a high competitive advantage on such sites but the heavy metals are in no sense directly beneficial to them. Many species (covering a wide range of flowering plant families, certain mosses and liverworts, lichens and microorganisms) occur in the flora of contaminated sites as tolerant ecotypes or races. Some of these ecotypes have evolved naturally over relatively short periods (often in less than 100 years); it is possible to select for heavy metal tolerance in *Agrostis tenuis* within one generation! These metal-tolerant races have no requirement for nickel or lead or for higher than normal levels of copper and zinc. In most cases they do not survive by exclusion (reduced absorption) of the toxic metal. Although in no case has the physiological basis of tolerance been fully elucidated there is evidence that a major factor is the affinity of the cell walls for the heavy metal and that a subsidiary factor may be ability to store within the vacuoles excess metal entering the cell protoplasts. In agreement with the tolerance, the binding sites in the cell walls are specific; thus races can be obtained tolerant to copper (but not to zinc), to zinc (but not to copper) or to both elements. The genetics of the tolerance have not been critically analysed but it seems that in most cases the inheritance of tolerance is polygenic. Metal-tolerant races may show associated morphological characters and sometimes are also adapted to growth under conditions of low fertility or adverse physical conditions (marked fluctuations in temperature and available moisture). In view of the speed of evolution of metal tolerance and the ease of experimental selection of tolerant genotypes their study is of particular interest for higher plant physiological genetics. On the practical side the developmennt and transplantation of metal-tolerant races of grasses and cover plants is already proving the key to the revegetation of many areas of industrial dereliction.

## The human viewpoint: breeding plants for stress resistance

As indicated by the foregoing sections, each of the environmental stresses discussed – water deficit, low temperature, high temperature, salinity, toxic metals – can seriously limit agricultural productivity, or completely prevent the exploitation of large areas of the earth's surface for crops. At the same time the expanding human population is faced with a food shortage. The study of the physiology of plants as affected by stress factors has therefore assumed practical importance over and above academic interest. Much thought is being devoted to the possibility of breeding stress-resistant plants. *If*, say, one could produce crops sufficiently salt-resistant to permit irrigation with sea water, cultivation would be possible in desert areas with an excellent record of sunshine, and no risk of frost. The selection of resistant varieties of established crop plants from the range already naturally present in a population has met with some success. In numerous instances, however, breeding programmes are still being hampered by lack of knowledge about the physiological basis of resistance. An alternative way of tackling the problem would be to look at species already native in extreme habitats, and to try to find among these plants species with a potential for agricultural use. No new major crop plants have so far been introduced in this way, but some progress has been made; e.g. oil-yielding species suitable for cultivation in semi-desert habitats have been found.

Since the possibility of gene transfer between species, 'genetic engineering', is becoming a reality, hopes have been raised that resistant plants might be produced by the transfer of 'resistance genes' into sensitive plants. This does not seem at the moment to be a promising proposition. Stress resistance is obviously a complex phenomenon and likely to be controlled by a large number of genes. One would also need to know much more about the molecular basis of stress resistance before genetic transfer of resistance can be seriously considered.

By selection and hybridization – and perhaps in the far-distant future, genetic engineering – we can hope to push the frontiers of cultivation some distance further into inhospitable habitats. But there may be an intrinsic limit to the level of productivity from extreme habitats. High stress resistance is associated with low average growth rates. Resurrection plants survive air dryness, but while dry they produce nothing. Cacti remain metabolically active over many months without an external water supply; but they manage this by conserving water in stems of very low surface/volume ratio, and they grow only slowly. Halophytes expend an appreciable amount of metabolic energy on osmoregulation – ion pumping across cell membranes, ion excretion by salt glands, and synthesis of organic cytoplasmic osmoregulatory solutes. One cannot have something for nothing; there is a price to be paid for stress resistance.

FURTHER READING

ANTONOVICS, J., BRADSHAW, A. D. and TURNER, R. G. (1971). Heavy metal tolerance in plants. *Advances in Ecological Research*, 7, 1–85.

BEWLEY, J. D. (1979). Physiological aspects of desiccation tolerance. *Annual Review of Plant Physiology*, 30, 195–238.

BURKE, M. J., GUSTA, L. V., QUAMME, H. A., WEISER, C. J. and LI, P.H. (1976). Freezing and injury in plants. *Annual Review of Plant Physiology*, 27, 507–28.

LEVITT, J. (1980). *Responses of Plants to Environmental Stresses.* Vol. I: *Chilling, Freezing and High Temperature Stresses.* Vol. II: *Water, Radiation, Salt, and Other Stresses,* 2nd edition. Academic Press, New York.

POLJAKOFF-MAYBER, A. and GALE, J. (eds) (1975). *Plants in Saline Environments.* Springer-Verlag, Berlin.

TURNER, N. C. and KRAMER, P. J. (eds) (1980). *Adaptations of Plants to Water and High Temperature Stress.* John Wiley, New York.

WAINWRIGHT, S. J. (1980). Plants in relation to salinity. *Advances in Botanical Research,* **8**, 221–61.

ZIMMERMANN, U. (1978). Physics of turgor – and osmoregulation. *Annual Review of Plant Physiology,* **29**, 121–48.

SELECTED REFERENCES

BECK, E., SENSER, M., SCHEIBE, R., STEIGER, H. M. and PONGRATZ, P.(1982). Frost avoidance and freezing tolerance in Afroalpine 'giant rosette' plants. *Plant, Cell and Environment,* **5**, 215–22.

CAMPBELL, N. and THOMSON W. W. (1976). The ultrastructural basis of chloride tolerance in the leaf of *Frankenia. Annals of Botany,* **40**, 687–93.

GAFF, D. F. and CHURCHILL, D. M. (1976). *Borya nitida* Labill. – an Australian species in the Liliaceae with desiccation-tolerant leaves. *Australian Journal of Botany,* **24**, 209–24.

HALLAM, N. D. and GAFF, D. F. (1978). Re-organization of fine structure during rehydration of desiccated leaves of *Xerophyta villosa. New Phytologist,* **81**, 349–55.

HSIAO, T. C., ACEVEDO, E., FERERES, E. and HENDERSON, D. H. (1976). Water stress, growth and osmotic adjustment. *Philosophical Transactions of the Royal Society B,* **273**, 479–500.

KROG. J. O., ZACHARIASSEN, K.E., LARSEN, B. and SMIDSRØD, O. (1979). Thermal buffering in Afro-alpine plants due to nucleating agent-induced water freezing. *Nature, London,* **282**, 300–1.

PALTA, J. P. and LI, P. H. (1980). Alterations in membrane transport properties by freezing injury in herbaceous plants: evidence against rupture theory. *Physiologia Plantarum,* **50**, 169–75.

SIMINOVITCH, D. (1979). Protoplasts surviving freezing to −196°C and osmotic dehydration in 5 molar salt solutions prepared from the bark of winter black locust trees. *Plant Physiology, Lancaster,* **63**, 772–5.

WOLFE, J. (1978). Chilling injury in plants – the role of membrane lipid fluidity. *Plant, Cell and Environment,* **1**, 241–7.

YEO, A. R., LÄUCHLI, A., KRAMER, D. and GULLASCH, J. (1977). Ion measurements by X-ray microanalysis in unfixed, frozen, hydrated plant cells of species differing in salt tolerance. *Planta,* **134**, 35–8.

# 8

# Growth: Progress and Pattern

## The definition and measurement of growth

Growth is one of the most fundamental and conspicuous characteristics of living organisms, being the consequence of increase in the amount of living protoplasm. Externally this is usually manifested by the growing system getting bigger and growth is therefore often defined as an irreversible increase in the mass, weight or volume of a living organism, organ or cell. The size increase must be permanent; the swelling of a cell in water is not growth, being easily reversed by returning the cell to a sufficiently concentrated solution. It is, however, possible to consider, as growth, changes not immediately involving increase in size. An amphibian embryo, or a *Selaginella* female gametophyte, for a long time utilizes the nutrient store with which it was released from the parent, to produce many new cells, without any size increase yet growing in the sense that protoplasm is increasing at the expense of stored nutrients.

Growth can be measured in a variety of ways. Since increase in the amount of protoplasm is difficult to measure directly, generally some quantity is measured which is more or less proportional to it: in higher plants or their organs, the four most commonly employed parameters are:

1) *Fresh weight* This measurement is usually easy. However, a plant organ must be detached from the plant to be weighed, and an entire plant which has been removed from its growth surroundings for weighing can rarely be replaced undisturbed. Thus growth measurements by weighing usually necessitate the taking of successive samples from a series of plants.

2) *Dry weight* This is sometimes considered more meaningful than fresh weight, because the increase in fresh weight may be largely the result of water uptake, and fluctuations in fresh weight may occur as a result of fluctuations in the plant's water content. In a germinating seedling, the dry weight decreases while the seedling is growing, and in such a case the fresh weight increase is a better indication of growth.

3) *Linear dimensions* For an organ growing predominantly in one direction, such as a root tip or a pollen tube, *length* is a suitable measure. Indeed linear elongation can be used to assess the growth of an entire shoot. An increase in *width* (diameter) may be relevant in other cases, e.g. for an expanding fruit, or a thickening axis.

4) *Area*   Used to assess growth in a system growing mainly in two dimensions, such as an expanding leaf.

Length, width and area measurements have the attraction that they can be carried out on the same organ over a period of time without destroying it. Growth in linear dimensions can now be measured with highly sophisticated *auxanometers*, which detect or magnify size increments far too small to be measured directly, and may be able to record the increments automatically. One class of auxanometers utilizes the *linear variable displacement transducer* (position-sensing transducer). The growing organ is fastened to a ferromagnetic armature bar which is movable within the transducer's conducting coils. The electrical output from the transducer varies according to the position of the bar within the coil. As the plant grows, it moves the core along in the coil, and the instrument is calibrated so that the voltage change in the coil is converted to a readout of the linear movement. With appropriate electronic equipment the data are automatically processed to the desired units and recorded as required, e.g. graphically. An example of such an instrument is illustrated in Fig. 8.1. Transducer auxanometers can be sensitive enough to record growth increments of less than 1 $\mu$m and have been very useful in detecting changes in growth rate within minutes of applying chemicals to an organ with an overall growth rate of no more than 1–2 mm h$^{-1}$. On the other hand they can be robust enough to withstand outside conditions. The auxanometer illustrated in Fig. 8.1 was designed for field use and can measure extension growth down to 0.1 mm h$^{-1}$, without interference from moderate wind, nor from temperature changes in the range of 1–24°C; it can be left recording in the field over periods of days.

Laser beams have been utilized in the construction of an auxanometer for sensitive measurement of growth over short time intervals. The growing organ is connected to a tiny mirror on which a laser beam impinges, to be reflected to a recording chart 10 m away. The growth of the organ tilts the mirror, deflecting the laser spot along the chart on which its position is recorded. The laser beam in effect acts as a giant but totally weightless pointer, which magnifies a minute growth movement to a visible deflection of the far end of the pointer beam; up to 2000-fold magnification of the growth increments has been obtained.

The drawback of both a displacement transducer and the laser beam auxanometer is that the growing object is somehow *fixed* to the apparatus and is subjected to some *tension* or *pressure*. Great care is taken to achieve the fastening without injury. Mechanical forces can be minimized, e.g. by counterweighting the transducer core (Fig. 8.1); or, the connexion between the plant tip and the transducer armature can be made via the delicately poised beam of a balance; as the plant grows, it tilts its end of the beam up, while the other end dips down, lowering the core. Small forces can be endured without affecting growth rate; in a series of experiments with lupin (*Lupinus angustifolius*) seedlings it was found that varying the weight producing a tension on the shoot from 0.5 to 4 g made no difference to the growth rate. When it is considered, however, essential to avoid all physical contact with the plant, growth can be estimated from successive photo-

**Fig. 8.1**   A displacement transducer auxanometer measuring the extension rate of millet in a glasshouse. The plant is connected to the armature by a white clip (lower right) and fine chain. The elongate tube contains the transducer coils, whilst terylene threads running over the pulley wheel at the top carry counterweights (not clearly visible) to balance the weight of the armature, so that the plant is kept under a slight tension. A perspex cover surrounds the auxanometer. At lower left is the electronic differentiator which records the growth rate. (From Saffell, Discoe and Gallagher, 1979, *Journal of Experimental Botany*, **30**, 199–204. By permission of the Clarendon Press, Oxford.)

graphs, with a camera electronically controlled to take exposures at set time intervals – the technique of *time-lapse photography*. Where growth curvatures are involved, auxanometers cannot be used, but photographic techniques are extremely useful.

For assessing the growth of tissue or cell cultures, and for microscopic unicellular organisms, other measurements are possible and indeed more suitable. *Volume* of a cell mass can be determined by centrifuging the cells in a graduated centrifuge tube, and for single cells volume can be calculated from measurements made under the microscope. Increase in the *turbidity* of a cell suspension is related to increase in the *number of cells* present. Cell number can also be estimated by direct counting using a special slide (haemocytometer) originally designed for counting blood cells. Other special criteria are estimates of *protein nitrogen* or *respiration rate* per unit of cell mass. The units in which growth is expressed are as diverse as the methods of measurement (increase in weight, area, length, cell number, protein nitrogen, per hour, day, week), and growth rates can be expressed in absolute or relative terms (see p. 144 *et seq.*).

Growth is always accompanied by change in form and physiological activity, by *differentiation*. The identical cells produced by cell division in an apical shoot meristem enlarge and at the same time become different from the meristematic cells, and from each other, forming for instance parenchyma or xylem or phloem. In their mature form these tissue cells are of very different structure and function. Nevertheless certain growth processes are common to all cells and, during the initial stages of a cell's development, these common processes predominate. In the meristem, the newly formed cells grow first by *plasmatic growth*, a synthesis of protoplasm. In cells destined to remain meristematic, a doubling of cell mass is followed by division; in those destined to undergo further growth there follows a phase of *expansion* (elongation) *growth*, characterized by rapid volume increase, water uptake accompanied by vacuolation, and cell wall synthesis. Then, as expansion slows down, divergences in cell development become dominant. While basic protoplasmic components, such as enzyme proteins and nucleic acids, increase in quantity in all cells during growth, the proportional increases in particular enzymes differ so that cells of varied metabolism are formed. Cell wall growth occurs in all cells, but as differentiation proceeds characteristic differences in the pattern, extent and chemical composition of the new wall material become apparent. The completion of differentiation leads to the formation of the mature living tissue cell or dead cell element. The mature tissue cells ultimately age and die. Such a sequence of changes, constituting the overall process of *development*, unfolds during the life history of every cell and is represented diagrammatically in Fig. 8.2. Similar developmental stages may be distinguishable in the life history of an organ or organism. Although developmental stages overlap in time it is still possible to separate growth from differentiation experimentally and conceptually. Thus it is possible to suppress differentiation by applying appropriate inhibitory chemicals while permitting growth to continue. Further these two aspects of development seem in many developing systems to be mutually competitive; conditions which favour rapid growth often suppress differentiation, and vice versa. For instance, plants whose growth is retarded by a deficient water supply

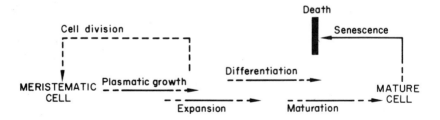

**Fig. 8.2**   Diagrammatic representation of the sequence of processes which constitute the development of a cell of a higher plant.

may show an enhanced degree of cellular differentiation (see also Chapter 9).

In the light of the above discussion, it is possible to advance a definition of growth as follows: 'Growth is a synthesis of protoplasm, usually accompanied by a change in form and an increase in mass of the growing system. The total mass increase may be many times that of the increase in the mass of the protoplasmic components proper.' Again, from the above it is clear that growth can be considered at several levels. In order of increasing complexity, one can distinguish growth at the level of the cell, the tissue, the organ and the organism. The growth of a system above the cell level is brought about by a combination of *cell multiplication* and *cell growth*. One single maize root tip can give rise to 17 500 new cells per hour, or 420 000 per day. During the development of a root tip meristem cell into a root parenchyma cell it expands by about 30 times; volume increases of several hundred-fold are not uncommon; in the water melon (*Citrullus vulgaris*) the volume of the initial cells increases by 350 000 times and, in extreme cases, development of water-storage cells may involve expansions of a million-fold.

## Localization of growth in space and time

Flowering plants continue growth throughout their life history by virtue of persistent localized growth centres or meristems. Elongation growth and the formation of organs result from the activities of meristems at the root and shoot apices and, in monocotyledons, also of additional intercalary meristems at internodes and leaf bases. Increases in the diameters of plant axes are brought about by the activities of vascular and cork cambia. New cells are formed in these meristems and the development of mature tissue cells by growth and differentiation takes place in close proximity to the meristems. Whilst still in the seedling stage the plant already contains cells at all stages of development – meristematic, expanding, mature, senescent and dead. The age of a plant and the age of its cells are quite different. Trees live for hundreds or even thousands of years, but their tissue cells have a much more limited life, rarely more than one or two years, exceptionally up to 100 years in tree pith. The life history of an entire plant of indeterminate growth cannot therefore be divided into stages corresponding to those depicted in Fig. 8.2. Frequently, however, during the development of organs of determinate growth, a phase of cell division comes first, with no or slight cell expansion and this is then followed by cell expansion growth not associated with increase in cell number. This growth pattern has been shown in the *Arum* spadix (Fig. 8.3) and in cucurbit fruits, tomato fruits and ripening bean cotyledons. In apical meristems a region of cell division is separate from a region of expansion; cells in the dividing region do not expand appreciably, and cells in the expanding region divide very infrequently. In fruits, separation of cell division and expansion is temporal rather than spatial. But this rule is not universal; in potato tubers, cell division and expansion proceed concurrently at least while the tuber grows from a weight

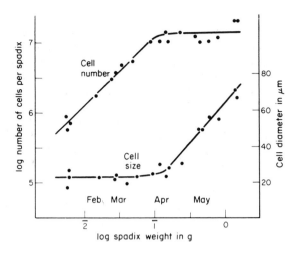

**Fig. 8.3** Development of *Arum maculatum* spadix. The stages of cell division and cell expansion are distinctly separate. (From Simon and Chapman, 1961, *Journal of Experimental Botany,* **12**, 414–20. By permission of the Clarendon Press, Oxford.)

of *c.* 40 mg to 200 g; during this time cell number increases 500 times, divisions continuing at a uniform rate throughout the period, while the average cell volume increases tenfold. Full size has not been reached in the tuber at 200 g, so that a later phase involving only expansion growth may be involved.

For leaves, it used to be thought that cell division was completed in the primordium stage, and then expansion followed. More recent work has shown that to be incorrect, at least in several species which have been investigated in detail. In the lupin (*Lupinus albus*) and sunflower (*Helianthus annuus*), cell division continues during 50–75% of the entire lifespan of the leaf, and most of the divisions take place after leaf emergence (Fig. 8.4). In the cucumber (*Cucumis sativus*) leaf, 70–98% of the cells are formed after unfolding. In marrowstem kale (*Brassica oleracea* var. *acephala*), cell divisions continue to the end of the life of a leaf. It is, however, true that the cell division rate is much higher in the primordium, as reflected by the observation that the ratio of number of cells dividing to total number of cells falls as leaf growth proceeds and, in the cucumber leaf, this fall in ratio occurs quite sharply, suggesting a change from one developmental phase to another. The earlier view that there was a clear distinction between division and expansion phases in leaf growth was based on counts and measurements of epidermal cells only, and it is now clear that these cells may cease dividing before all the rest of the leaf cells as in the tobacco leaf. Again, all species do not behave in an identical manner and, in the leaves of strawberry (*Fragaria vesca*), the cell division and expansion stages are much more distinct than in lupin, sunflower and cucumber.

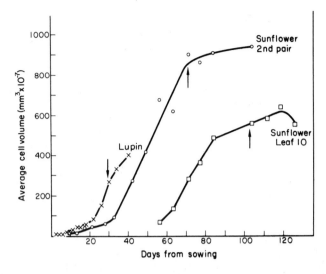

**Fig. 8.4** Development of sunflower and lupin leaves. Arrows denote the cessation of cell division; the phases of cell division and cell expansion overlap. (From Sunderland, 1960, *Journal of Experimental Botany*, **11**, 68–85. By permission of the Clarendon Press, Oxford.)

Size differences between organs of similar basic anatomy may be traceable to differences in cell volume or cell number or both. Table 8.1 presents a selection of data from an investigation of fruit growth in 12 varieties of cucurbits drawn from 4 genera and in which the largest fruits had a volume more than a thousand times that of the smallest. The final fruit size

**Table 8.1** The relationship between cell size and fruit size in cucurbit fruits. (After Sinnott, E. W., 1939, *American Journal of Botany*, **26**, 179–89.)

| Species | Diameter of mature fruit (mm) | Average diameter of inner cells of the fruit wall ($\mu$m) | | |
|---|---|---|---|---|
| | | When fruit diameter = 2 mm | At time of last cell division | At maturity |
| *Cucurbita pepo* | 39 | 14.5 | 27.0 | 80 |
| (7 varieties) | 45 | 15.5 | 26.8 | 98 |
| | 62 | 17.1 | 25.8 | 142 |
| | 120 | 17.5 | 30.7 | 125 |
| | 300 | 16.8 | 38.0 | — |
| | 315 | 12.4 | 27.8 | 200 |
| | 360 | 15.8 | 36.0 | — |
| *Lagenaria* | 105 | 14.5 | 34.0 | 140 |
| *vulgaris* (3 varieties) | 170 | 13.2 | 26.2 | 165 |
| | 100 | 15.5 | 29.0 | 300 |
| *Cucumis anguria* | 31 | 17.7 | 18.5 | 150 |
| *Citrullus vulgaris* | 310 | 15.2 | 29.6 | 700 |

was determined by a combination of three factors: cell number, the degree of cell expansion during the period of cell division, and the amount of expansion after cell division had stopped. On the whole, the larger-fruited races had a more extended period of cell division, thus producing more cells, and a greater amount of expansion both during the division and expansion phases. However, the same fruit size could be reached by different balances between cell division and expansion. Several authors have concluded that, in leaves of any one species, size differences are due to differences in cell number rather than cell size; in the sunflower and the morning glory (*Ipomoea hederacea*) the larger leaves actually have the smaller cells.

## Conditions necessary for growth

A number of conditions are necessary for plant growth. The plant system must be in a potentially growing state: a mature tissue is no longer capable of growth except in response to special stimuli such as result from wounding. It must also receive or be able to synthesize the growth hormones which control cell development (see Chapter 9).

Its environment must provide a supply of water, oxygen and nutrients, and a suitable temperature. Plants can grow only when their cells are turgid and, in nature, water supply often limits growth. Most flowering plants require oxygen for growth, although the seedlings of some aquatic plants can pass through the stages of germination and early seedling growth under anaerobic conditions, examples being rice (*Oryza sativa*) and the water plantain (*Alisma plantago-aquatica*). In their natural environment these plants normally germinate under water or in mud, where there is very little oxygen.

The flowering plant is characteristically autotrophic, obtaining the mineral nutrients for its growth from soil or water and manufacturing all its essential organic constituents from carbon dioxide, water and inorganic ions. However its separate organs are not self-sufficient with respect to organic nutrients; even young leaves that already are photosynthesizing continue to require to import organic nutrients or particular growth hormones. The dividing and growing cells of a plant are nurtured by metabolites and growth hormones (or their precursors) synthesized in the specialized tissues of the organism.

The temperature range compatible with plant growth varies from species to species (Table 8.2). Within this range there is an optimum temperature whose value will depend also upon the other conditions controlling growth. Plants native to warm habitats require higher temperatures for growth than those of cooler regions. The optimum growth temperature for winter wheat, a cereal of cool temperate climate, is 20–25°C; for maize, a cereal from warmer climate, it is 30–35°C. Higher plants are less tolerant of extreme temperatures than some micro-organisms (Table 8.2). The effect of temperature on growth is complex. An alternation of a lower temperature at night and a higher temperature during the day is frequently better for growth

**Table 8.2** Temperature limits for the growth of some flowering plants and microorganisms. (Data from Pfeffer, W., 1903, *The Physiology of Plants,* II. Clarendon Press, Oxford; Ingraham, J. L. and Stokes, J. L., 1959, *Bacteriological Reviews,* **23,** 97–108; Stiles, W., 1950, *An Introduction to the Principles of Plant Physiology.* Methuen, London.)

| Species | Minimum °C | Maximum °C |
|---|---|---|
| Flowering plants: | | |
| *Pisum sativum* (roots) | −2 | 44.5 |
| *Sinapis alba* | 0 | > 37 |
| *Triticum vulgare* | 0–5 | 42 |
| *Lepidium sativum* | 2 | 28 |
| *Acer platanoides* | 7–8 | 26 |
| *Zea mays* | 9 | 46 |
| *Cucurbita pepo* | 14 | 46 |
| *Cucumis sativus* | 15–18 | 44–50 |
| | | |
| Psychrophilic bacteria | −10 | Up to 45 |
| | | |
| Thermophilic micro-organisms: | | |
| *Aspergillus* | 15 | 60 |
| Bacteria | 38–45 | 75 |
| Blue-green algae | ? | 85–89 |

than any one constant temperature. The optimum growth temperature varies between organs of a plant and changes as a plant ages; what emerges as the optimum depends also on whether one considers the total growth over a long time period or the growth rate during a short interval. The temperature limits for growth of a plant are generally narrower than the temperature limits for individual physiological processes. Respiration for instance continues at a temperature at which growth is inhibited (Fig. 8.5).

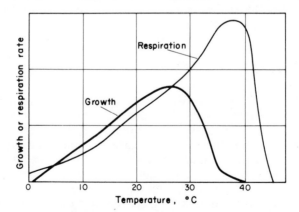

**Fig. 8.5** The effect of temperature on growth and respiration in the bean, *Phaseolus vulgaris.* At extreme temperatures, growth ceases before respiration stops. (From Bünning, 1953, *Entwicklungs- und Bewegungsphysiologie der Pflanze,* Springer-Verlag, Berlin.)

Growth follows from a harmonious interaction of all physiological processes and the required balance is disturbed at temperatures not inhibitory to the component processes, or immediately damaging to the cells.

Light was not listed above as a growth requirement because although flowering plants normally need light for photosynthesis, and hence for growth, they can grow in the dark if a supply of organic nutrient is available and may be able to complete their life history in darkness. Maize has been grown from seed to seed setting in the dark on an artificial sugar supply and *Arisaema triphyllum* (Araceae) has been raised in the dark from its large corm for four successive seasons. If light is thus not absolutely necessary for growth, it nevertheless profoundly influences growth and development (see Chapters 11 and 12). The phenomenon of etiolation indicates that light tends to suppress elongation growth, to promote leaf expansion (particularly in dicotyledons), and to enhance differentiation. The effect that light has on growth depends on the species, its age, previous growth conditions, and on irradiance and wave-length. Young plants are more sensitive to light inhibition of their elongation than older ones. The inhibition of elongation by light may involve an effect on the supply of growth hormones and on the sensitivity of the cells to these hormones.

Under natural conditions, plants often grow more in height during the night than during the day provided that the temperature at night is not too low. The growth of the date palm is stopped completely by direct sunlight. Bamboos in various tropical regions have been shown to grow in total height 1.8–2.8 times more during the night than during the day. However in this instance the controlling factor seems to be water supply rather than light. In general, the water contents of plants are higher by night than by day (Fig. 4.1) due to the development of a water deficit from transpiration. Many studies indicate that growth is greater by day if temperature is the limiting factor and greater by night if water is limiting. The highest growth rates therefore often occur in the early morning (temperature rising, and water content still high) and early evening (temperature still high, and water content beginning to increase).

## Growth rates

Quantitative comparisons between the growth of living systems can be made from two viewpoints. One can measure and compare their *absolute growth rates*, i.e. the total growth of each per unit time; or one can compare their *relative growth rates*, the growth of each per unit time expressed on a common basis, e.g. per unit initial weight. To estimate plant yields the absolute amounts of growth are required; to compare the growth activities of two systems their relative growth rates must be measured. If two leaves with initial areas of 5 and 50 cm$^2$ respectively both expand by a further 2 cm$^2$ in a day, then the absolute growth of each leaf is the same, but the smaller leaf has a ten times higher relative growth rate than the larger.

The relative rates of linear elongation growth of a number of plant organs

**Table 8.3** Rates of linear elongation growth of some higher plant organs and fungal hyphae expressed as percentage increase per minute per unit length of the growing zone. (After Pfeffer, W., 1903, *The Physiology of Plants,* II. Clarendon Press, Oxford; and Stiles, W., 1950, *An Introduction to the Principles of Plant Physiology,* Methuen, London.)

| Organ | Species | Growth rate |
|---|---|---|
| Pollen tubes | *Impatiens Hawkerii* | 220 |
| | *Impatiens balsamia* | 100 |
| Staminal filaments | *Triticum* and *Secale* | 37.5 |
| Shoot growing zones | *Bambusa* | 1.27 |
| | *Bryonia* | 0.58 |
| Root, fastest zone | *Vicia faba* | 0.36 |
| Fungal hyphae | *Botrytis cinerea* | 83–200 |
| | *Mucor (Rhizopus) stolonifer* | 118 |

are compared in Table 8.3. The values are given as percentage increases per minute, i.e.

$$\frac{\text{Length increase per min, in mm}}{\text{Length of the growing zone, in mm}} \times 100$$

This table is compiled from results obtained under varied experimental conditions, but the differences in growth rates recorded are far greater than accounted for by differences in external conditions. However favourable the conditions of growth, the growing zone of a *Vicia* root will not double its length in a minute, as will the growing zone of some fungi. Very high steady rates are maintained by fungal hyphae; similar but only short-lived rates are developed by pollen tubes and staminal filaments. The elongation of the latter is completed in a matter of minutes and results from a very rapid expansion of the constituent cells, during which wall extension greatly outstrips synthesis of new cell wall material.

Bacteria are frequently described as having the highest growth rates of all living organisms; bacterial cells can double their mass and divide to form daughter cells in 20–30 minutes. The duration of the cell cycle in the meristems of higher plants is much longer than this. In pea root tips, the duration of mitosis is 3 hours at 15°C and about 1 hour at 30°C, while the average interval between two successive cell divisions is about 22 hours (see Chapter 9 for further discussion of the cell cycle). The high growth rates of microorganisms, and their high rates of physiological activity in general, are believed to be related to their small size, which allows rapid diffusion of metabolites and gases into and out of the cells.

Table 8.3 depicts the relative growth rates per unit growing zone. The absolute growth rates will depend on these values, and on the sizes of the growing zones. In the hyphae of *Botrytis*, the growing zone is only 0.018 mm long, so that even with a 200% increase in length per minute, the total extension growth made in 24 hours (the daily absolute growth rate) is about 5 cm. In the bamboo shoot, with a relative growth rate of only 1.27% per minute, the growing zone is 5 cm long, and hence the daily absolute growth rate is about 90 cm. Differences in absolute growth rates of plants are thus largely the result of differences in the length or volume of the growing zones.

The extent of total growth achieved during a given time period varies enormously between plants. A marrow plant (*Cucurbita*) or a hop (*Humulus*) grows from seed to a length of 12 metres in a summer; an oak seedling will grow to 12 cm in this time; since mass varies as (length)$^3$, the differences in mass are much greater still. Marrow and hop are annual plants, with large growing zones, and produce much soft, thin-walled tissue; the oak, a perennial, has a small growing zone and quickly becomes woody. The oak plant can be regarded as showing a higher degree of differentiation and, as already indicated, growth and differentiation are often mutually antagonistic.

### Mathematical analysis of growth

Growth lends itself to mathematical analysis, and formulae have been devised to express various types of growth. The simplest case is that of *constant linear* or *arithmetic growth* shown by a root elongating at a constant rate. Then:

$$L_t = L_0 + rt \qquad (8.1)$$

where     $L_0$ = length at zero time

$L_t$ = length at time $t$

$r$ = growth rate, or elongation per unit time.

If $L$ is plotted against time we obtain a straight line (Fig. 8.6); the intercept of the graph gives $L_0$, and the slope, $a/b$, gives the growth rate $r$. If $r$ is known, the length $L_t$ can be obtained from the formula for any value of $t$, the time; or alternatively, the value of $r$ can be obtained from successive measurements of $L_t$.

Constant linear growth is observed only in relatively few growing systems. More frequently, the growth rate does not remain constant. In an entire root system any one tip may be elongating at a steady linear rate; but more and more new tips are continually formed; hence the rate of elongation of the root system as a whole keeps increasing. What may still remain constant is the relative growth per unit growing mass and, in that case, we have *constant exponential* or *constant logarithmic growth*. Denoting the relative growth rate by $r'$, for constant exponential growth we have:

$$L_t = L_0 e^{r't} \qquad (8.2)$$

where $e$ is the base of natural logarithms. This is the formula for continuous compound interest. In this case a plot of $L$ against time will give a curve as

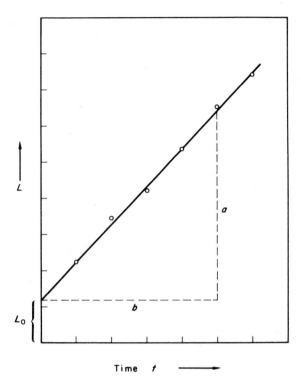

**Fig. 8.6** Constant linear growth in length (diagrammatic): a plot of total length, $L$, against time, $t$, gives a straight line. See text for further explanation.

shown in Fig. 8.7; and to obtain a straight line, the logarithm of $L$ must be plotted against $t$ (Fig. 8.8). From *equation 8.2*

$$\ln L_t = \ln L_0 + r't \qquad (8.3)$$

or, converting to logarithms of base 10,

$$\log L_t = \log L_0 + \log e^{r't} \qquad (8.4)$$

In Fig. 8.8, the intercept gives $\log L_0$ and the slope $a/b$ gives $\log e^{r'} = 0.43\, r'$.

A period of such exponential growth is found in the development of many living systems but, since it implies growth at an ever-increasing rate, it can proceed only for a limited period in the development of an organism. If mass increase of the plant is plotted against time from the commencement to the cessation of growth we obtain an *S*-shaped or *sigmoid curve* as illustrated in Fig. 8.9. This curve shows that growth is slow initially but it then speeds up and for a time approximates to exponential, later slowing down and finally stopping; with the onset of senescence there may occur an actual loss of mass. This sequence covers what is often called the **grand period** or **grand curve** of growth of the organism. It is not difficult to see why the period of high relative growth rate is limited. In the very young plant, mass increase

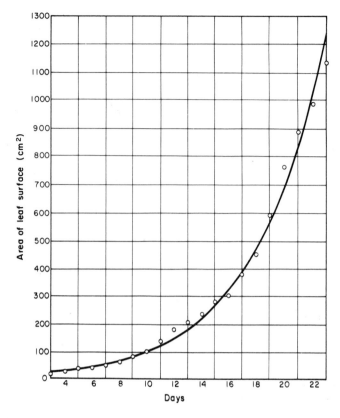

**Fig. 8.7**   Constant exponential growth illustrated by the increase in total leaf area of a cucumber (*Cucumis sativus*) plant; a plot of the leaf area (equivalent to *L* of equation *8.2*) against time gives a curve of ever-increasing steepness. (From Gregory, 1921, *Annals of Botany*, **35**, 93–123. By permission of the Clarendon Press, Oxford.)

means an increase in growing points and in photosynthetic area, the growth potential expands as the mass expands and growth is exponential. However not all the new tissue remains growing nor does it all add to the synthetic capacity of the plant. Thus the proportion of non-growing and non-photosynthetic tissue soon increases, and leads to a decline in relative growth rate. The loss of mass during senescence results from an excess of respiration over photosynthesis and from abscission of organs.

There is one formula which in some cases (though by no means universally) can be used to describe the grand curve of growth; this corresponds to the formula for an autocatalytic monomolecular reaction:

$$\frac{\mathrm{d}x}{\mathrm{d}t} = kx(a-x) \qquad (8.5)$$

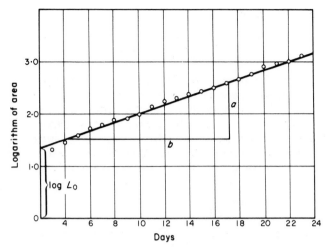

**Fig. 8.8** A plot of the logarithm of cucumber leaf area (data of Fig. 8.7) against time: a straight line is now obtained. See text for further explanation.

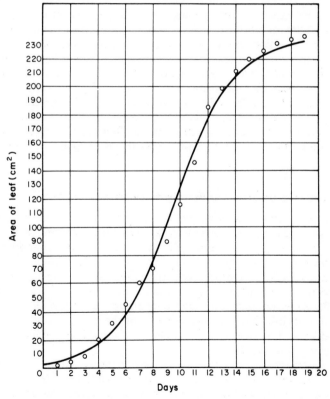

**Fig. 8.9** The grand period of growth of a single cucumber leaf (source as for Fig. 8.7). Curve for a whole plant would be similar.

which on integration gives:

$$\log \frac{x}{a-x} = k(t-t_1) \qquad (8.6)$$

where $a$ is the final maximum size, $x$ is the size at any time $t$, $t_1$ is the time when half the growth is completed (i.e. when $x = a/2$), and $k$ is a constant. This relationship holds for growth when the curve for the grand period of growth is quite symmetrical, the decrease from exponential growth occurring at exactly the same rate as the increase to exponential growth. It can be applied to describe the leaf growth illustrated in Fig. 8.9 by substituting the following numerical values in the formula:

$$\log \frac{x}{236-x} = 0.200\,(t-9.46) \qquad (8.7)$$

the maximum leaf area being 236 cm$^2$, and the time for half growth being 9.46 days.

Though in some cases *equation 8.6* describes the growth sequence almost perfectly, no particular conclusions can be drawn from the mathematical fit. Growth is not governed by one master reaction, but involves directly or indirectly all the metabolic reactions of cells. Growth and a monomolecular autocatalytic reaction are expressed by the same formula for different reasons. In the chemical reaction, rate initially increases because the more product is formed, the more catalyst there is; subsequently the rate decreases because the substrate of the reaction is exhausted. The initial increase in growth rate, in so far as it is due to an increase in the enzymes and templates that catalyse the growth reactions, can be considered as an autocatalytic process. The fall-off in rate is, however, brought about by a complex of factors not comparable with those involved in the chemical reaction.

**The progress of growth during development: growth rhythms**

Smooth growth curves as shown in Figs 8.7 and 8.8 are obtained only when the experimental material is grown under uniform conditions, and measurements are taken at fairly long time intervals. In nature, irregular fluctuations are superimposed on the growth curve by chance fluctuations of environmental factors such as temperature, irradiance and water supply. Figure 8.10 shows the course of maize growth in the field; the weekly growth rate is seen to fluctuate considerably and temperature is apparently one factor responsible for this. But the general shape of the grand period curve is still recognizable, and a smooth curve could be obtained by putting the line of best fit through the points.

If measurements are made at short time intervals such as hourly, then rhythmic changes in growth rate become apparent. Growth has a **diurnal rhythm**, with maxima and minima occurring at definite times of day. This is only to be expected under natural conditions, because of the regular diurnal

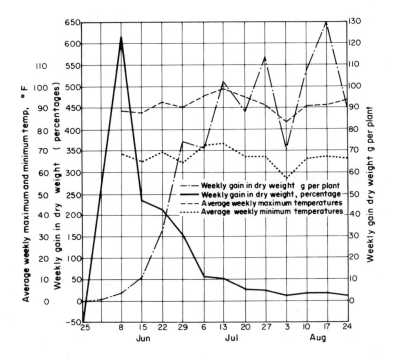

**Fig. 8.10** The growth rate of maize (*Zea mays*), expressed as the weekly dry weight gain, under natural conditions. (From Miller, *Plant Physiology,* Copyright 1931, McGraw-Hill Book Co. Used by permission of McGraw-Hill Book Co., New York and Maidenhead.)

alternation of day and night conditions, in particular the alternation of light and darkness. A plant grown from seed under constant conditions and in darkness does not show a diurnal growth rhythm. But once a rhythm has been initiated, and a single period of illumination may suffice for this, the rhythm will persist in continuous darkness for 2–3 days, or even longer. Thus growth conforms to the general pattern of diurnal rhythms in metabolic activity shown by many physiological processes which are partly controlled by the diurnal changes in external conditions and partly endogenously by a 'biological clock' mechanism of unknown nature. The growth of tomato plants is inhibited if the plants are subjected to light:dark cycles of 6:6 or 24:24 hours and this has been explained on the basis that the endogenous rhythm of the plants cannot be adjusted to these light:dark cycles. The diurnal growth cycle may show one or several maxima and minima of growth rate. In the roots of a number of species, 2 to 4 maxima of elongation occur in a 24-hour period, the maxima of elongation growth coinciding with minima of cell division activity (see also Chapter 12, p. 241).

Growth rhythms not related to the 24-hour period are also known. Growing shoot tips rotate in a circle (as viewed from above). This movement or **nutation (circumnutation)** is most pronounced in climbers, in which it aids the plant to twine round a support. Nutation is brought about by a wave of growth moving round the axis tip; at any instant, growth is most intense in one limited region of the tip, and the wave of activity completes a cycle round the tip in anything from 1.25 to 24 hours according to species and environmental conditions. The nutational rhythm is not dependent upon any rhythm in the external conditions, and will continue in a constant environment, though the speed of rotation is affected by changes in the environmental conditions, such as temperature. The direction of nutation is fixed and, in most species, is anti-clockwise as viewed from on top.

The grand curve of growth represents the life history of ephemerals and annuals and the course of growth during one growing season in perennials. In temperate habitats, most of the shoot growth of perennials is completed within a short period in the spring and early summer; some trees for example complete 90% of a year's growth in a 30-day period, starting 7–14 days after the commencement of growth, these first days representing the slow phase of the grand period. Such an early cessation of growth must be the result of some internal control mechanism and sometimes there does occur a second brief flush of growth later in the season. Root growth continues longer into the summer. In wet tropical habitats, growth in a plant community as a whole continues all the year round and with equal intensity, but in individual plants periods of high and low growth activity alternate, with a periodicity measured in months. Perennials therefore appear to show an endogenous growth rhythm in which each cycle may extend to several months. In habitats where climatic conditions vary over the year, such growth rhythms have become synchronized with the climatic rhythms, but in the more constant tropical climate where the environment does not exert a synchronizing influence, each plant or even separate branch of a plant grows according to its own innate rhythm.

Normally a phase of vegetative growth is followed by one of reproductive growth, and this transition is usually marked by the shoot apex changing quite abruptly from the production of vegetative organs to the production of flowers. This is usually reflected in the growth curve of the plant as a new surge of growth after the grand period curve of vegetative growth has levelled off. This transition is discussed more fully in Chapter 12.

## Morphogenesis

Plant development involves increase in size (growth) and the emergence of its organs and their arrangement in space (morphogenesis). The process of morphogenesis is inseparable from growth; the location of the apical meristems determines the branching and overall shape; a lobed leaf is produced by unequal growth along different radii; a fruit that grows equally in all directions becomes round, one which elongates preferentially along one axis becomes oblong. Shaping of organs by a morphogenetic movement of cells, which plays an important part in animal morphogenesis, is not possible in a flowering plant, where the cells are immobile. Thus, in plants, shape results from differential growth, in different regions and in different directions.

In morphogenesis we are concerned with the primary initiation of organs, and their subsequent development. Organ initiation establishes the basic pattern, the branching of shoot and root, the positioning of leaves and so on. Each organ is then moulded by its pattern of growth and differentiation.

Leaves and axillary buds are initiated close to the shoot apex. Commonly the primordia arise singly on the apical dome of the stem and the position of initiation moves in a circle round the apex. Since the apex elongates between the formation of every two successive primordia, the leaves get pulled out into a spiral round the stem. The precise positioning of the leaves, their *phyllotaxis*, is characteristic of the species. A very common arrangement is a 2/5 spiral, in which the sixth leaf lies directly above the first, and the spiral joining the leaves in order passes twice round the axis in moving from the first leaf to the sixth. Sometimes leaves arise in pairs or in whorls. Buds are initiated a little below the shoot apical meristem in most leaf axils.

The branching of the shoot could therefore potentially reflect the phyllotaxis, but in fact many of the buds initiated never grow out, being inhibited by the apical bud (*apical dominance*). The inhibition is usually thought to be mediated by growth regulators secreted by the apex, but monopolization by the apical bud of growth factors transported upwards from the roots has also been suggested as a causative factor. Branch angles are also under apical control. Apical dominance is much more marked in some species than in others. When the apical meristem is cut off, laterals below it often grow out in profusion; this reaction is utilized when pruning is applied to encourage bushiness in a shrub or tree (see also Chapter 11).

In roots, laterals are formed some distance back from the tip meristem, in a region where the primary tissues are mature. Initiation takes place in the pericycle, opposite or lateral to the protoxylem points, so that root branching is related to vascular anatomy. Roots do differ in the extent of lateral initiation and in the duration of growth of individual root axes but nevertheless exhibit less range of form than do shoot systems.

After organ initiation and outgrowth have mapped out the overall growth pattern, each organ develops by its own pattern of growth and different-iation. As an example of how leaf shapes result from differential growth, leaves of three strains of *Tropaeolum* are illustrated in Fig. 8.11. At an early developmental stage, the leaves have identical shapes in all three strains (leaf 1). Shape differences in the mature leaves are brought about by different amounts of growth along the main veins relative to growth in the intervenal areas. In leaf 2, approximately equal growth along both directions has preserved the outline of the young leaf; in the others, greater relative growth in the intervenal regions has served to obliterate the lobing, partly in leaf 3 and completely in leaf 4.

It is frequently observed that a constant ratio is maintained between the growth rates of different parts of a plant. This type of relationship is known as *allometric* (heterogonic) growth, and the relationship can be expressed mathematically by the allometry formula. If $x$ and $y$ represent the sizes of

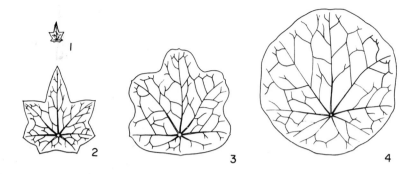

**Fig. 8.11** Morphogenesis in leaves of three strains of *Tropaeolum*. The juvenile leaf shape is identical for all (1), but through differences in the relative growth rates in the main vein and intervein directions, the adult shapes come to differ. (From Whaley and Whaley, 1942, *American Journal of Botany*, **29**, 195–200.)

two parts growing allometrically, and $k$ equals the ratio of their growth rates, then

$$y = bx^k \qquad (8.8)$$

where $b$ is a constant ($b$ = the value of $y$ when $x$ is taken as unity). Taking logarithms,

$$\log y = \log b + k \log x \qquad (8.9)$$

A plot of $\log x$ against $\log y$ gives a straight line and the slope of the graph gives $k$. If both parts are growing at equal rates, the slope of the graph is at 45° and $k = 1$. Figure 8.12 shows allometric growth exhibited by the stems and roots of some cultivated plants. An allometric relationship has also been observed in many cases between growth rates of leaf parts, and between the growth rates of the organs in a developing flower. Allometric growth plays an important part in the morphogenesis of all organisms.

Primary organ initiation results from cell divisions oriented so as to produce a tissue mass standing out from the initiating tissue. Subsequent differential growth involves both directional and localized cell divisions and directional cell elongation. In the *Tropaeolum* leaves shown in Fig. 8.11 the growth differences in the intervenal regions of the three types are effected through cell divisions, more cells being produced in these regions in leaves 3 and 4 than in leaf 2. Fruit shape in the Cucurbitaceae similarly depends on the direction of cell divisions. In round-fruited species, cell divisions in the fruit occur at random in all directions; the more elongate the fruit is to be, the more strictly are the planes of cell divisions orientated so as to produce new cells mainly along the axis of elongation. On the other hand, leaf shape in the aquatic plant *Callitriche intermedia* is determined both by the direction of cell division and cell expansion. This plant produces linear submerged leaves and ovate aerial leaves. Orientated cell divisions even while the leaf primordia are small produce shape differences for the two types and, when

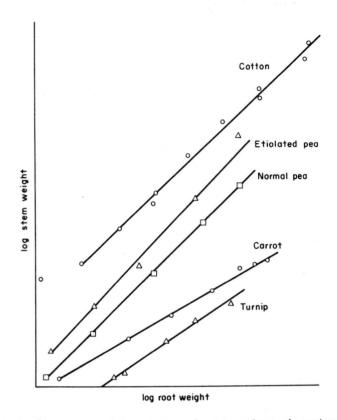

**Fig. 8.12** Allometric growth shown between the stems and roots of a number of species. The vertical scale is reduced to one-half in the case of etiolated peas. (From Pearsall, 1927, *Annals of Botany,* **41**, 549–56. By permission of the Clarendon Press, Oxford.)

elongation commences, cells of the linear leaves expand primarily in the direction of leaf elongation whereas the cells of the ovate leaves expand isodiametrically.

Certain aspects of morphogenesis are discussed in more detail in Chapters 11 and 12.

FURTHER READING

BONNER, J. T.(1963). *Morphogenesis.* Athenaeum, New York.

CUTTER, E. G. (1971). *Plant Anatomy: Experiment and Interpretation. Part II. Organs.* Contemporary Biology Series. Edward Arnold, London.

DORMER, K. J. (1972). *Shoot Organization in Vascular Plants.* Chapman & Hall, London.

HUNT, R. (1978). *Plant Growth Analysis.* Studies in Biology, no. 96. Edward Arnold, London.

KRAMER, P. J. and KOZLOWSKI, T. T. (1979). *Physiology of Woody Plants.* Academic Press, New York.

MOREY, P. R. (1973). *How Trees Grow.* Studies in Biology, no. 39, Edward Arnold, London.

WHALEY, W. G. (1961). Growth as a general process. In *Encyclopedia of Plant Physiology*, ed. Ruhland, W., **14**, 71–112. Springer-Verlag, Berlin.

SELECTED REFERENCES

BROWN, R. and BROADBENT, D. (1950). The development of cells in the growing zones of the root. *Journal of Experimental Botany*, **1**, 249–63.

GALLAGHER, J. N., BISCOE, P. V. and SAFFELL, R. A. (1976). A sensitive auxanometer for field use. *Journal of Experimental Botany*, **27**, 704–16.

SINNOTT, E. W. (1939). A developmental analysis of the relation between cell size and fruit size in cucurbits. *American Journal of Botany*, **26**, 179–89.

SUNDERLAND, N. (1960). Cell division and expansion in the growth of the leaf. *Journal of Experimental Botany*, **11**, 68–85.

# 9
# Cell Growth and Differentiation

Consideration of the problems of growth and differentiation at the cellular level in depth centres around aspects of cell physiology, biochemistry and ultrastructure outside the central theme of this text, the physiology of the whole plant. However, the delineation of the visual patterns of tissue and organ development – the formation of the whole plant – depends on the behaviour of the individual cells in various parts of the plants, as evident from the preceding chapter. The position of cells within the multicellular plant body is one of the controlling factors determining their behaviour: cells become what they are because of where they are. When the question is posed why the primary plant meristems are restricted to the root and shoot apices, and how, from such meristems, cellular expansion and different-iation proceed to give the patterns of tissue distribution characteristic of roots, stems and leaves, one must inevitably start at the cellular level. We must seek to describe the environment within the plant – the distribution within the developing plant of organic and inorganic nutrients, $O_2$, $CO_2$ and growth regulating substances. Then we must endeavour to understand how the interaction of such factors impinges on the metabolism of the cells to initiate and maintain cell division, to stimulate and later limit cell expansion, and to determine the various pathways of differentiation. It is therefore essential to have some appreciation of the biochemical events involved in cell division, cell expansion and cell differentiation.

The controlling factors regulating plant growth and development are still poorly understood. We shall, however, have cause to refer to hormonal control repeatedly in this chapter, and indeed subsequently. Before discussing cell growth and differentiation, we shall therefore make a preliminary survey of the plant growth hormones. Their involvement in various aspects of growth and development, and the controversial nature of their mode of action, are discussed in later sections.

## Plant growth hormones: a brief survey

### Auxins

The auxins can be defined as compounds which, in very low concentrations ($<10^{-3}$M to below $10^{-12}$M), promote the elongation growth of certain sensitive test organs, e.g. segments of coleoptiles or hypocotyls of seedlings. Historically they were the first of the plant hormones to be discovered, during studies on the growth and tropisms (see Chapter 10) of the grass coleoptile (Fig. 9.1). The elongation of this coleoptile (from a length of 10 mm or less) to its final length (of 80 mm or more) is a consequence of the expansion, and predominantly elongation, of its constituent cells. This

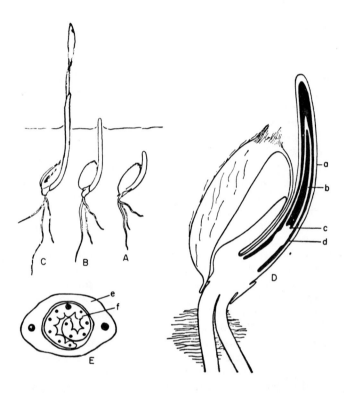

**Fig. 9.1**  Structure of the grass seedling. A, B, C; stages in germination; stage C shows emergence of first leaf from apex of the coleoptile. D: section taken at stage A; a, coleoptile; b, first leaf; c, stem apex; d, node; the region below d, mesocotyl. E: transverse section through coleoptile; e, coleoptile; f, first leaf; vascular strands in coleoptile and first leaf shown shaded. (From James, 1943, *An Introduction to Plant Physiology.* Clarendon Press, Oxford.)

elongation can be prevented by removal of the apical 2 mm of the coleoptile and restored by replacing the coleoptile tip back in position or replacing the excised tip by a block of agar jelly which has previously collected diffusible material from the cut base of an excised tip (Fig. 9.2). These findings can be taken to indicate that a substance is passed back from the tip, and promotes elongation of the cells behind the tip. The substance was named auxin and

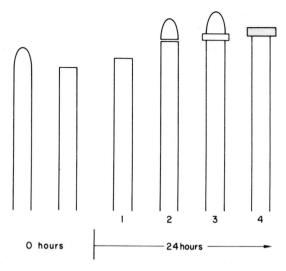

**Fig. 9.2** Diagram showing the effects on growth of an etiolated coleoptile of (1) decapitation (growth ceases then slowly resumes); (2) placing the tip back in position; (3) placing the tip back in position but with a thin sheet of agar or gelatin between tip and stump; (4) replacing the tip with a block of agar which has previously been enriched with auxin by diffusion from an excised tip.

eventually identified as indol-3yl acetic acid, IAA (Fig. 9.3). The cereal coleoptile has served for many years as a favourite test object for studies of auxin movement, auxin action, and auxin bioassays, i.e. assays of the amount of auxin in a plant extract, based on measuring the magnitude of the response produced by the extract in a test organ. The classical Went curvature test for auxin activity involves measuring, under standardized conditions, the curvature induced in the etiolated decapitated oat coleoptile by supplying the auxin in an agar block to one side only of the cut apical surface. The assay depends on the polarized downward movement of the auxin, with very little lateral diffusion. For straight growth tests, coleoptile segments are floated on auxin solutions. The responses are linear with respect to the logarithm of the auxin concentration, within a wide concentration range, depending on the tissue. As Fig. 9.4 shows, organs that respond to auxin with elongation growth have optimal IAA concentrations, above which the hormone becomes inhibitory and finally toxic. The toxic effects of excess auxin are utilized in auxin herbicides.

Auxin that diffuses out of tissues, the *diffusible auxin* is free IAA. Plant tissues also contain several forms of IAA associated with cellular components: the *extractable* auxin extracted with organic solvents, and *hydrolysable* auxin removed only after hydrolysis. Some of the bound auxin may represent the active form, bound to auxin receptors; other fractions may act as auxin stores, inactive in growth and inaccessible to enzymic degradation.

In addition to IAA, plant tissues contain other indole compounds having

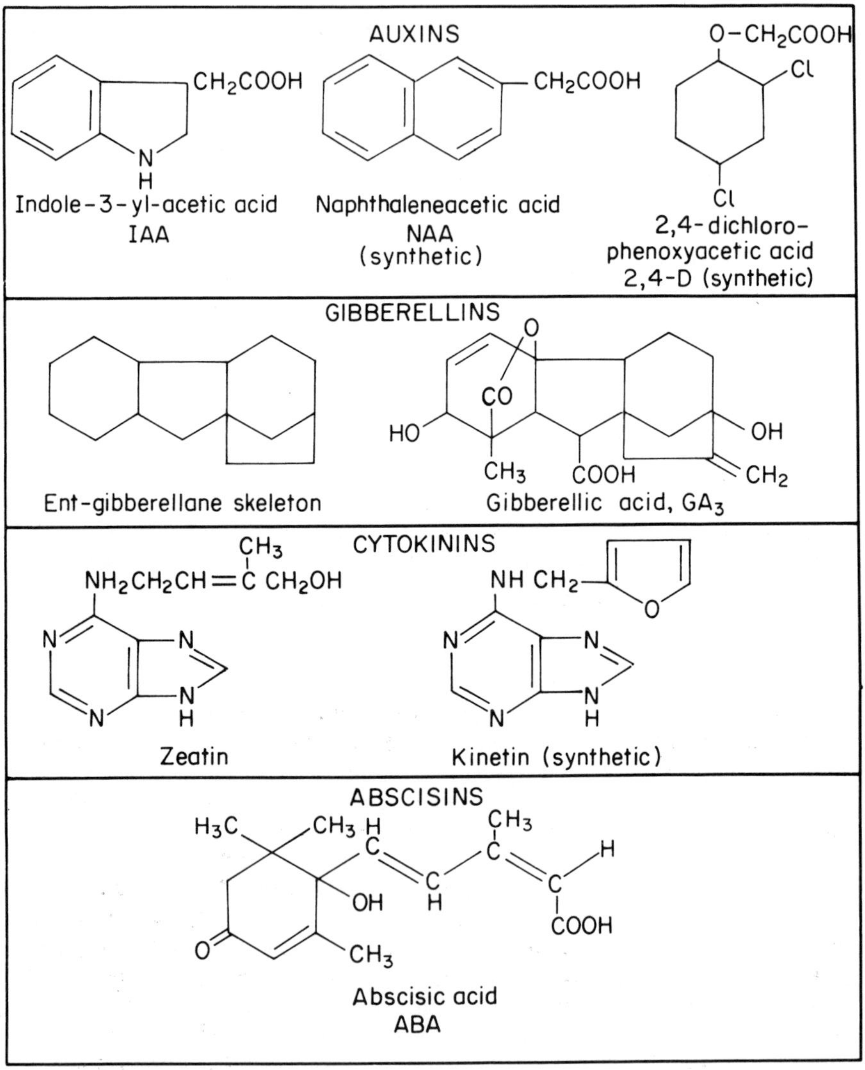

**Fig. 9.3** The structural formulae of some plant hormones.

auxin activity. It is disputable whether these are natural auxins, or IAA precursors, or IAA breakdown products. A wide range of synthetic compounds also have properties which justify their classification as auxins on the basis of their effects on growth and differentiation, and their polar transport within the plant. Examples of such compounds are the substituted phenoxyacetic acids (such as 2,4-D = 2,4-dichlorophenoxyacetic acid), the naphthalene and naphthoxyacetic acids, and the phenylaliphatic acids, such as phenylacetic and phenylpropionic acids. As Fig. 9.3 shows, auxin activity

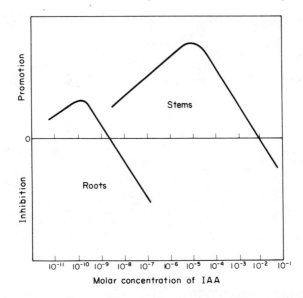

**Fig. 9.4** Generalized representation of the influence of indol-3yl acetic acid (IAA) on the linear growth of roots and stems. Note the log. scale of molarity of IAA. (After Thimann, 1938, *American Journal of Botany*, **24**, 407–12.)

can be exhibited by chemically diverse compounds. A ring structure and an acid group are imperative, but attempts to work out 'rules' which determine whether a compound should have auxin activity have not met with much success. Too little attention may have been paid to the importance of physical properties (e.g. surface tension effects, lipid solubility) as opposed to chemical structure, in relation to auxin activity.

Auxins are defined on the basis of their ability to promote cell elongation. They have many other effects on plants. Auxins can promote cell division; promote parthenocarpic fruit development in some species; promote rooting of cuttings (Fig. 9.5); they are active in apical dominance (Chapter 11); sometimes they inhibit leaf abscission and senescence; influence the sex of unisexual flowers, and they play a part in morphogenesis.

### The gibberellins

On the basis of biological activity, the gibberellins can be defined as compounds which, in very low concentrations, promote the elongation of certain genetic dwarf varieties of pea and maize (Fig. 9.6). Chemically, the gibberellins are a very well defined group, all having the 5-ring gibberellane skeleton (Fig. 9.3), and individual gibberellins differ only in the side groups; they are numbered as $GA_1$, $GA_2$, $GA_3$ ... and so on. Over 40 gibberellins are known from plants. One tissue commonly yields several gibberellins; e.g. in immature seeds of *Phaseolus coccineus*, $GA_1$, $GA_3$, $GA_5$ and $GA_6$ have been identified. The significance of this bewildering array of

**Fig. 9.5** Rooting of lemon cuttings. Treatments: *upper row,* basal end of cuttings in tap water for 8 hours; *lower row* in indol-3yl acetic acid (500 ppm) for similar period. Cuttings photographed 17 days after treatment. (From Cooper, 1935, *Plant Physiology, Lancaster,* **10**, 789–94.)

gibberellins is not yet clear. Particular gibberellins may perhaps be specific to particular functions, but although the activities of different gibberellins in different bioassays vary, it has not been possible to associate a particular gibberellin with one growth or differentiation process. One of the most widespread gibberellins, and the most potent in numerous assays, is $GA_3$, gibberellic acid, the first gibberellin to be discovered. Some gibberellins may

**Fig. 9.6** The influence of gibberellic acid (GA₃) on the growth of variety 'Meteor' dwarf pea. The plant on the left received no GA₃ and shows the typical dwarf habit. The remaining plants treated with $GA_3$; the dose per plant in micrograms is shown. With doses up to 5 $\mu$g there is increased linear growth of the stems with increase in GA₃ dosage. This is the principle of the dwarf pea assay for gibberellins. (Photograph from Marshall Cavendish *Encyclopedia of Gardening,* 1968. By Dr L. C. Luckwill, Long Ashton Research Station.)

be precursors or degradation products of others.

The gibberellins, like the auxins, have multiple effects. They can: promote cell elongation and division; promote parthenocarpy in some species; delay leaf senescence and inhibit leaf abscission; break dormancy, and stimulate flowering in long day plants (Chapter 12). In excised aleurone cells of cereal grains, as related in Chapter 2, gibberellins may act as stimuli for enzyme synthesis. The gibberellins differ from auxins in that high concentrations can be tolerated by plants without toxic symptoms. The gibberellins tend to be more effective at the whole plant level than on isolated tissues; until their effect on the cereal grains was discovered, the most reliable bioassays for gibberellins were based on their promotion of elongation of intact dwarf plants. Gibberellins are stated to have very little effect on roots.

### The cytokinins

The cytokinins were first discovered as agents necessary for cell division, and this group of hormones is defined as compounds that, in very low concentrations, stimulate cell division in tissue explants and cell cultures. The cytokinins are 6-substituted adenine derivatives (Fig. 9.3). Some fifteen natural cytokinins are known from plants. Many synthetic 6-substituted adenine compounds have cytokinin activity; the artificial compound kinetin

(Fig. 9.3) is commonly utilized in laboratory experiments when a cytokinin is required.

Almost all the effects exhibited by auxins and gibberellins can also be duplicated by cytokinins: promotion of cell elongation, promotion of parthenocarpy and of flowering, breakage of dormancy, control of apical dominance, delay of senescence. Comparison of the lists of effects of the three classes of hormones, auxins, gibberellins and cytokinins, shows them to be very similar. For a particular effect on any one species, however, only one of these hormones may be active. Fruit set in tomato is stimulated by auxins; in citrus fruits auxin is ineffective, but gibberellins are highly active. This should not be taken as indicative of a fundamental difference in the hormonal requirements of the two kinds of fruit. More probably, both auxins and gibberellins are required for fruit set in tomatoes and citrus fruits alike; but these plants may differ in their sensitivity (quantitative requirements), or the fruit of one species may be able to produce enough auxin without pollination, whereas the other is better able to synthesize its own gibberellins. Much evidence points to the interactions of several hormones in any developmental process. Where a specific quantitative response to one single hormone is obtained, the system giving that response is limited by the particular hormone.

### Abscisins

In contrast to the three previously discussed classes of hormone, abscisins are growth-inhibiting; the best known representative of this group is abscisic acid, ABA (Fig. 9.3). The natural and active compound is the D(+) isomer. It was first discovered as a compound which promoted dormancy in buds and abscission in cotton fruits. The recorded activities of ABA in plants include inhibition of growth and seed germination, promotion of fruit abscission, and induction of bud and seed dormancy. In spite of its growth-inhibitory action, it can be present in plant tissues in relatively high concentrations without adverse effects. ABA differs from other plant hormones in having effects quite separate from those on growth and differentiation: it causes stomatal closure, and is possibly generally involved in stress resistance. Other compounds, e.g. xanthoxin, with structure resembling that of ABA and with similar effects in bioassays, have been isolated from plants, but less is known about their physiological activity.

### Ethylene

The gas ethylene, $C_2H_4$, was the last member to be added to the catalogue of plant hormones. It had been known since the first decade of the century that ethylene had marked effects on plant growth, and by the mid-1930s it was recognized that ripening fruits produced ethylene and fruit ripening was speeded up by the gas. The fact that almost any plant tissue evolved at least small amounts of ethylene was, however, not established until much later (*c.* 1959), when the application of gas chromatographic techniques enabled trace amounts of ethylene to be identified reliably. Another ten years elapsed before it was accepted as a 'respectable' plant hormone, there being much reluctance to accord the status of a hormone to a chemically very simple gas. There is, however, no reason to segregate ethylene from other plant hormones. Its physiological effects are comparable to those of other hormones. Elongation growth is inhibited by ethylene (except at extremely low concentrations); plumular hook opening and polar transport of auxin are inhibited. On the other hand, ethylene promotes fruit ripening, seed germination (some species) and bud outgrowth, and it accelerates leaf senescence and abscission. The synthesis of ethylene is stimulated by auxin and some auxin effects, e.g. the promotion of flowering in pineapple, which has been utilized in cultivation, are now known to be mediated by the auxin-induced ethylene production.

### Problems of extraction, identification and quantitative estimation of plant hormones

Hormones are present in plant tissues usually in very low concentrations. The level of auxin in vegetative tissues (including bound auxin) varies from a few $\mu$g kg$^{-1}$ to about 350 $\mu$g kg$^{-1}$. Gibberellin concentrations in vegetative tissues go up to 1 mg kg$^{-1}$. In seeds, higher levels may be found, e.g. maize grain has yielded nearly 80 mg kg$^{-1}$ of auxin, while the gibberellin content of the liquid endosperm of *Echinocystis* has been estimated at 470 mg kg$^{-1}$. The low concentrations pose tremendous problems for the isolation and assay of hormones. In one series of experiments, 1300 kg of pea fruits were used to obtain a final yield of 9 mg ABA! Commercially grown seeds and fruits can be obtained in kg quantities, but the laboratory worker usually has only gramme or even mg quantities of experimental material at his disposal. Moreover, processing of bulk material necessitates large scale processing facilities. For many years therefore the identification and quantification of hormones depended almost entirely on **bioassays**, which relied on the high sensitivity of living plant material to minute amounts of hormone. The plant extract was purified by paper chromatography, (subsequently by column and thin layer chromatography); the more or less purified extract was then applied to test plant material for observation of qualitative and quantitative effects e.g. as described for auxin (p. 159). Even crude extracts have been bioassayed. While bioassays are very sensitive, they are time-consuming to set up and may need much test material; e.g. the dwarf pea test for gibberellins (Fig. 9.6) would need a large number of uniform-sized pea plants. Also the results, depending on growth effects, may take a long time to obtain. An even more serious drawback is the lack of specificity in bioassays: the same biological effect can be elicited by more than one hormone. There is also the possibility of interference by impurities and the effect of a hormone may be boosted or antagonized by the presence of another hormone, or non-hormonal chemical, in an impure extract.

Sensitive and reliable physico-chemical methods for the identification and assay of plant hormones are now available. One of these is the use of *gas-liquid chromatography* (GLC). Whereas in paper, thin layer or column chromatography, a compound is carried in a liquid along an adsorbing solid surface, in gas chromatography, a gaseous compound is carried in a gas stream over an adsorbing liquid surface, and the compound is detected as a pulse passing over an appropriate detector. The pulse pattern from an unknown extract can be compared with the pulses from known marker compounds. Although the plant hormones are not gaseous with the exception of ethylene, they form volatile derivatives which are suitable for GLC analysis. This is a quick and accurate method detecting from as low as 50 pg (some gibberellins) to 10–20 ng (IAA) of hormones. Bioassays may detect lower amounts (down to 1 pg for IAA is claimed), but with much less accuracy. Methods of *high performance liquid chromatography* (HPLC) using columns, have been used to separate gibberellins (Fig. 9.7). For final absolute identification, GLC can be followed by *mass spectrometry*, where

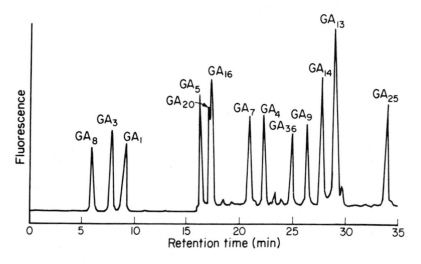

**Fig. 9.7** Separation of gibberellins (as esters) by HPLC; the gibberellin esters were detected by their fluorescence. (From Crozier, 1981, *Advances in Botanical Research*, **9**, 33–149.)

the separated fractions are injected into a mass spectrometer for determination of molecular mass; mass spectrometry can be used also to authenticate hormone samples purified by other means (Fig. 9.8).

The latest technique introduced for plant hormone detection is the *radioimmunoassay*. An antiserum is prepared by injecting rabbits with a hormone-protein complex. The antibodies in that serum then have a high and specific affinity for the hormone; the specificity is so high that an antiserum to $GA_3$ showed reactivity in addition to $GA_3$ only with $GA_7$ from amongst ten gibberellins and related compounds tested. The amount of hormone in a plant extract is obtained by measuring the efficiency with which the extract competes with pure radioactive hormone for the anti-

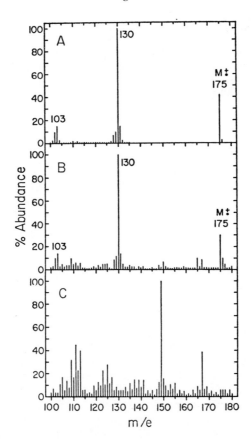

**Fig. 9.8** Mass spectra of (A) authentic IAA; (B) the spot from the IAA Rf after the chromatographic purification of the diffusate from 15 000 *Zea* coleoptile tips; and (C) the same zone of a chromatogram as used in (B), except that the extracts were made of blank agar blocks. (From Greenwood, Shaw, Hillman, Ritchie and Wilkins, 1972, *Planta*, **108**, 179–83.)

bodies. IAA, gibberellins, cytokinins and ABA have all been immunologically assayed. With the detection limit as low as e.g. 2 pg for gibberellins, very little plant material is needed, and the assay has such high specificity that there is no need for a rigorous purification of the extract, with possible loss of hormone. Weiler and Ziegler (1981) were able to divide 0.25ml of phloem sap between assays for ABA, three cytokinins, $GA_3$ and IAA!

As such techniques become more widely used, plant hormone analysis will be put on an accurate quantitative basis.

## Cell division

Cells undergoing repeated division go through a repeating cell cycle during which there occurs an increase in cell mass, a replication of the genetic material and division of the cell mass and of the duplicated genetic material to give rise to daughter cells. The sequence can be represented thus:

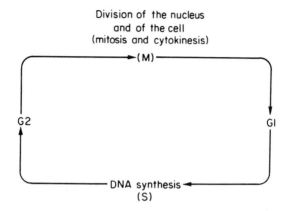

The two periods of interphase which respectively precede (G1, the first gap) and follow (G2, the second gap) the S phase are periods during which events essential to the initiation of chromosome replication and mitosis occur. Cell growth proceeds throughout interphase. In plant cells this cycle can be completed in 16–24 hours but can take much longer owing to extensions of interphase (particularly of G1). Both the whole cycle and mitosis itself are sensitive to temperature. At near optimum temperatures mitosis takes from 30 minutes to 3 hours depending upon species and cell size. Prophase nuclei contain twice the diploid DNA content (4C) and telophase nuclei the diploid content (2C); a doubling of nuclear DNA content occurs during a particular period (the S phase) of interphase. The duration of the S phase and its initiation can be determined by studying the incorporation of radioactive phosphorus ($^{32}$P) or tritium ($^{3}$H)-labelled thymidine into chromosomal DNA. This phase in which duplication of the genetic material of the chromosomes takes place is therefore a critical step in the cell cycle. Without DNA duplication there can be no cell division.

These considerations naturally lead us to ask what initiates the S phase of the cell cycle and determines its duration. There are many cases when nuclei within the same cytoplasm show very high synchrony in their division, suggesting the operation of a control mediated via the cytoplasm.

During interphase there occurs not only replication of DNA, but the synthesis of the other chromosome components, and of material involved in the mitotic spindle, increase in the number of mitochondria and other organelles, and increase in cell mass. While DNA synthesis is confined to the S phase, the synthesis of other cytoplasmic and nuclear constituents occurs over a longer time (in G1 and G2). Thus RNA is synthesized from late telophase, through the entire interphase and up to the mid-prophase of mitosis. Synthesis of histones (basic nuclear proteins), begins in G1 and continues through G2. It is also during interphase that the energy source is built up for the contraction, condensation and movement of chromosomes and for the reconstruction of the interphase nucleus during mitosis. Studies with plant cell cultures (see Chapter 12), synchronized in cell division, have shown that the rates of physiological processes and the syntheses of individual enzymes and metabolites during interphase follow a sequential and repeating pattern during successive cell cycles.

Mitosis does not proceed without prior DNA replication but DNA replication alone is not sufficient to ensure that mitosis will follow as evidenced by the development of polytene (many-stranded) chromosomes. The nature of the signal which terminates G2 and initiates mitosis is not known. The hypothesis that the achievement of a critical cell mass by cell growth during interphase precipitates mitosis cannot be upheld. It is true that mean cell size is often very stable in active meristems and that the growth phase can be prolonged and mitosis prevented by repeated amputation of the cytoplasm (for example in *Amoeba*). However in contrast to this there are many instances where cells go through a succession of divisions without further growth so that daughter cells become smaller with successive divisions; this is seen during plant embryology, in the behaviour of cultured plant cells and in the growth, under appropriate conditions, of certain unicellular algae. In such cases interphase is usually shorter than where cell mass is being maintained by interphase growth but it still occupies a high proportion of the cell cycle and exceeds in duration the S phase.

This description of the cell cycle has been developed to emphasize the many points at which inter- and intra-cellular controls could operate. To talk of some experimental treatment promoting cell division is only the very first step; to understand its mechanism of action it is necessary to be able to describe its specific impact on some component process in a very complex sequence of integrated biochemical events.

The discovery of the unique properties of coconut milk (the liquid endosperm of *Cocos nucifera*) in stimulating cell division in isolated tissue fragments raised the question of the natural occurrence in plants of specific cell-division factors. Thus pieces of the stem pith of tobacco (*Nicotiana tabacum* var. Wisconsin 38) when placed on a medium supplying sugar, inorganic salts, various vitamins and an auxin such as indol-3yl acetic acid (IAA), naphthalene acetic acid (NAA) or 2,4-dichlorophenoxyacetic acid (2,4-D) (Fig. 9.3) remained alive for some time but failed to grow. Increasing the auxin concentration in the culture medium induced swelling of the cells but did not initiate cell division in the explanted tissue. However addition of coconut milk to the medium led to continuing cell division throughout the explant and to the development of a growing callus tissue. The coconut milk could be replaced by an appropriate concentration of an extract of yeast or of aged or autoclaved herring sperm DNA. Further examination of the extract of autoclaved DNA led to the isolation of its active principle and its chemical identification as 6-furfurylaminopurine.

This compound was termed *kinetin* because of its activity in promoting cell division (cytokinesis). Subsequently it was shown that compounds similar in physiological activity and chemical composition to kinetin occur universally in higher plants. These are the natural cytokinins, which obviously are essential for cell division in higher plants.

Under certain conditions the cells of a tissue culture derived from pea roots show doubling of the chromosomes without division of the nucleus (endomitosis). Mitosis and cytokinesis in such cells can be initiated by supplying kinetin and its effect appears to be quite specific. This and other observations indicate that cytokinin is not implicated in the DNA duplication occurring during the S phase of the cell cycle.

A number of workers have shown that applications of kinetin or of other active cytokinins are quickly followed by changes in the content of cellular RNA and in certain such studies evidence has been obtained by simultaneous feeding of radioactive phosphorus ($^{32}$P) and orotic acid labelled with $^{14}$C that the cytokinin either changes the rate of RNA synthesis or its rate of turnover (synthesis versus degradation). In a number of cases where cytokinin application has raised the RNA level this has been shown to be accompanied by enhanced protein synthesis. Indirect evidence that cytokinins can promote mRNA synthesis comes from experiments involving the use of inhibitors such as actinomycin-D which inhibit this DNA-dependent RNA synthesis and from the observation that kinetin can preserve polyribosomes under conditions otherwise leading to their breakdown. Other studies point to cytokinins promoting nucleolar activity and increasing or maintaining the number of ribosomes per cell (promoting rRNA synthesis). Although increased synthesis of some enzymes (for instance the enzyme, tyramine methylpherase, in seedling barley roots) has been reported to follow kinetin application, in other cases cytokinin-mediated repression of enzyme synthesis has been observed (for example repression of the nucleic acid degrading enzymes, ribonuclease and deoxyribonuclease). The physiological activity of cytokinins appears to arise from their controlling influence in RNA metabolism, but we cannot yet define how they exert this control nor how the changed RNA metabolism leads to the observed physiological responses. Cytokinins are found as integral parts of some tRNA molecules, where they occupy a position adjacent to the anticodon. But this also occurs in tRNAs from other organisms where cytokinins have no hormonal activity, and there is no conclusive evidence that the hormonal nature of cytokinins in higher plants is related to the presence of these bases in tRNA.

Cytokinins are not the only hormones implicated in the control of cell division. Auxins also act as cell division factors, so much so that cytokinins have been defined as substances promoting cell division 'in association with the auxins'; e.g. the formation of adventitious roots, which is stimulated by auxin application (p. 162) begins with the stimulation of cell division to form a new root meristem. From studies with tissue cultures, evidence has been adduced for the necessity of auxin for DNA doubling, mitosis, and cytokinesis. In explants of artichoke tuber tissue, auxin is an absolute requirement for DNA replication. In this tissue there is a lag phase of at least 16 h between auxin application and DNA synthesis and during this period about 40 new proteins become detectable in the nuclei. In other tissues, too, effects of auxin on nuclear proteins have been noted, so that auxin may act in cell division by affecting synthesis of nuclear proteins involved in DNA replication. Cell and tissue cultures do not require exogenously added gibberellins, but the elongation of dwarf pea, maize and some rosette plants under the influence of gibberellins results largely from an increased number of cell divisions in the shoot subapical meristem.

## Cell expansion

The transition of the cells of plant meristems into differentiated cells almost invariably involves increase in cell size, often a many hundred-fold increase and this cell expansion is the first visible sign of differentiation. Accompanying this expansion process there is a massive uptake of water and associated with this the appearance of liquid vacuoles and their coalescence into a central vacuole. As indicated earlier (p. 138) the cell wall does not thin out as it increases in area; synthesis of new cell wall material takes place continuously throughout the period of increase in cell volume. Further although cells remain turgid during cell expansion, the uptake of water is not the consequence of an increased solute content in the vacuoles but arises through changes in the plasticity of the cell wall and the maintenance of an effective (although sometimes decreasing) solute concentration within the protoplast. Again although much of the increase in protoplast volume is accounted for by growth of vacuoles there is during cell expansion an increase in the volume of cytoplasm and often in the number of cytoplasmic organelles per cell. Thus either permanently or transiently (according to the subsequent course of differentiation) the expanding cell has a higher protein, lipid and RNA content than the meristematic cell from which it is derived (Fig. 9.9). The period of cell expansion is one characterized by active metabolism and net biosynthesis. The cell wall undergoes chemical changes. Increases in the numbers of mitochondria, and increased development of mitochondrial cristae have been observed, correlated with rises in the $O_2$ uptake per cell. There are increases in the extent of the endoplasmic reticulum and in the proportions of ribosomes bound to the endoplasmic reticulum and organized into polysomes. Associated with the rise in protein content there occur changes in the pattern of proteins as detected by gel electrophoresis, and changes in the absolute and relative amounts of enzymes. Changes take place also in the patterns of cellular RNA and, from the known roles of RNA in translating genetic information into cellular activity, it may be postulated that these RNA changes are critical in initiating, sustaining and terminating the phase of cell expansion.

For many years, auxin has been regarded as *the* hormone that promotes and controls cell expansion. However, it is now recognized that auxin does not hold a monopoly in this matter and that gibberellins and cytokinins, too,

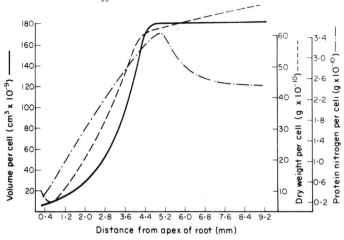

**Fig. 9.9** Changes occurring during cell expansion in seedling pea roots. The progress of cell expansion parallels the scale of distance from the apex of the root (i.e. cells examined at different distances from the apex are equivalent to stages in the expansion process through which each cell passes). Cell expansion is completed at c. 4.8 mm from the apex. As the cells expand in volume there is a parallel increase in the dry weight per cell and the protein content per cell. (Data from Brown and Broadbent, 1950, *Journal of Experimental Botany*, **1**, 249–63.)

may stimulate cell enlargement. In leaf and cotyledon tissue, for example, cytokinins can stimulate expansion, and gibberellins lead to cell elongation in germinating embryo radicles. Nevertheless the effects of auxin have been studied most intensively; it has become almost a dogma that cell elongation is primarily under auxin control.

### The acid-growth theory

Quite early observations revealed that auxin has an effect on the wall properties of elongating cells. As previously stated, during cell expansion the turgor pressure remains constant or declines and there is an increase in cell wall materials so that wall thickness is maintained or increased as cell surface area rises. During cell expansion there occur, particularly early in the process, changes in the physical properties of the wall and middle lamella which permit increase in cell area without increase in turgor pressure and irrespective of simultaneous synthesis of wall material. Soon after cell expansion is initiated, there however also occurs a continuing synthesis of the cell wall polysaccharides including cellulose. Heyn in 1931 first reported values for the elasticity (reversible stretching) and plasticity (irreversible stretching) of the plasmolysed etiolated oat coleoptile calculated from measuring the bending of the horizontal coleoptile under applied weights and its recovery on removal of the weights and obtained evidence that immediate pretreatment with auxin increased the value of both properties and particularly of the plastic component. Such softening of the cell wall by auxin has been confirmed by subsequent studies (Fig. 9.10).

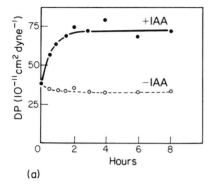 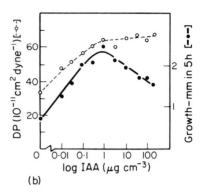

**Fig. 9.10** (a) Time course of development of auxin effect on plastic extensibility (DP) of *Avena* coleoptile sections pretreated for 90 min in water, then incubated 0–8 h in buffer and sucrose, with or without IAA. (b) Effect of auxin concentration on growth and DP. Sections were incubated for 5 h in buffer and sucrose with varying levels of IAA; growth and DP were then determined. (From Cleland, 1967, *Planta,* **74**, 197–209.)

The growing cell wall is a very complex system. The structural framework is provided by cellulose microfibrils, which are embedded in a matrix of hemicelluloses, pectic substances, and some protein. The wall constituents are held together largely by H bonds; the mixture of polymers has been considered to form a supermacromolecule, of ordered structure, but how much is held in covalent linkage is uncertain. During cell extension bonds must be broken to permit the microfibrils to slide past each other, and then the bonds must be reformed in new positions. The precise molecular events in the loosening of the wall are not known. The plasticizing of the cell wall is associated with the secretion of $H^+$ ions from the cells into the wall, resulting in an acidification of the wall. Auxin stimulates, in the plasmalemma, a $H^+$ pump which moves $H^+$ ions out across the plasmalemma while splitting ATP (an $H^+$ ATPase); the ATP acts as the energy source. The acidification of the wall is believed to mediate the loosening of the structure, possibly via effects on enzymes in the cell wall. In agreement with this hypothesis, excised plant organs elongate more rapidly, and exhibit increased wall plasticity, at low pH. The optimal pH is around 4.8. Perhaps even more convincingly, transfer of tissue to a low pH stimulates growth in the absence of auxin, though transiently (Fig. 9.11). The fungal toxin fusicoccin, which is a very potent stimulator of the plasmalemma $H^+$ ATPase, is an even more potent stimulator of elongation growth than auxin (Fig. 9.11). The postulate that auxin acts via lowering cell wall pH is known as the *acid-growth theory*. It is generally presumed that auxin must bind with a receptor site on the plasma membrane.

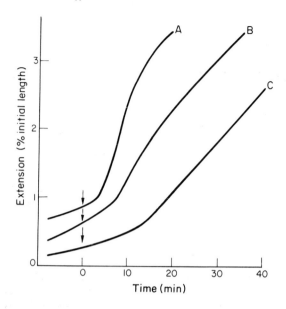

**Fig. 9.11**    Kinetics of elongation response of *Avena* coleoptiles to auxin, fusicoccin and $H^+$ (i.e. acidity). Sections were incubated for 30 min in an optical growth recorder (Evans and Ray, 1969) with 10 $\mu$mol $1^{-1}$ auxin (C), 10 $\mu$mol $1^{-1}$ fusicoccin in pH 7 buffer (A), or 10 $\mu$mol potassium phosphate-citrate buffer, pH 3.0 without hormones (B). (From Cleland, 1977, *Symposium of the Society for Experimental Biology,* **31**, 101–15.)

An alternative view is that the primary action of auxin is exerted at the genetic level, auxin stimulating the transcription of mRNAs which lead to the synthesis of proteins needed for the expansion process. The protein pattern of soybean hypocotyl segments changes in response to auxin. Nuclear material isolated from auxin-treated tissues has a higher RNA synthesizing activity than nuclear material from corresponding control tissues. Auxin-induced growth can be inhibited by actinomycin D, an inhibitor of mRNA synthesis, or by cycloheximide, which inhibits protein synthesis. Cell wall loosening is inhibited in parallel to the growth inhibition produced by actinomycin D and cycloheximide.

The acid-growth theory can be reconciled with the idea that auxin controls RNA and protein synthesis by postulating a dual action for auxin: first a rapid effect on the plasmalemma, mediating the wall loosening; secondly action on the genetic system. Growth stimulated by acid (in the absence of auxin) ceases after 60–90 min: apparently the acid can achieve the plasticizing of the wall, but other actions of auxin are required for sustained growth. An even closer combination of the two theories is incorporated in the hypothesis that the stimulation of the $H^+$ ATPase is brought about by a protein whose synthesis is controlled by a mRNA produced under the influence of auxin.

Yet while biochemical analyses are delving ever deeper into the effects of auxin on membrane pumps and nucleic acid metabolism, doubts are arising as to whether auxin is such an important controlling agent of cell elongation under natural circumstances. The great stimulation of growth by auxin is observed in *excised* tissue segments but auxin fails to promote normal elongation growth of *intact* whole plants, where excess applications result in overproduction of RNA and protein and recommencement of cell divisions in mature regions. The classical interpretation was that excision deprives the isolated tissues of their natural auxin source (e.g. the coleoptile tip), and therefore auxin now stimulates growth. An alternative interpretation is that excision is accompanied by a shock reaction, entailing among other effects loss of activity of the membrane $H^+$ ATPase, and the application of auxin merely speeds up the recovery from shock. Excised tissues do tend to resume growth spontaneously within a few hours, as if after a recovery process. In studying the effects of auxin on the growth of excised plant segments, the system may be so far removed from the whole plant that extrapolation from the results to the intact organism may not be justified. The dilemma is that data gained on isolated segments may be invalid for the whole plant, but an entire plant, even a young seedling, is too complex a system for the analysis of cellular control mechanisms.

## Cell differentiation

Cellular differentiation occurs in multicellular organisms and is the process leading to the development of the specialized structures and functions which characterize mature tissue cells. These cells are derived in higher plants from cells initiated by division at the primary and secondary meristems.

Cell expansion is an aspect of differentiation involving not only change in shape and volume but in physiological function (as evidenced by the changing activity of enzymes). Further some measure of cell expansion is involved in the differentiation of all tissue cells.

Differentiation into incipient shoot and root poles occurs prior to division in the fertilized ovum of flowering plants, gymnosperms and pteridophytes. Following the establishment of this polarity, the walls which effect cell cleavage are appropriately orientated as development of the embryonic sporophyte or gametophyte is initiated. Such polarity can be regarded as a spatial differentiation within the cytoplasm of a single cell. However tissues derived from such polarized cells themselves appear to show polarity and often for some particular physiological activity it is, in consequence, possible to demonstrate polarity in the whole organ concerned. One example of such polarity has already been given in describing auxin transport (p. 107). Vöchting in 1878 described how willow shoots form roots at their physiologically basal ends and buds at their apical ends irrespective of their orientation during the rooting period. A similar phenomenon can be readily demonstrated with segments of *Taraxacum* roots (Fig. 9.12). The establishment of such polarity in cell aggregates and growing organs appears to be involved in the normal differentiation of cells to give the tissue patterns

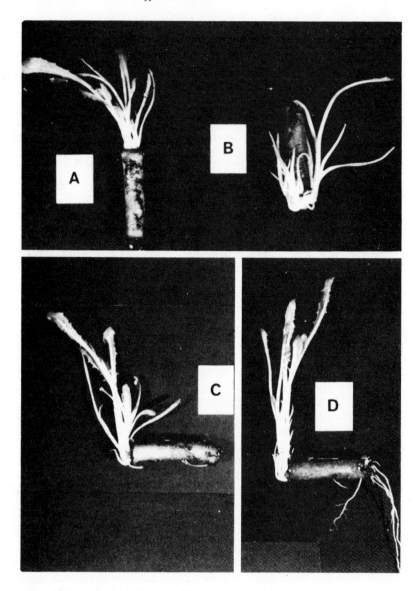

**Fig. 9.12**   Polarity of segments of roots of *Taraxacum officinale,* as shown by the development of shoots at their apical ends (ends adjacent to junction of root and shoot) irrespective of orientation during the sprouting period. A: upright; B: inverted; C and D: horizontal. Lateral root development is enhanced by application of a lanoline paste containing IAA to the basal end of the segment (D). Application of IAA to the apical ends of the segments will suppress shoot bud initiation, causing roots to develop at both ends of the segments. (Photo. by G. Asquith.)

characteristic of plant organs. Longitudinal and lateral gradients of major nutrients and metabolites, of dissolved gases and of growth-regulating substances are characteristic of growing organs and may be visualized as a three-dimensional physico-chemical pattern established or modified by tissue polarity. The forces operating on each cell are determined by its position (albeit changing with time), in this micro-environment pattern; cells are what they are because of where they are within the plant.

Tissue cells can always in theory (and sometimes in practice) be traced back to a daughter cell arising within the group of permanently meristematic cells and subsequently displaced from this group into a region of cell expansion and differentiation. Cells are what they are not only because of where they are but because of where they originated within the region where cells are being initiated. This can be illustrated by drawing upon studies on the differentiation of epidermal (piliferous layer) cells in roots such as those of *Phleum pratense*. The seedling roots of this grass are transparent and have a rudimentary root cap. This has enabled growth and differentiation to be followed in continuous files of epidermal cells by a photomicrographic technique. The epidermis shows two kinds of cells; short cells (trichoblasts) which produce root hairs and long cells which remain hairless. Ultimate, or occasionally penultimate, divisions in the epidermis differentiate the trichoblasts from the hairless cells. About one-third of these cell divisions are equal and both daughter cells of the division become hairless cells. The remaining two-thirds of the divisions are very unequal and, of the pair of cells, the short more apical ones develop into trichoblasts and the longer basal ones develop into hairless cells. The initiation of a trichoblast occurs as a result of an asymmetric mitosis (for further discussion of this aspect of differentiation the reader is referred to the volume in this series by Professor Elizabeth Cutter entitled *Plant Anatomy: Experiment and Interpretation, Part I*). Similar unequal divisions are involved in the differentiation of the sieve tube unit cell and the companion cell from a common mother cell and in the differentiation of stomatal guard cells and supporting cells in the leaf epidermis.

Studies on the giant chromosomes of insect cells and nuclear transplantation experiments in amphibians can be quoted as evidence that as cells differentiate more or less permanent changes take place in their nuclei. Examination of nuclei isolated from seedling pea roots has revealed differences in nuclear volume and in DNA, RNA and protein contents according as to whether the nuclei were from meristematic, expanding or mature cells. Such differences would however be expected between nuclei from different cytoplasmic environments. Further the capacity of certain mature plant tissue cells to return, under the influence of appropriate stimuli such as wounding or treatment with growth-regulating substances, to the meristematic state and subsequently to display their capacity to function like the fertilized egg (see p. 255) or to embark upon new lines of differentiation argues that the nuclear changes associated with differentiation in plants are reversible. Cells with very different phenotypes may be of identical genotype. The term *totipotency* is used to describe this capacity of plant cells to display at different times and under appropriate stimuli different aspects of their genetic potentialities, including their potentiality to initiate a new multicellular individual. Nevertheless it is important to recognize that it has not been demonstrated that *all* plant tissue cells, whilst still alive, retain this totipotency.

Doubt regarding the universal retention of totipotency during the differentiation of plant cells seems to be justified on various grounds. Callus cultures derived from a number of plants will, when first isolated, give rise to shoot buds and roots either spontaneously or in response to the incorporation into or omission from their culture media of appropriate concentrations of growth-regulators, like auxin and cytokinins (see Chapter 11). In most cases, the callus if maintained in culture loses this capacity for organogenesis. This could be due to progressive depletion of some unknown growth-regulator carried over in the initial explant and in certain cases this may prove to be the explanation. However in some instances a correlation has been observed between loss of the capacity to initiate organs and the development of polyploidy (both euploidy and aneuploidy) and of chromosomal aberrations. Such cytological changes could be regarded as resulting from the abnormal conditions of *in vitro* culture. However it has been observed: (*i*) that separate callus isolates from the same plant may differ markedly in their initial capacities to give rise to organs and (*ii*) that polyploidy is a widespread concomitant of normal cellular differentiation. Thus in roots it has been found that certain tisues are characterized by a predominant level of polyploidy. Cortical cells are frequently tetraploid, vessel unit cells in the metaxylem are tetraploid or octoploid or of even higher ploidy and endodermal cells are frequently tetraploid. A close correlation has been found between the presence of tetraploid cells and the origin of root nodules in leguminous plants. However the cells of such tissues are rarely if ever *all* at the predominant level of ploidy suggesting that polyploidy arises during differentiation and is not essential to its initiation and normal progress. The physiological consequences of this development of polyploidy in tissue cells are uncertain although studies on volumes reached by metaxylem vessel unit cells and polyploidy of the vessel unit cells suggest that each increase in ploidy of the nucleus permits a further increment of cell expansion (Fig. 9.13); nuclear volume (which is related to DNA content) and cell volume at maturity are related. That a given amount of nuclear material may be able to support a given extent of cell expansion is also suggested by studies on the development of fibres. Thus often the long primary phloem fibres become multinucleate in plants whereas the shorter secondary phloem fibres remain uninucleate. In explaining what happens in the cell expansion phase of differentiation we have to account not only for its initiation and orderly continuation but for its limitation.

Even as during cell extension cell wall growth plays an important part, during differentiation there occur profound changes in cell wall thickness, chemical composition, and changes in cell shape which imply localized growth of the cell wall: during the differentiation of fibres their apices grow and intrude among associated cells. Numerous plant cell types are defined

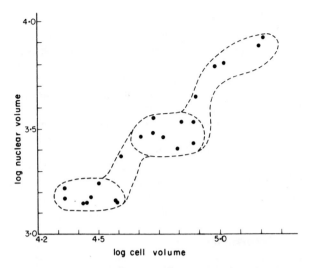

**Fig. 9.13** Plot of log. nuclear volume against log. cell volume for developing metaxylem unit cells in *Arisaema triphyllum*. The clustering of the groups (indicated by enclosure in the broken lines) indicates a periodicity in the growth of cell and nucleus. (From List, Jr., 1963, *American Journal of Botany,* **50**, 320–9.)

on their cell wall characteristics (collenchyma, sclerenchyma). Highly localized depositions of cell wall material in some cell types result in uneven wall thickenings in organized patterns. The obvious examples of such patterns are the annular, spiral and reticulate thickenings of the longitudinal walls of proto- and meta-xylem vessels. Microscopists working in the second half of the nineteenth century, including the botanist Strasburger, observed an aggregation of cytoplasm, more dense and granular than the rest, which was streaming in a definite pattern and in a topographical relation to the wall thickenings such as rings and reticulation. A cytoplasmic organization preceded the appearance of a pattern in the secondary cell wall. More recent studies with the electron microscope have revealed a close correlation between the organization of the cell wall, and the arrangement of cytoplasmic *microtubules*. These microtubules are composed of protein subunits (tubulin), arranged helically into tubules 25 nm in diameter and up to a few $\mu$m long. They are found in the peripheral layers of the cytoplasm, just to the inside of the plasmalemma, and the cellulose microfibrils are laid down parallel to the microtubules, which thus seem to control the microfibrillar pattern of the cell wall. The arrangment of the microfibrils is characteristic for each cell type; in secondary walls there are frequently several wall layers with different orientations of microfibrils. How the microtubules inside the plasma membrane can control the orientation of microfibrils which are synthesized on the outside of the membrane remains a fascinating puzzle. The microtubules may also map out the pattern of wall thickenings: in developing xylem, the future wall thickenings are marked by bands of microtubules, which continue to overlie the growing wall deposits.

Furthermore, in cells about to divide the positions of future cell divisions become marked by the *pre-prophase bands* of microtubules, which run around the cell in the position where the new cell plate will eventually join the mother cell wall. This has been noted in apical meristems, in the unequal divisions of trichoblasts (p. 177), and during the highly asymmetric divisions leading to the formation of stomatal complexes. Microtubules obviously play an important role in cell differentiation. They are, however, highly labile; the pre-prophase bands appear some hours before nuclear division, and disappear before division commences. In meristems with a very regular pattern of division, the pre-prophase bands appear in a fixed relation to the surrounding cells. Thus microtubule arrangement may reflect some preceding invisible differentiation, some polarization, in the outer layers of the protoplast – possibly in the plasmalemma.

## Patterns of differentiation

The arrangement of the various tissues of roots and stems is not only characteristic of the organ but in its details is characteristic of each species. The view has been advanced that this pattern is determined by the pre-existing pattern, that the formative influences are transmitted from the mature to the young cells. This hypothesis however faces the immediate difficulty that the original development of the characteristic tissue patterns, both in the embryo and whenever new shoot or root meristems arise, must be explained. The alternative hypothesis that the pattern is determined by the apical meristem itself is supported by many lines of evidence. Thus by the technique of culturing isolated root tips it can be shown that the very small root tips (in certain species tips of only 0.5 mm) which do not include any cells showing visible differentiation will give rise to roots having the anatomy typical of the radicle of the species. Again using the same technique, decapitated roots (apex removed) can be induced to form a new apical meristem. The vascular pattern in the new growth then does not line up with that in the root stump. Seedling and cultured pea roots are triarch. When cultured pea roots initiate a new apex following decapitation, the number of protoxylem groups in the new growth can be increased to six if the new apex is developed in presence of $10^{-5}$ M IAA (Fig. 9.14). The hexarch condition persists when growth is continued in this auxin medium, but on transfer to auxin-free medium linear growth is accelerated and the xylem becomes triarch. In the auxin-free medium not only is growth accelerated but the root apex decreases in diameter. Size and growth rate of the root apex determine the pattern of vascular tissues differentiated below the apex.

   The concept that a distribution pattern of growth regulators emanating from the apical meristem may predetermine the differentiation pattern is supported by some studies with callus cultures. Under appropriate conditions of culture such callus masses are free of vascular tissues or contain only a few scattered tracheid-like cells. Experimental induction of vascular tissue development in such callus masses has been achieved by making a

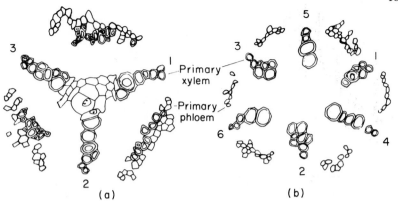

**Fig. 9.14** Vascular pattern in cultured pea roots as seen in transverse section. (**a**): triarch root normally observed (protoxylem points numbered); (**b**) hexarch pattern arising during culture in a medium containing $10^{-5}$M indol-3yl acetic acid. (From Torrey, in *Cell, Organism and Milieu,* ed. Rudnick, 1959, 189. Ronald Press, New York.)

V-shaped incision in the upper surface of the callus and inserting into this incision either a shoot bud or a supply of an auxin and a sugar (Fig. 9.15a), which must be either sucrose or one of a few other α-glycosyldisaccharides. In these experiments both xylem and phloem were differentiated and the balance between them could be controlled by the concentration of sugar applied. When the callus masses were sectioned, transversely to their vertical orientation during the experiment, the vascular strands were seen to be arranged in a roughly circular pattern and to be linked by an 'interfascicular cambium' giving rise to additional vascular tissue (Fig. 9.15b). The diameter of the circle and distance below the incision at which differentiation was initiated were influenced by the auxin concentration applied; the circle was inflated and the region of differentiation lowered at high auxin concentrations. The incision is acting as an artificial apex from which substances exerting a controlling influence on cambial initiation and cell differentiation are diffusing.

Studies on vascular cambium initiation and the development of secondary vascular tissues in cultured roots give further support to the concept enunciated above but indicate that, within the organ, distribution patterns of substances additional to auxin and sugar may be important. As normally cultured, excised roots show little or no secondary thickening. Professor Torrey and his co-workers at Harvard noted that excised pea or radish roots showed very limited secondary thickening during the first culture passage following initiation of the root cultures from seedling root tips. No secondary thickening was observed on further subculture. Attempts to enhance the secondary thickening in the first culture passage by introducing growth-regulating substances (auxin was tested exhaustively) or enhancing the sucrose concentration of the medium were unsuccessful. It is however possible to supply all or part of the nutrients to the excised root tip via its

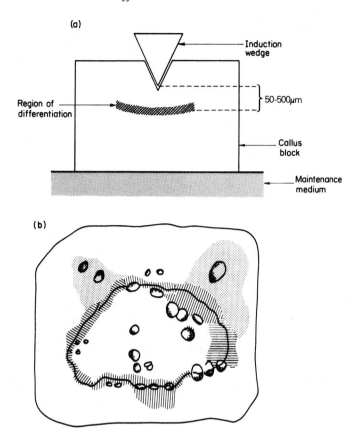

**Fig. 9.15** (**a**) Diagrammatic section drawn through a cylindrical block of callus with an induction incision cut in the centre of the upper surface. Introduction of auxin and sucrose via this induction wedge resulted in differentiation of vascular tissue in the region indicated. (From Jeffs and Northcote, 1967, *Journal of Cell Science*, **2**, 77–88.) (**b**) Diagrammatic transverse section through a callus cylinder of *Syringa vulgaris* taken 450 μm below the surface 54 days after insertion of an induction wedge (see diagram **a**) containing 4% sucrose and 0.5 ppm of the synthetic auxin, naphthalene acetic acid. Note the ring-like distribution of vascular nodules and the formation of a cambium (continuous line) involving the nodules and becoming more or less continuous across the interfascicular regions. Cellular regions laid down by cambial activity are shown by radial parallel lines. (From Wetmore and Rier, 1963, *American Journal of Botany*, **50**, 418–30.)

basal cut end (Fig. 9.16). Torrey used this technique to supply sucrose and auxin via the base to excised seedling root tips of pea and then found that there was enhanced cambial activity and development of secondary xylem well beyond the confines of the vial in which the IAA and sucrose were contained. On subculture, after growth for 14 days the root tip continued growth but no cambium was initiated. Subsequently the same workers

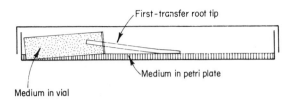

**Fig. 9.16** Technique to enable sucrose and growth regulators to be fed to a growing excised root tip via the basal cut end. (From Torrey, 1963, *Symposium of the Society for Experimental Biology,* **17**, 285–314.)

induced a high level of vascular cambium activity and good differentiation of secondary xylem and phloem in first passage radish (*Raphanus sativus*) roots supplied via their basal ends with sucrose, auxin (IAA, α-naphthalene acetic acid and 2,4-D were all effective), a cytokinin (kinetin and a number of other substituted aminopurines were effective) and *myo*-inositol (Fig. 9.17).

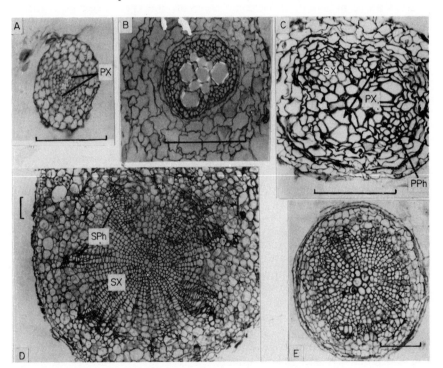

**Fig. 9.17** Influence of an auxin, a cytokinin and *myo*-inositol on the formation and functioning of vascular cambium in cultured radish roots (*Raphanus sativus*). Sections cut in the region of maximum root diameter. All scales equal 100 μm. PX, primary xylem; SX, secondary xylem; PPh, primary phloem; SPh, secondary phloem. A: No growth factors supplied via the vial (see Fig. 9.16); B: No cytokinin supplied; C: No auxin supplied; E: No *myo*-inositol supplied; D: All three growth factors supplied. (Photograph supplied by J. G. Torrey, Biological Laboratories, Harvard University.)

However here as with the pea roots the above treatment did not induce cambial activity when the root tips were taken into a second culture passage. There is presumably critical depletion of some unidentified factor(s) as the roots grow in culture. It was concluded that the induction of secondary thickening in roots depends upon the root apex receiving sucrose and an appropriate mixture of growth-regulating substances by transport from the shoot systems. The early (1920) observation by Garner and Allard that radish forms thickened roots when the shoot is under short day conditions, but only fibrous roots if the shoot is in long days seems to support Torrey's hypothesis. However the extent to which growth regulators synthesized in the root may be involved in the promotion of secondary thickenings may vary from species to species. Roots are known to synthesize auxins, cytokinins and gibberellins. Cambium initiation and activation and the differentiation of the new cells into xylem and phloem cells are under multifactor control and we cannot yet specify all the factors involved. It can be postulated that the root tip meristem is the centre from which these factors move in a basipetal manner to create the necessary concentration patterns and that the root tip receives these factors or their precursors by transport (possibly via the phloem and procambial cells) from mature root tissue cells or from the shoot system or from both simultaneously. At the same time the sensitivity towards hormones varies among cells in different positions, as discussed below (p. 185).

## The problem of the mode of hormone action

A vast body of data has been collected about the effects of hormones on plant development. But we are still far from comprehending how plant growth hormones work. Not for one single hormone do we know one single biochemical reaction involving the hormone. In discussing the effect of auxin on cell extension, two possible modes of action were considered: either direct action on the genetic level, the hormone acting as a gene repressor or derepressor; or action on a membrane, probably the plasmalemma. The same two basic modes of biochemical action have been proposed for all classes of plant hormones. The best evidence for action at the gene level is provided by the gibberellin-induced synthesis of $\alpha$-amylase and other enzymes in barley endosperm (p. 21). In this system, $GA_3$ does stimulate the appearance of the specific mRNA for the enzyme, though even here there is uncertainty about whether the mRNA is newly transcribed, or just released from an inactive form; the proof of direct interaction between the hormone and the genetic material is lacking. The time between the application of the gibberellin and the first detectable appearance of the enzyme is long enough to allow a chain of reactions to intervene between the primary action of the hormone and the derepression of the genes. Other evidence for hormones as gene regulators is even more indirect. Many examples can be quoted of a hormonal effect on development accompanied by changes in protein patterns, levels of RNA, and rates of synthesis of RNA and protein. But any factor that influences development is bound to affect these features even if indirectly. The very essence of development is differential gene activation in space and time; control of development involves, at some step, control of gene expression.

The alternative idea, that hormones act on cellular membranes, receives its strongest support from observations that hormone effects can be very rapid. Auxin-induced change in membrane potential of bean roots is measurable within 1 min of hormone application; protoplasmic streaming is stimulated within 2 min, respiration rate begins to rise after 5 min. The stimulation of cell elongation by auxin is generally quoted as having a lag period of 10–15 min, but the lag is due to slow penetration: with the methyl ester of IAA, which penetrates faster, growth stimulation has been detected within 1 min. These times are believed to be too short for effects mediated via gene activation. In bacteria, transcription takes 2–10 min, translation a further 1.5–5 min. Many attempts have been made to isolate the postulated membrane receptors for hormones, and membrane proteins capable of auxin binding have been found. There is, however, no proof that these proteins are involved in mediating growth or control of development.

It has also been suggested that hormones might act as allosteric effectors of enzymes; no direct evidence of this is available. The possibility of several primary actions for hormones must not be precluded.

**Multiplicity of hormone effects: the competence of target tissue**

Each plant growth hormone has a multitude of effects, the result of hormone application depending on the tissue to which it is applied. When $GA_3$ is applied to barley endosperm, $\alpha$-amylase synthesis is stimulated, without any cell division; the application of $GA_3$ to a dwarf pea apex initiates cell proliferation. The effect of the hormone depends on the *competence* (or potentiality, sensitivity, reactivity) of the target tissue. This is beautifully illustrated in an experiment by Gotô and Esashi (1974), as shown in Fig. 9.18. These workers excised 4 cm lengths of young etiolated bean hypocotyl, with hook; marked it into 3 mm segments, and then treated replicate segments with different hormones. The effect of the hormones on elongation of the segments is shown in Fig. 9.18: every segment reacted differently. As the cells passed from the stage of cell division at the hook to stages of expansion and differentiation further back, their receptivity towards the hormones changed profoundly. So far we have discussed the tissue–hormone interaction from the (conventional) angle, namely that hormones control development. It is equally true that development controls the reactions of tissues with hormones.

The classical view has been to regard plant hormones as very like animal hormones: compounds synthesized in one part of an organism, transported to another part, where the rise in hormone concentration elicits a specific response. Superficially, a similar case can be made for plant hormones, but detailed examination shows discrepancies with this classical view, as pointed out particularly by Trewavas (1981) and Hanson and Trewavas (1982) (see

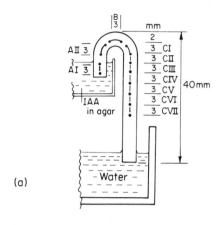

(a)

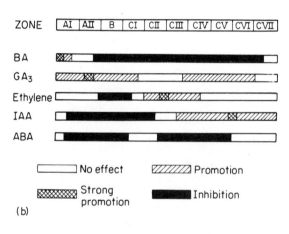

(b)

**Fig. 9.18**   The variation in sensitivity to hormones of the growth of different zones of etiolated bean (*Phaseolus vulgaris*) hypocotyl. (**a**) The hypocotyl segment with the zones marked on it, set up for the application of IAA. The other hormones (except ethylene) were applied via the basal end. (**b**) The effect of the hormones on elongation of the successive zones. BA = benzyladenine, a cytokinin. The hormones were applied for 6 h and the growth measurements were made 3 days later. (Adapted from Goto and Esashi, 1974, *Planta,* **116**, 225–41.)

Further Reading, p. 187). These authors note, for instance, that effects of plant hormones are usually proportional to the *logarithm* of hormone concentration, i.e. plant tissues are very insensitive to small changes in hormone levels. Changes in growth rate, e.g. in intact plants or organs on intact plants, are associated with natural changes in hormone concentration too small to bring about reactions of the observed magnitude, judging by the dose-response curves obtained in bioassays. Numerous authors have commented on this inconvenient fact from time to time, and have invoked subsidiary factors, e.g. interactions with other hormones, to explain it. Hanson and Trewavas suggest that tissue competence, rather than hormone concentration, is the limiting factor. According to that view, bean hypocotyl zone AI (Fig. 9.18) is not stimulated strongly by benzyladenine because it is lacking in cytokinins (being excised), but because the cells in that zone are at a stage particularly sensitive to cytokinins. Another fact, often neglected, is that hormone synthesis is *not* highly localized in plants. In the much-studied cereal coleoptile, the tip is the source of auxin for the elongating region (though much of that auxin may have come originally from the seed reserves, not from tip-located synthesis). In other young growing shoots auxin synthesis proceeds throughout the extension zone, and there is little transport – a fact which makes the idea of control by the tip difficult to sustain. The time courses of developmental processes may correlate poorly with the time courses of change in the concentrations of the hormones believed to control them. For instance, the fruit climacteric is often stated to be triggered off by increased ethylene production by the fruit; but the climacteric may start *before* there is any rise in ethylene production by the tissues!

Plant hormones are essential factors in all aspects of growth and differentiation. Hormones should not, however, be facilely accepted as triggering certain processes, or controlling a particular process by virtue of their limiting concentration. Our understanding of the action of plant hormones is still very incomplete, both at the biochemical level and at the level of the whole plant.

FURTHER READING

HANSON, J. B. and TREWAVAS, A. J. (1982). Regulation of plant cell growth: the changing perspective. *New Phytologist*, **90**, 1–18.

HESLOP-HARRISON, J. (1967). Differentiation. *Annual Review of Plant Physiology*, **18**, 325–48.

HILL, T. A. (1980). *Endogenous Plant Growth Substances*. Studies in Biology. No. 40. Edward Arnold, London.

JACOBS, W. P. (1979). *Plant Hormones and Plant Development*. Cambridge University Press, Cambridge.

MOORE, T. C. (1979). *Biochemistry and Physiology of Plant Hormones*. Springer-Verlag, New York.

RAPPAPORT, L. (1980). Plant growth hormones: internal control points. *Botanical Gazette*, **141**, 125–30.

STREET, H. E. (1969). *The Repertoire of Plant Cells*. Leicester University Press, Leicester.

STREET, H. E. (1969). Growth in organised and unorganised systems: knowledge gained by culture of organs and tissue explants. In *Plant Physiology*, ed. Steward, F.C. Vol. 5B, 3–224. Academic Press, New York and London.

TREWAVAS, A. J. (1981). How do plant growth substances work? *Plant, Cell and Environment*, **4**, 203–28.

## 188   Cell Growth and Differentiation

WAREING, P. F,. and PHILLIPS, I. D. J. (1981). *Growth and Differentiation in Plants*, 3rd edition. Pergamon Press, Oxford.
WEILER, E. W. (1982). Plant hormone immunoassay. *Physiologia Plantarum*, **54**, 230–4.

SELECTED REFERENCES

BOYER, J. S. and WU, G. (1978). Auxin increases the hydraulic conductivity of auxin-sensitive tissue. *Planta*, **139**, 227–37.
EVANS, M. L. (1974). Rapid responses to plant hormones. *Annual Review of Plant Physiology*, **25**, 195–223.
GUNNING, B. E. S., HARDHAM, A. R. and HUGHES, J. E. (1978). Pre-prophase bands of microtubules in all categories of formative and proliferative cell division in *Azolla* roots. *Planta*, **143**, 145–60.
MELANSON, D. and TREWAVAS, A. J. (1982). Changes in tissue protein pattern in relation to auxin induction of DNA synthesis. *Plant, Cell and Environment*, **5**, 53–64.
ROLAND, J. C. and VIAN, B. (1979). The wall of the growing plant cell. *International Review of Cytology*, **61**, 129–66.
STREET, H. E. (1966). The physiology of root growth. *Annual Review of Plant Physiology*, **17**, 315–44.
STREET, H. E. (1968). Factors influencing the initiation and activity of meristems in roots. In *Root Growth*, Proc. 15th Easter School, Univ. Nottingham, ed. Whittington, W. J., pp. 20–41. Butterworths, London.
VANDERHOEF, L. N. and DUTE, R. R. (1981). Auxin-regulated wall loosening and sustained growth in elongation. *Plant Physiology, Lancaster*, **67**, 146–9.
WEILER, E. W. and WIECZOREK, U. (1981). Determination of femtomol quantities of gibberellic acid by radioimmunoassay. *Planta*, **152**, 159–67.
WEILER, E. W. and ZIEGLER, H. (1981). Determination of phytohormones in phloem exudate from tree species by radioimmunoassay. *Planta*, **152**, 168–70.
ZURFLUH, L. L. and GUILFOYLE, T. J. (1980). Auxin-induced changes in the patterns of protein synthesis in soybean hypocotyl. *Proceedings of the National Academy of Science, U.S.A.*, **77**, 357–61.

# 10

# Growth Movements

## Tropisms

### Introduction

In flowering plants, growth not only results in mass increase and morphogenesis, but is involved in various responses to external stimuli. Plants alter the orientation of their parts in space in relation to external stimuli through either growth movements, or localized turgor changes. Growth movements, which occur in response to unidirectional stimuli and result in the positioning of the plant part in a direction related to the direction of the stimulus, are termed *tropisms*. If the growth movement is directly towards the source of the stimulus, the reaction is described as *positively orthotropic*; if directly away from it, as *negatively orthotropic*. If the organ becomes orientated at an angle to the stimulus we speak of a *plagiotropic* response and, if this angle is a right angle, of a *diatropic* response.

Pioneer studies on the orthotropic reactions towards light and gravity, phototropism and gravitropism, led to the discovery of plant hormones. Charles Darwin, in 1880, reported upon his studies of the positive phototropic curvature of grass coleoptiles and of the hypocotyls of dicotyledonous seedlings and, finding that the bending failed to occur when extreme tips of the organs were shaded by metal foil 'hats', concluded that some stimulus must be transmitted from the tip to the curving region. Rothert (1894) extended these observations to a wider range of plant material. Then came the studies of Fitting (1907), Boysen Jensen (1910–1911) and Páal (1914–1919), which established that when the tip of a coleoptile is cut off and placed back in position the stimulus is not interrupted but passes across a moist cut surface or a gelatin layer. The stimulus failed to be transmitted when a thin sheet of mica, platinum foil or cocoa butter was placed between the tip and the responding part of the organ. Such observations made a chemical nature of the stimulus probable (Páal), though explanations in terms of 'induced polarity' were also advanced (Fitting). Appropriate transverse incisions indicated that the conduction of the stimulus occurred along the shaded side of an illuminated organ, and replacing the coleoptile tips asymmetrically on their stumps by itself produced curvatures without unilateral illumination (Páal, 1918). The straight growth of *Avena* coleoptiles decreased severely or stopped when they were decapitated and could be restored by placing the tips back in position (Söding, 1925; Went, 1928) (Fig. 9.2, p. 159). The stimulus involved in the phototropic reaction was thus apparently also active in controlling normal growth. Parallel observations were made in the study of gravitropism, namely, that the tip of a stimulated organ produced a growth-controlling stimulus which under the influence of gravity was transmitted along the lower side.

Clear demonstration that a chemical capable of controlling growth was transmitted from the tip was achieved in 1928 by Went. He demonstrated that the chemical agent could be collected in agar blocks by diffusion from the cut surfaces of coleoptile tips, and that when these blocks were placed unilaterally on decapitated coleoptiles, they caused a curvature away from the side receiving the chemical messenger (the growth hormone) by diffusion from the block. Further, by means of collecting the hormone from the lighted and shaded sides of an illuminated coleoptile tip into separate agar blocks, Went was able to prove that the total amount of hormone released from the tip was not altered significantly, but more hormone passed to the shaded than the lighted side (Fig. 10.1). Between the years 1926 and 1928, Went and Cholodny independently put forward a hormonal theory of gravitropism and phototropism, whilst the discovery of auxin in these early studies led on to the recognition that plant growth in general is regulated by plant hormones (see Chapters 9 and 11). For excellent detailed accounts of these classical investigations reference should be made to Went and Thimann (1937).

As indicated in subsequent sections of this chapter, advancing knowledge has required considerable modifications of the Cholodny–Went theory of tropisms, and now even its basic tenets are being challenged. Nevertheless this theory has held the field for such a long time, and has so greatly influenced thought and experimental work, that one can scarcely avoid making it the starting point for any discussion on the physiology of tropisms.

### Survey of tropisms

The most important tropisms manifested by flowering plants are *gravitropism (geotropism)* in response to gravity and *phototropism* in response to light; *haptotropism* (thigmotropism) in response to touch is also quite widespread. In addition to these, some plant organs can also show tropic responses towards chemicals (chemotropism), temperature gradients (thermotropism), presence of water (hydrotropism), flowing water (rheotropism), oxygen (aerotropism) and even towards electric fields (galvanotropism), but these reactions are of more limited occurrence. Only gravitropism and phototropism will be considered further here.

The capacity to react to certain stimuli can be lost by mutation. Mutants of rice, pea and maize are known which have lost the ability to react to gravity, i.e. are agravitropic (non-gravitropic), and the maize mutant is also

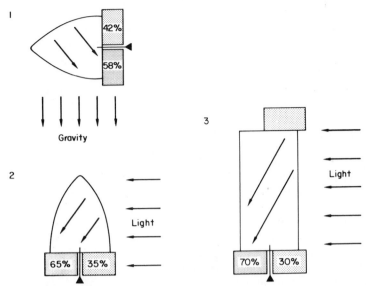

**Fig. 10.1**   1. Diversion of diffusible auxin to the lower side when a coleoptile tip is placed horizontally, demonstrated by collecting the auxin released from the cut surface in two agar blocks (stippled) separated by a razor blade. The total amount of auxin leaving the coleoptile tip is not significantly altered by submitting the tip to the gravitropic stimulus, but now 58% of the total is released from the lower half. (Experiment by Dolk, 1929.) 2. Diversion of diffusible auxin to the shaded side when a coleoptile tip is unilaterally illuminated. Technique as in 1 above. Again the total amount of auxin collected is not significantly altered by the phototropic stimulus but 65% of the total is released from the half furthest from the light source. (Experiment by Went, 1928.) 3. An agar block containing auxin is placed upon the upper cut surface of a segment of hypocotyl from a radish (*Raphanus*) seedling. The segment is correctly orientated (upper part uppermost). The hypocotyl is subjected to unilateral illumination. The auxin as it is transported through the hypocotyl is diverted to the shaded side as shown by the two-block technique described above. (Experiment by van Overbeek, 1933.)

aphototropic. These tropic responses of plant organs resemble the tactic movements of motile unicells (phototaxis, chemotaxis and thermotaxis) in so far as similar unidirectional stimuli are involved and in that the motile organisms move in a direction related to the direction of the environmental stimulus. Plants moreover show the phenomenon of autotropism, in that the orientation of organs can be fixed positionally in relation to other organs. Leaf positions with respect to the stem, the angles of lateral roots (higher than first order) in relation to the axis that bears them, and the angles between axillary branches and the main stem are autotropically determined. In reacting to a unidirectional stimulus a plant organ may overshoot the equilibrium position and return to it by an autotropic counter-reaction; a plant in which an over-curvature is produced by gravity will straighten this out on a klinostat (p. 194) in the absence of any further gravitational stimulus.

All tropisms are effected by differential growth rates on opposite sides of the reacting organ. A vertically orientated main root grows equally on all sides and therefore grows straight; when placed at right angles to the field direction of gravity by being laid horizontally, the upper side in the region of curvature grows faster than the lower and a curvature is produced, directing the root again along the field direction of gravity. The curvature results from an enhanced growth rate on one side, or a reduced growth rate on one side, or a combination of both, and hence only a growing organ can show a tropic growth reaction. During development, the sensitivity of an organ towards a particular stimulus increases to a maximum, then decreases and finally is lost when the organ reaches maturity and ceases to grow. Mature nodes of grasses and some other jointed plants, e.g. *Tradescantia* and *Dianthus*, react gravitropically while mature nodes of the Commelinaceae also react phototropically but, in such nodes, growth is resumed as a result of stimulation. The differential growth rates involved in tropisms are very largely confined to regions where growth is by cell expansion, the cells on the convex side expanding faster and becoming longer and thinner-walled than on the concave side; cell division has been implicated in only a few cases.

Tropisms are of supreme importance in establishing and maintaining the correct orientation (the liminal direction) of plant organs (Fig. 10.2). Rice mutants which have lost gravitropic reactivity through X-ray treatment collapse flat on the ground and become lodged. Interactions between different tropisms occur: leaf orientation, for instance, is determined by a combination of tropic (or nastic, see p. 209) responses towards light and gravity, in conjunction with autonomous tendencies for unequal growth on the upper and lower sides. Light modifies the reactivity of plants towards gravity; the orthogravitropic sensitivity of shoots is lowered by light, but etiolated shoots of *Sinapis alba* show a poor gravitropic reaction. The plagiogravitropic runners of *Circaea* and *Fragaria*, which normally grow along the soil surface, turn upwards as if negatively gravitropic when the plants are kept in the dark; plagiogravitropic rosette leaves and shoots may turn upwards even in weak light. Some plagiogravitropic subterranean organs, on the other hand, behave as if positively gravitropic on illumination (see p. 208).

The sign (positive or negative) of the tropic response of an organ may change during its growth; this happens frequently in reproductive organs, where the flower bud, the open flower and the fruit may each have a characteristic orientation due to a different gravitropic response of the pedicel (see also Chapter 12 and Fig. 12.6, p. 244). The flower stalk of the peanut *(Arachis hypogaea)* is at first negatively gravitropic, but after fertilization it becomes positively gravitropic and buries the fruit in the soil. The phototropic reaction of the flower stalk of the ivy-leaved toadflax (*Cymbalaria muralis*) changes from positive to negative after fertilization; this results in the fruits being pushed into crevices of the rocks or walls on which the plant grows.

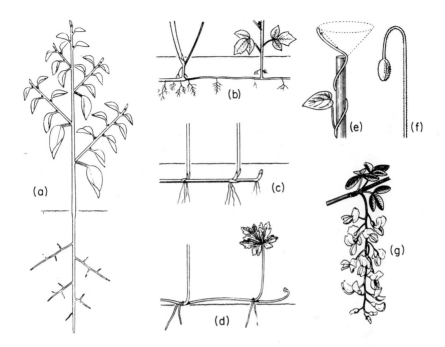

**Fig. 10.2** Various types of growth habit controlled by gravitropic responses. Schematic. a: dicotyledonous plant with orthotropic main root (positively gravitropic), main stem (negatively gravitropic), secondary roots and side shoots (plagiotropic) and tertiary roots (non-gravitropic). b: diagravitropic root of *Rubus idaeus,* raspberry. c: diagravitropic rhizome of *Eleocharis.* d: diagravitropic runners of *Ranunculus repens,* creeping buttercup. e: twining shoot of *Ipomoea purpurea* showing strong circumnutation and lateral gravitropism. f: young pedunde of *Papaver,* poppy, showing positive gravitropism. g: inflorescence of *Laburnum anagyroides*; the axis is non-gravitropic, and the inflorescence hangs down under its own weight. (From Larsen, 1962, in *Encyclopedia of Plant Physiology*, ed. Ruhland, **17**/2, 34–73. Springer-Verlag, Berlin.)

# Gravitropism

Suitable experimental material for the investigation of gravitropism is provided by primary seedling roots (positive response), and by young seedling shoots and cereal coleoptiles (negative response). The tropic curvature in a seedling root laid horizontal becomes visible after 30–60 minutes and the apex becomes directed downwards; the bending is produced in the region of most rapid growth. Gravitropic root curvature often results from a combination of some growth stimulation on the upper side with strong inhibition on the lower side (Fig. 10.3); sometimes there is strong growth inhibition on the lower side and a weak inhibition on the upper. Hence during curvature root growth as a whole slows down; but a combination of growth stimulation on the upper side, with no change on the lower, has also been reported. In shoots, inhibition of growth on one side is usually associated with stimulation on the other, so that the overall growth

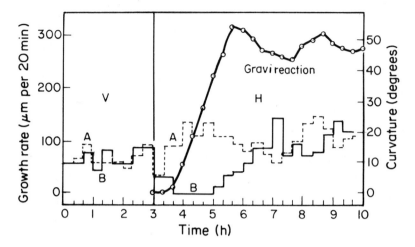

**Fig. 10.3** The growth rate (in μm per 20 min) of the two sides of *one* primary root (initial length 17 mm) of maize (cv. LG 11), first kept 3 h in the vertical (V) position, and then horizontal (H) for 7 h; and its gravireaction (downward curvature in degrees) with time. A = upper side; B = lower side. (From Pilet and Ney, 1981, *Planta,* **151**, 146–50.)

rate remains unchanged, though even in shoots there may be an overall inhibition of growth. Continuous stimulation is not necessary; after a period of stimulation curvature will occur if the test object is subsequently rotated on a klinostat (slowly, to avoid development of any appreciable centrifugal force). The minimum time during which the stimulus must be allowed to act to lead to subsequent curvature is the **presentation** or **perception time**; the time required for an observable response to appear is the **latent time** or **reaction time**, and this is usually measured from the end of presentation time, so that the time for the response to appear is presentation time + latent time. In the organs of flowering plants, the minimum presentation time with natural gravity and at room temperature varies from 10–30 s to 25 min, and the latent time from 10 to 120 min. The duration of presentation time and reaction time is increased by lowering the temperature, and availability of oxygen is needed for completion of both processes.

The intensity of gravitropic stimulation can be varied in two ways. One is to place the material at different angles to the vertical; the force acting on it is then **g** sin $\alpha$, where $\alpha$ is the angle with the vertical and **g** is the force of gravity. The other method depends on the fact that a centrifugal force has the same effect on plants as natural gravity, and involves rotation at various speeds on a klinostat or centrifuge, thereby subjecting the plant to a centrifugal force which can be expressed as a fraction or multiple of **g**. It has been found that a certain minimum or threshold stimulation is needed to produce a gravitropic curvature; a long presentation time is needed with a weak force, and vice versa. The product of presentation time and force required to produce the first perceptible curvature remains constant over a wide range of force, i.e. gravitropism obeys the **reciprocity law**: stimulus for given response = (time of stimulation) × (intensity of stimulation). With the oat coleoptile, this has been found to hold true for forces of 58.4 to 0.0014 **g**, with corresponding presentation times of 5 seconds to 68 hours. The threshold for the oat coleoptile is thus about 300 **g** s, at room temperature. Under continuous stimulation, maize roots have responded to a force as low as 0.0005–0.001 **g**. Several periods of sub-threshold stimulation can be additive and result in curvature, provided the intervals between the periods of stimulation do not exceed a value which depends on the organ investigated. With short intervals the value of the threshold may actually be reduced when stimulation is applied intermittently rather than given continuously. This indicates that the primary stages in gravitropic perception can be saturated at a very low level of stimulation and that some system must revert to the original (unstimulated) state before it can again be affected by the stimulus.

As the amount of stimulus is raised above threshold value, the response becomes stronger, the latent time decreasing and the curvature increasing. When a physiological response is directly proportional to the amount of stimulus which produces it, it is said to obey the ***stimulus-quantity law***. For gravitropism, the law holds within a limited range of stimulus quantities; the response rises to a maximum with increased stimulation; then further increases in the stimulus give no further increase in reaction, which now must be limited by internal factors, and very strong stimulation becomes inhibitory; in one series of experiments with pea roots, inhibition was found to result from forces in excess of 111 **g**.

Gravitropic curvature occurs in the region of cell elongation, a little distance back from the root or shoot apex . In roots, perception may be confined to the tip and in a number of species the root cap has been identified as the site of perception, decapping leading to loss of response; these species include *Zea mays, Pisum sativum*, the cress *Lepidium sativum*, and the lentil *Lens culinaris*. This implies that, following stimulation, a message is transmitted from the perception site to the reacting region.

It might be argued that decapitation causes loss of the capacity to react through the shock effect of wounding, but in a number of instances it has been shown that growth is not inhibited, indeed removal of the root cap while abolishing gravitropic sensitivity may give a transient increase in growth rate. Further, if decapitation is delayed for some period after presentation time but carried out before reaction is complete, a normal curvature develops, and deliberate wounding of root tips without removal of the cap does not prevent the development of curvatures. In some maize varieties the cap can be teased off without any cutting but there is complete loss of gravitropic reaction. Again curvatures usually, although admittedly not always, take place when a stimulated tip is placed on an unstimulated stump, and the two need not be of the same species, nor from the same organ: a root stump can, for instance, be induced to curve by a stimulated coleoptile tip. Evidence for the importance of the root tip in the perception of gravity can also be obtained in experiments not involving its excision. By suitable orientation on a klinostat, the root tip and the elongating region can

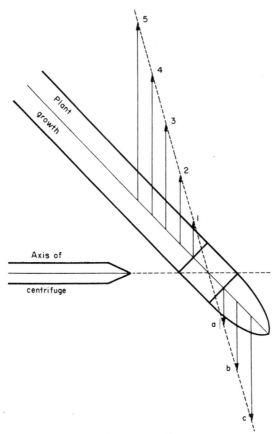

**Fig. 10.4** Centrifugal stimulation of root tip and base in opposite directions. The arrows a, b, c, show the direction of forces acting on the tip which projects beyond the centrifuge axis; the arrows 1 to 5 represent the forces acting on the basal part; the lengths of the arrows are proportional to the forces. If 1.5 to 2 mm of the tip is allowed to project beyond the axis, the root behaves as if the tip alone were stimulated, though the forces acting on the base in the opposite direction are much stronger. (From Larsen, 1962, in *Encyclopedia of Plant Physiology,* ed. Ruhland, **17**/2, 34–73, Springer-Verlag, Berlin.)

be subjected to centrifugal forces in opposite directions (Fig. 10.4). The resulting curvature is then determined by the direction of the force acting on the tip, even when this is considerably weaker than that applied further back.

In shoots, perceptivity is not confined to the tip. Decapitation of coleoptiles and of shoots of dicotyledonous seedlings does not abolish their reactivity. The reaction is often slowed, or the threshold of stimulation raised, by decapitation. From this it has been argued that the shoot tip is the most sensitive region of perception. An alternative view is that any disturbance of the shoot response to gravity brought about by detipping results from a general growth disturbance. In contrast to decapped roots, detipped coleoptiles do show inhibition of growth, and normal sensitivity towards gravity can be restored by application of auxin, which restores growth (see p. 159). In grass nodes the gravity stimulus is perceived by the reacting region. Even in shoots, however, some cell to cell transfer of stimulus must occur, for perception is believed to be confined to certain cells only.

The gravitropic response thus consists of a number of stages separable in space and/or time: (1) *perception* of the external stimulus (gravity), resulting in (2) *transduction* of the stimulus (*induction* of a metabolic change) in the sensitive region; (3) *transmission* (*conduction*) of an internal stimulus to the region of (4) the **gravitropic reaction**, expressed through differential growth on the two sides of the organ.

### Perception of gravity: the statolith theory

The primary perception of gravity must involve the movement of some entity in the sensitive cells; this is the only way in which the physical force of gravity can react on a cell; and one must search for an organelle at least 0.5 μm in diameter. In 1900, Haberlandt and Neměc independently put forward the *statolith theory* of gravitropism. This supposes that in the perceptive cells, the *statocytes*, mobile starch grains (the *statoliths*) move under gravity to lie in the lower parts of the cells (Fig. 10.5). Thereby the upper and lower parts of the cells come to differ and the organ acquires an 'up-and-downness', a polarization. Not all starch grains in cells are mobile under gravity and capable of acting as statoliths. On the whole, organs which are gravitropically sensitive contain statolith starch, and non-gravitropic organs lack it. In root tips, statolith starch is abundant in the root cap; in stems, the endodermis contains movable starch grains, and it has been noted that these statolith starch grains persist even in extreme starvation. Disappearance of

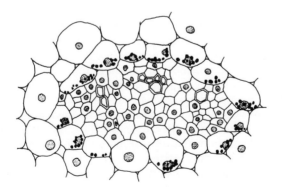

**Fig. 10.5** Transverse section of young onion (*Allium cepa*) cotyledon base; the statocytes form a sheath round the vascular bundle and the statolith starch grains (in black) have been made to accumulate on the lower sides of the cells. (From Hawker, 1932, *Annals of Botany*, **46**, 121–57. By permission of the Clarendon Press, Oxford.)

the starch grains can be induced by low temperature treatment or by treatment with gibberellins and this leads to loss of gravitropic sensitivity; when the cells are allowed to recover their starch, gravitropic sensitivity reappears. There is evidence that in some organs bodies such as crystals of calcium oxalate may also function as statoliths. The statolith theory faces the difficulty that some organs known to be sensitive to gravity do not appear to contain statoliths; fungal sporangiophores and roots of some seedlings are in this category. In these cases, one can postulate that the statoliths have not yet been identified.

If the displacement of statoliths is the primary event in graviperception, it must next be considered how the cells sense this displacement: what is the primary event in transduction? The three main possibilities currently considered are that (*a*) the *position* of the statoliths is critical; (*b*) the *movement* of the statoliths through the cytoplasm is sensed; (*c*) the *pressure release*, as the statoliths fall away from their original site, is detected. Possibility (*b*) has not received much support; the experimental evidence for it is tenuous and open to alternative interpretations. If the position of the statoliths is critical (possibility *a*), they might act by exerting pressure on the plasmalemma, thereby altering its permeability or enzymatic properties. The plastids never actually make direct contact with the plasmalemma, but pressure could be transmitted through the relatively gelled outer layers of cytoplasm. Being metabolically active, the amyloplasts could also carry out localized enzymatic reactions; or they might have the purely mechanical function of displacing upwards lighter organelles, whose position is important. The hypothesis that the final positioning of the amyloplasts is decisive in graviperception has been criticized on the grounds that the presentation times are shorter than the times of sedimentation of the amyloplasts to their new equilibrium position. Sedimentation rates of no more than 1 $\mu$m s$^{-1}$ have been widely accepted and, at such speeds, an amyloplast would travel only a fraction of the way towards the lower wall within a perception time of 10 s. These velocities are, however, based on indirect measurements, e.g. examination of fixed cells. More recent studies, in which living cells of dandelion flower stalk and mung bean hypocotyl have been examined by cinematographic techniques, have demonstrated that the speed of movement of individual amyloplasts varies tremendously. Whereas complete sedimentation requires at least some minutes, a few amyloplasts do reach the new bottom wall within 10–20 s. It is not known, however, how large a proportion of amyloplasts must sediment for an effect to be registered. Moreover, sub-threshold stimulations of 0.5 s duration are additive in producing a response, and no appreciable redistribution can occur in so brief a period.

The idea that pressure release on the endoplasmic reticulum might be the critical event (possibility *c*) has been proposed by Sievers and Volkmann largely on the basis of studies on the graviperceptive root cap cells of cress (*Lepidium sativum*). The sensitive cells contain near their walls layers of endoplasmic reticulum arranged in a precise way (Fig. 10.6) and the shape of the cells and the arrangement of the membranes is such that, with the root in

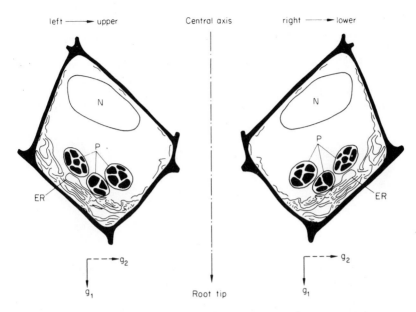

left ──▶ upper      Central axis      right ──▶ lower

Root tip

**Fig. 10.6** Diagrammatic representation, based on tracings from electron micrographs, of two cells on either side of the root axis of a *Lepidium sativum* root cap. In the vertical position as here drawn (gravity acting in the direction $g_1$) the amyloplasts P press equally on the endoplasmic reticulum, ER, in the two cells. When the root is placed horizontally (gravity acting in the direction $g_2$), sedimentation of the amyloplasts will greatly relieve pressure on the ER in the cell remaining uppermost, the left-hand cell in the drawing, but in the lower (right-hand) cell, the amyloplasts will move less and still remain largely in contact with the ER. N = nucleus. (From Sievers and Volkmann, 1972, *Planta*, **102**, 160–72.)

the vertical position, the amyloplasts press equally against the endoplasmic reticulum in cells on both sides of the root axis. An actual pressing of the amyloplasts against the endoplasmic reticulum is indicated by visible deformation of the membranes beneath the plastids. When the root is placed horizontally, the plastids slide off the endoplasmic reticulum and considerably relieve pressure on it, in the *upper* cells only, i.e. the cells above the central axis of the root; in the *lower* cells the pressure scarcely changes (Fig. 10.6). There is not only a change in pressure on the membranes, but the pressure after stimulation is now different in the upper and lower cells – a polarization has been induced. The proponents of this idea point out that when a root is placed horizontally, pressure relief is instantaneous, and very short presentation times would be feasible. In root cap cells of other species, endoplasmic reticulum complexes near cell walls have been observed, though the arrangement of organelles is not as regular in all plants as in cress. In a non-gravitropic pea mutant (*Pisum sativum ageotropum*), the mutant differs from the normal in lacking a distal endoplasmic reticulum complex in the root cap statocytes, pointing to the importance of this complex in graviperception. In some root caps, however, endoplasmic reticulum is displaced by the amyloplasts and, in statocytes of shoot organs, regular endoplasmic reticulum complexes have not been detected. In such cases, pressure relief on the plasma membrane might be sensed.

According to another hypothesis, the opening and closing of plasmodesmatal 'valves' by amyloplasts impinging on the endoplasmic reticulum are postulated to result in changes in the direction of intercellular transport, and thereby to effect the upper-lower polarization of the organ (Fig. 10.7).

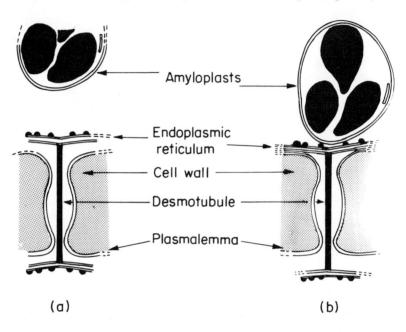

**(a)**                                                   **(b)**

**Fig. 10.7**  A model for the proposed amyloplasts/plasmodesmatal valves. (**a**) open, (**b**) closed. For clarity, the size of the plasmodesmata is much enlarged in relation to the amyloplasts. (From Juniper, 1976, reproduced, with permission, from *Annual Review of Plant Physiology*, **27**, 385–406, © 1976, by Annual Reviews Inc.)

In the root tip of *Lepidium sativum*, gravistimulation causes a rapid change (within 30 s) in the pattern of electric currents associated with the organ. If this phenomenon should prove to be of universal occurrence, it could be highly significant in the transduction phase of gravitropism. A sequence may be visualized, with the sensing of a pressure change by cellular membranes first affecting ion flow through them, and this leading to changes in electric currents in the tissues; the current flow in turn could establish gradients of charged compounds, i.e. effect a polarization. The idea is attractive in – possibly – explaining how polarization is achieved in an entire organ where the number of statocytes is low relative to the total number of cells involved in the growth reaction. Electric current flow would not be confined to individual cells.

To summarize so far, the primary transduction event in gravitropism most probably involves the sensing of a pressure change, brought about by statolith movement, on a membrane or membrane system. The enzymatic properties or and permeability of the membranes could vary according to whether they are under pressure or not. This, however, is only the first step in a chain of reactions which culminates in the differential growth response. The two sides of a stimulated organ develop differences in pH, osmotic potential, concentration of reducing sugars, enzyme activities and respiration rate. The lower side of a horizontally placed organ typically becomes electropositive with respect to the upper by 5–30 mV – the geoelectric effect. The greatest significance, however, has for many years been attributed to the differences in the concentration of growth hormones that can be detected between the upper and lower sides.

## The Cholodny–Went theory and gravitropism

The classical Cholodny–Went theory for the development of the differential growth rates proposed the induction of differences in auxin concentration on the two sides to be the key event. In a root or a shoot growing vertically, the concentration of auxin reaching the growing cells is equal on all sides. When a root or shoot is, however, placed horizontally, more auxin is found on the lower side than the upper (Fig. 10.1). In shoots, if auxin is synthesized in the apical meristem and transported basipetally to the elongating region, the auxin concentration difference is achieved by a redistribution, or lateral transport; the lower side gains at the expense of the upper. For roots, there is controversy about sites of synthesis: at least some of the auxin found in roots is thought to be transported acropetally from the shoot and there have been also suggestions of some additional synthesis of auxin in a stimulated root, increasing the amount on the lower side. The difference in reaction of roots and shoots is explained by the very different auxin sensitivities of root and shoot cells (Fig. 9.4): concentrations of auxin which are stimulatory to shoot growth are inhibitory to root growth. It is accordingly postulated that the increased auxin concentration in the lower half of a horizontally placed shoot stimulates growth in that half, causing the shoot to curve upwards while, in a horizontal root, the auxin increase on the lower side results in growth inhibition and a downward curvature. The effect of the statolith movements must, in accordance with the Cholodny–Went theory, be a stimulation of auxin movement in the lateral direction.

### Modifications of the Cholodny–Went theory

The Cholodny–Went theory has the attraction of great simplicity and of providing a common basis for root and shoot gravitropism and indeed for phototropism as well. An increasing body of observations has, however, necessitated considerable modifications of this hypothesis.

Auxin redistribution in gravistimulated shoot organs does occur; this has been confirmed by the application of radioactive IAA to coleoptiles of *Zea* and *Avena*, and to some dicotyledonous seedlings. The changes in growth rate are, however, greater than would be expected from differences in auxin distribution as judged by bioassays of extractable auxin, so it has been suggested that stimulation could also change the sensitivity of the two sides of the organ towards auxin. Hormones other than auxin have been implicated in tropic movements. For roots, a number of authors have adduced strong evidence that the inhibitory hormone accumulating on the lower side is not IAA but ABA, produced by the root cap; the ABA has been identified chromatographically. But this is not universally accepted; radioactive IAA is transported laterally in roots, IAA is also present in root caps, and roots are far more sensitive to growth inhibition from IAA than from ABA. When ethylene synthesis is inhibited in stems, their gravitropic reaction is slowed and hence a role for ethylene in the response has been suggested. In roots, too, the growth inhibition on the lower side has been attributed to ethylene synthesis in response to an increased auxin level on that side. Gibberellins have been reported to be asymmetrically distributed in gravistimulated shoots and roots, with the higher gibberellin content in each instance on the side with the higher growth rate. In sunflower shoots, the gibberellin content on the lower side has been claimed to be ten times higher than on the upper – a concentration gradient much steeper than for auxin, and one which can scarcely exist without an effect on growth. It must be noted, however, that the claims of differences in gibberellin concentration are based on bioassays only and have not been verified by more reliable methods.

The hormonal relations of gravitropism are obviously not yet clear, especially in the case of roots. Several hormones may be implicated and roots and shoots may differ in the hormones involved. It is to be hoped that application of techniques such as radio-immunoassays will eventually yield reliable data on the hormonal contents of various parts of stimulated organs. On the other hand, changes in hormone content should not be over-emphasized as growth-controlling factors, as discussed in Chapter 9. Changes in the growth competence of cells in the upper and lower halves may be induced by gravistimulation. Perhaps more attention should be paid to the other physiological gradients that are established across gravi-stimulated organs. It could be that research, originally stimulated by the Cholodny–Went theory, has become so preoccupied with the hormonal relations of gravitropism that other important growth-controlling factors are being missed.

## Phototropism

Phototropism has been studied mainly in etiolated coleoptiles of cereals, particularly those of oats, wheat and maize, and in the epicotyls and hypocotyls of dicotyledonous seedlings, including light-grown material, particularly pea and sunflower. The threshold of stimulation is $c.$ $10^{-2}$ to $10^{-1}$ J m$^{-2}$. Shoots are generally stated to be positively phototropic whilst roots are negatively phototropic or indifferent to light. The direction of curvature of shoots can however depend on the light dosage. The response of coleoptiles changes with increasing dosage from a first positive to a first negative and a second positive reaction (Fig. 10.8); with yet higher energies a second positive reaction occurs and finally a third positive, but this last is probably a damage manifestation caused by turgor loss on the intensely illuminated side. But not all organs show the same response; e.g. in etiolated epicotyls of *Lens culinaris*, the first and second positive curvatures are separated by an indifferent zone (Fig. 10.8); light-grown radish seedlings are insensitive to stimulation at low dosages and begin to curve (positively) only in the second positive dosage range. Under natural conditions, the second positive curvatures would prevail. Reactions beyond the second positive curvatures have scarcely been studied.

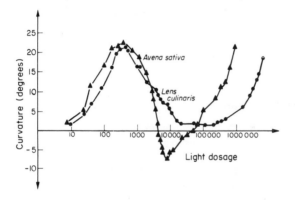

**Fig. 10.8**  Phototropic dose-response curves for etiolated coleoptiles of *Avena sativa* (▲) and etiolated epicotyls of *Lens culinaris* (●). Note that the light dosage is on a logarithmic scale; the units can be converted *approximately* to J $m^{-2}$ by multiplying by $4 \times 10^{-3}$; thus the threshold lies at around $4 \times 10^{-2}$ J $m^{-2}$ for these organs. (Adapted from Steyer, 1967, *Planta*, **77**, 277–86.)

The phototropic reaction has a number of features in common with gravitropism. There is a presentation time and a reaction time; at high irradiances, the presentation time may be a fraction of a second and the reaction time a few seconds; subthreshold stimulations are additive. The stimulus–quantity law holds however only for the first positive and first negative curvatures and the reciprocity law is valid only over limited dosages.Hence a certain dosage of energy given as high irradiance over a brief time can produce a different degree of response compared with the same dosage applied as low irradiance over a prolonged period and this can make comparisons between different sets of data difficult.

Phototropism is dependent on the establishment of a gradient of illumination across the reacting organ. Although a grass coleoptile is a delicate translucent object, it is hollow and light entering it is subject to considerable internal reflection. Unilateral illumination therefore produces in it an illumination gradient of from 1.5:1 to 35:1 depending on factors such as irradiance level and the region of the coleoptile under illumination. A 1%

irradiance difference between the illuminated and shaded sides has been stated to evoke a phototropic curvature. In more robust plant organs there is sufficient light absorption by the tissues to produce a steep illumination gradient.

In coleoptiles the tip has for many years been regarded as the most sensitive region of perception, for unilateral illumination of only the tip suffices to call forth the response, although illumination of the lower regions has indicated also some sensitivity throughout the length of the organ; decapitated coleoptiles need an auxin supply to react. Investigations by Mandoli and Briggs (1982) have now shown that a coleoptile can conduct light in a manner analogous to a synthetic optical fibre bundle, so that light applied to the tip can be transmitted down the organ; in view of these findings, claims of the high sensitivity of the coleoptile tip need re-examination. In a green sunflower seedling the growing region of the hypocotyl is both the perception and the reaction site and removing or covering its tip does not affect curving.

## Light perception in phototropism

The primary step in any light-mediated reaction must be the absorption of the light by, and hence activation of, a photoreceptor chemical. Theoretically one should be able to identify the photoreceptor from the action spectrum of phototropism. The action spectrum of a reaction is obtained by plotting the amount of response obtained at different wavelengths of incident light against the wavelength. Usually the response to the same number of incident quanta at each wavelength is plotted (a quantized action spectrum) but sometimes the response is measured for the same amount of total incident energy (e.g. $J\ m^{-2}\ s^{-1}$) at each wavelength. The action spectrum so obtained can then be compared with the absorption spectra of known pigments, i.e. the plots of light absorption by these pigments against wavelength. Coincidence of the action spectrum of a reaction with the absorption spectrum of a specific pigment is taken to mean that this pigment absorbs the light utilized for the reaction. In phototropism, blue light is active and the action spectrum has three peaks in the blue, at $c$. 436, 440 and 480 nm (Fig. 10.9); there is also a peak in the near ultraviolet around 370 nm. This action spectrum does not correspond exactly with the absorption spectrum of any known pigment, but resembles the absorption spectra of both flavins and carotenoids. Accordingly there is controversy about which of these is the phototropic photoreceptor.

In the visible region the correspondence of the action spectrum with carotenoid absorption spectra is very good (Fig. 10.9), but extracted carotenoids do not absorb near 370 nm. Flavins do absorb in the ultraviolet, but in usual solvents they show only single peaks in the blue (Fig. 10.9 E). In certain solvents, or at low temperature, flavins do exhibit triple peaks in the blue. In the cell the absorption spectra of pigments may be modified by the combination of pigments with other molecules; e.g. a protein-carotenoid

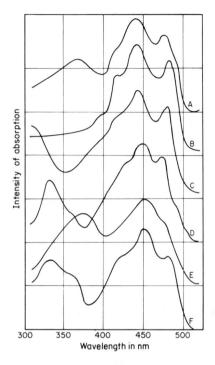

**Fig. 10.9** The action spectrum of the phototropism of the *Avena* coleoptile (A), compared with absorption spectra of: α-carotene in hexane (B); hexane extract of 50 coleoptile tips (C); 9-9' *cis-β*-carotene in hexane (curve of Inhoffen *et al.,* 1950). (D); riboflavin in water (E); 3-methyl-lumiflavin in benzene (curve of Hanbury *et al.,* 1959) (F). (From Thimann and Curry, 1961, in *Light and Life,* ed. McElroy and Glass, 646–72. Johns Hopkins Press, Baltimore.)

complex might have an ultraviolet-absorbing peak. Present evidence from action spectra is equivocal.

Carotenoids are present in organs showing phototropism, and several mutants with a lowered carotenoid content also show a lowered phototropic sensitivity (the flavin content of such mutants being normal). On the other hand, some phototropically reactive mutants have a very low carotenoid content indeed; thus a maize mutant, in which the coleoptile contains not more than 0.1% of the normal carotenoid content, is still 50–80% as reactive as the normal type. From calculations of light absorption and quantum yield (a measure of the number of quanta required for the response), it has however been concluded that the effective pigment concentration needs to be only about $10^{-9}$ M. Since carotenoids are photoreceptors in phototaxis and in animal vision, their involvement in phototropism would not be unexpected. Those who favour flavin as the photoreceptor point to the universal occurrence of flavins in sensitive organs and to the involvement of riboflavin in animal vision. The supporters of flavins also stress the significance of the low-carotene mutants with high phototropic reactivity, and of the failure of some inhibitors of carotenoid biosynthesis to inhibit phototropism. Changes in the phototropic sensitivity of maize coleoptiles have been correlated with changes in the amount of membrane-associated flavins. Some authors have attempted to resolve the controversy by suggesting that riboflavin is the photochemically active pigment, but that carotenoids act as light screens controlling the amount of light received by the riboflavin.

**Transduction and transmission: the Cholodny–Went hypothesis and phototropism**

The Cholodny–Went hypothesis of phototropism is based on the classical experiments as illustrated in Fig. 10.1, indicating that auxin is diverted to the shaded side of a unilaterally illuminated organ. In a shoot this would increase the growth rate of the shaded side and the organ would bend towards the light. In those roots which show phototropism, the growth rate would be retarded on the shaded side by a supraoptimal auxin level and a negative curvature would result. In coleoptiles most of the auxin redistribution seems to occur in the extreme tip; this implies transmission of stimulus from the tip. Very little work has been carried out on roots and the following discussion applies to phototropism of the shoot.

As with gravitropism, difficulties have arisen in accepting this simple hypothesis, both as regards the quantitative relationships between the levels of auxin and growth rates, and the precise nature of the hormones involved. The measured differences in auxin concentration on the two sides again appear to be insufficient to account for the differences in growth rates. The original hypothesis was based entirely on bioassays of the auxin activity of agar blocks in contact with stimulated plants (Fig. 10.1). On occasions it has been possible to demonstrate the transport of auxin to the shaded side by application of radioactive IAA to unilaterally illuminated maize and wheat coleoptiles, but workers in this field have had to concede that phototropic curvatures can occur in coleoptiles in the absence of conclusive evidence for lateral transport of auxin. In green sunflower seedling hypocotyls, there is no evidence for lateral redistribution of endogenous auxin, but an abscisin, *xanthoxin*, is redistributed, with more moving to the illuminated side. It is suggested that xanthoxin is the active hormone in that tissue and inhibits growth on the illuminated side; or inhibits auxin transport down that side.

The discussion above indicates that neither the primary perception event in phototropism, nor the physiological basis for the differential growth rates on the opposite sides is known with certainty. Any hypotheses about the intervening transduction events must of necessity be equally uncertain. Some of these hypotheses are discussed below. They all concentrate on accounting for differences in hormone levels on the opposing sides. The possibility that other physiological differences may be equally important should be kept in mind.

Considering flavin as the photoreceptor, it was first suggested that the phototropic reaction results from a riboflavin-mediated destruction of auxin on the illuminated side, for Galston (1949) had found that, *in vitro*, riboflavin catalysed auxin breakdown in the light. The lower concentrations of auxin on illuminated sides of organs were, however, shown fairly conclusively to result from auxin diversion to the opposite side and not from auxin breakdown. The hypothesis was then modified to the suggestion that the light absorbed by riboflavin causes a temporary inactivation of an enzyme involved in auxin synthesis, on the illuminated side, and as a result more auxin precursor reaches the shaded side and is there converted to auxin. In support of this view is the fact that riboflavin has been shown to catalyse the photo-inactivation of some enzymes. However, this fails to explain the observation that with light treatments producing negative curvatures there is more auxin on the lighted than on the shaded side. Neither does this hypothesis account for the diversion of externally applied radioactive IAA across a coleoptile.

Carotenoids are located mainly in plastids and the perceptive region of coleoptiles is rich in yellow plastids. Thimann and Curry in 1961 suggested that light absorbed by the carotenoids may cause the plastids to move to the cell walls which are at right angles to the direction of illumination, and thus favour auxin transport across these walls, i.e. laterally. It is well known that the chloroplasts in some cells orientate themselves according to the direction and intensity of illumination along certain walls, but it has not yet been demonstrated that the etiolated plastids of organs commonly used in experiments on phototropism move in this way. If the plastids change their direction of movement with changing irradiance, congregating at the walls nearest to or furthest from the light source, then more auxin could be transported to the lighted side or to the shaded side. An attraction of this hypothesis is that it proposes a mechanism of phototropic perception similar to that postulated in the statolith theory of gravitropism, involving the movement of cell inclusions in response to the unilateral stimulus. Light and gravity stimuli can be made to act additively, or to cancel each other out, although this in itself does not imply a common mechanism of perception of the stimuli. An alternative suggestion regarding carotenoids is that photolysis of carotenoids gives rise to xanthoxin (and perhaps other abscisins) on the illuminated side. This implies a net synthesis of xanthoxin; at present it is uncertain whether any accumulation of xanthoxin on the illuminated side results from a synthesis of xanthoxin or a lateral redistribution.

## Plagiotropic, diatropic and nastic reactions towards gravity and light

The discussion developed above has been confined to a consideration of mechanisms which might account for the orthotropic reactions to gravity and light. It is clear that the mechanisms underlying these orthotropic reactions are by no means elucidated. We are still less able to explain the plagiotropic reactions which keep the natural direction of growth at a fixed angle to the controlling stimulus rather than parallel to it. In nature, such plagiotropic growth reactions are much more numerous than orthotropic reactions and are more difficult to investigate because the growth angle is usually controlled by an interaction between several factors.

Leaves of many species are stated to be diaphototropic. In some instances (e.g. *Tropaeolum*) the orientation of the leaf blade at right angles to the incident illumination is due to a positive orthotropic response of the petiole, the curvature of the petiole towards light bringing the blade to lie transversely to the incident light. Here the organ of perception is the petiole, although the presence of the blade is necessary, since this is the source of the auxin reaching the petiole. A debladed petiole behaves like a decapitated coleoptile: it is able to react phototropically only when auxin is supplied to its apical cut surface. In other cases of diaphototropic leaves the lamina actively participates in the orientation movement, as in rosette plants such as species of plantain (*Plantago*), species of hawkweed (*Hieracium*), and the shepherd's purse (*Capsella bursa-pastoris*).

In many instances the natural position of an organ is achieved by the balanced interaction of several influences. The apparent plagiotropic response of side shoots and some leaves has been found to result from a negative gravitropism tending to make the organ grow up, balanced by an autonomous *epinasty* (inherent tendency for greater growth on the upper surface, see p. 209) tending to bend it down. The inherent epinasty can be revealed by eliminating the influence of gravity on a klinostat, in which case the continued excess growth on the upper side causes the organ to coil up. The epinasty of side shoots can be eliminated by removal of the main axis growing point (a source of auxin), when the negative gravitropism becomes apparent and the shoots grow upwards. Auxin is apparently involved in all plagiotropic reactions, but how it acts is obscure. One notable difference between organs reacting orthotropically and those reacting plagiotropically is that the former are radially symmetrical, while the latter are dorsiventrally symmetrical, with distinct upper (dorsal) and lower (ventral) surfaces. This suggests very strongly that the symmetry of an organ is of basic importance in determining the type of growth movement executed.

Rhizomes of many plants grow horizontally at a more or less fixed level below the soil surface and, if the level is disturbed, they will grow up or down at an angle till they reach the normal depth, and then proceed to grow horizontally once more. The horizontal natural direction is generally regarded as a gravitropic response, but it is not clear how the constant level is maintained. Rhizomes of *Aegopodium podagraria* react to light by a positive gravitropism (the response, which is elicited even by weak red light, is not a negative phototropism); this causes the rhizomes to turn down when they chance to come too near the surface. Plagiotropic roots and rhizomes of some other species are also known to react similarly towards illumination. Less is known of the factor(s) preventing such organs from penetrating to excessive depths; in the case of *Aegopodium* rhizomes, the controlling factor may be the increased carbon dioxide concentration at greater depths, since the rhizomes react to high carbon dioxide concentrations by turning upwards. For *Polygonatum multiflorum* rhizomes, the control has been suggested to reside in a balance between negative gravitropism and photoepinasty, the rhizomes growing below the surface at a level where the irradiance is just sufficient to induce a nastic reaction balancing the inherent negative gravitropism.

Nastic growth movements also involve unequal growth on opposite sides of an organ in response to environmental stimuli. They are, however, not elicited by directional stimuli, and the direction of the response is determined by the organ. Characteristically nastic reactions take place in organs with bilateral symmetry whereas radially-symmetrical organs execute tropic movements. A higher growth rate on the upper surface is termed *epinasty;* a higher growth rate on the lower, *hyponasty.* Gravinastic, photonastic, thermonastic, haptonastic (thigmonastic) and chemonastic movements are known. Autonasty also occurs, i.e. unequal growth on two sides of an organ independent of external conditions. The physiological basis of nastic movements is still puzzling; the hormones auxin and ethylene are known to be involved. The view has been expressed that the only fundamental difference between organs showing a tropic or a nastic reaction is that anatomical or physiological differentiation in the organ showing a nastic reaction determines that curvature must occur in a fixed direction, irrespective of the direction of the stimulus.

## FURTHER READING

DENNISON, D. S. (1979). Phototropism. In *Encyclopedia of Plant Physiology,* New Series, **7**, ed. by Haupt, W. and Feinleib, M. E., pp. 506–66. Springer-Verlag, Berlin.

FIRN, R. D. and DIGBY, J. (1980). The establishment of tropic curvatures in plants. *Annual Review of Plant Physiology,* **31**, 131–48.

JACKSON, M. B. and BARLOW, P. W. (1981). Root geotropism and the role of growth regulators from the cap: a re-examination. *Plant, Cell and Environment,* **4**, 107–23.

JUNIPER, B. E. (1977). The perception of gravity by a plant. *Proceedings of the Royal Society B,* **199**, 537–50.

VOLKMANN, D. and SIEVERS, A. (1979). Graviperception in multicellular organs. In *Encyclopedia of Plant Physiology,* New Series, **7**, ed. by Haupt, W. and Feinleib, M. E., pp. 573–600. Springer-Verlag, Berlin.

WENT, F. W. and THIMANN, K. V. (1937). *The Phytohormones.* Macmillan, New York.

WILKINS, M. B. (1977). Light- and gravity-sensing guidance systems in plants. *Proceedings of the Royal Society B,* **199**, 513–24.


## SELECTED REFERENCES

BEHRENS, H. M., WEISENSEEL, M. H. and SIEVERS, A. (1982). Rapid changes in the pattern of electric current around the root tip of *Lepidium sativum* L. following gravistimulation. *Plant Physiology, Lancaster,* **70**, 1079–83.

CLIFFORD, P. E. and BARCLAY, G. F. (1980). The sedimentation of amyloplasts in living statocytes of the dandelion flower stalk. *Plant, Cell and Environment,* **3**, 381–6.

FIRN, R. D., DIGBY, J. and HALL, A. (1981). The role of the shoot apex in geotropism. *Plant, Cell and Environment,* **4**, 125–9.

FRANSSEN, J. M. and BRUINSMA, J. (1981). Relationships between xanthoxin, phototropism and elongation growth in the sunflower seedling *Helianthus annuus* L. *Planta*, **151**, 365–70.

HEATHCOTE, D. G. (1981). The geotropic reaction and statolith movements following geostimulation of mung bean hypocotyls. *Plant, Cell and Environment*, **4**, 131–40.

MANDOLI, D. F. and BRIGGS, W. R. (1982). Optical properties of etiolated plant tissue. *Proceedings of the National Academy of Science, U.S.A.*, **79**, 2902–6.

OLSEN, G. M. and IVERSEN, T-H. (1980). Ultrastructure and movements of cell structures in normal pea and an ageotropic mutant. *Physiologia Plantarum*, **50**, 275–84.

PILET, P. E. and NEY, D. (1981). Differential growth of georeacting maize roots. *Planta*, **151**, 146–50.

ZIMMERMANN, B. K. and BRIGGS, W. R. (1963). Phototropic dosage-response curves for oat coleoptiles. *Plant Physiology, Lancaster*, **38**, 248–53.

# 11

# Vegetative Development

## Morphogenesis

Morphogenesis may be defined as the genesis or initiation of form. The study of plant morphogenesis is directed to identifying the physiological events which have their visible manifestation in the appearance of new structures in the plant body. The term development embraces, at the level of the individual organ, all the changes which intervene between the initiation of form and the achievement of the mature morphology and, at the level of the organism, the whole sequence of changes in form and function which intervene between the fertilized ovum and the death of the plant. Morphogenetic events may therefore be regarded as the qualitative changes through which patterns of development unfold. Further, only if we concentrate our attention on some particular criterion of change do we talk of developmental morphology, anatomy or physiology. Only if we concern ourselves with particular levels of organization do we distinguish between cellular differentiation, organization of tissues, organ development, and the ontogeny of the individual organism. Clearly a vast field of biological enquiry can properly be regarded as directed towards an understanding of development. However this chapter and the following chapter will concentrate upon certain selected facets of plant development considered from the stand-point of physiology with particular reference to their control by natural growth-promoting and growth-inhibiting substances.

## Initiation of lateral roots and buds

Lateral root primordia arise in the pericycle at positions related to the primary xylem poles of the central cylinder. Initiation takes place at a level where cell differentiation is proceeding in the primary root tissues and it may be that the substances inducing localized divisions in the pericycle are released by differentiating vascular elements. There is also evidence that the shoot system influences root branching and it has been postulated that both lateral root and adventitious root development are controlled by substances *(rhizocalines)* synthesized in the shoot. The observations that auxin may greatly enhance lateral root initiation in cultured roots and adventitious roots formed on stem cuttings points to auxin of shoot origin as a rhizocaline.

With cultured roots, an initial application of auxin to roots in their first or second culture passage may greatly enhance lateral development. However, a second application to such roots or an initial application to roots which have already been in culture for some time may completely fail to result in lateral initiation. When auxin is the limiting factor its application results in lateral initiation; at other times absence of other essential factors renders auxin ineffective. In particular instances other factors have been shown to enhance lateral initiation; these include the amino acids arginine, ornithine and lysine, the vitamins thiamine and nicotinic acid, the growth hormones kinetin and gibberellic acid, and certain concentrations of $CO_2$ in the solution bathing the roots. Hence it is likely that root initiation is determined by the balance between acropetal transport from the shoot or older regions of the root of auxin, vitamins and amino acids and basipetal transport from the root apex of cytokinin(s) and gibberellin(s) which are believed to be synthesized in the root apex.

Various workers have attempted to study the factors involved in bud and leaf initiation in flowering plants and ferns by experiments involving excision, removal and culture of primordia or isolation of segments of the stem apex from adjacent tissues by incisions at right angles to the surface of the apical meristematic dome. Studies of this kind have revealed that the lateral centres of cell division arising at the shoot apex are for a time capable of giving rise to either a bud or a leaf; there is a developmental interval between initiation of the lateral growth centre and its determination as a leaf or bud primordium. Such investigations have also shown that each lateral growth centre is surrounded by a zone in which the immediate origin of further growth centres is inhibited and that such interacting 'spheres of influence' are concerned in the determination of the characteristic leaf and bud arrangement of the growing shoot. It is however from studies with callus cultures that some insight has been gained into the chemical factors which may control bud initiation.

As mentioned in Chapter 9 (p. 169), it was found that the growth in culture of tobacco callus required the presence in the culture medium of both an auxin and a cytokinin. The callus remained parenchymatous when grown in a medium containing 2.0 mg $1^{-1}$ IAA and 0.2 mg $1^{-1}$ kinetin. However if the ratio of auxin to kinetin was decreased either by raising the concentration of kinetin (to 0.5–1.0 mg $1^{-1}$) or lowering the concentration of auxin (to 0.03 mg $1^{-1}$) the cultures initiated shoot buds which grew into leafy shoots (Fig. 11.1). With a medium containing an auxin:kinetin ratio favourable to bud initiation it was further observed that addition of casein hydrolysate or of the single amino acid, tyrosine, enhanced bud initiation and promoted subsequent growth of the buds (Fig. 11.2). Manipulation of the levels of auxin and kinetin to give a high ratio of auxin to kinetin also initiates organogenesis but now roots and not shoot buds are initiated (Fig. 11.1). The auxin: cytokinin balances appear to exert a controlling influence upon morphogenesis in the tobacco callus. Further work on organogenesis in

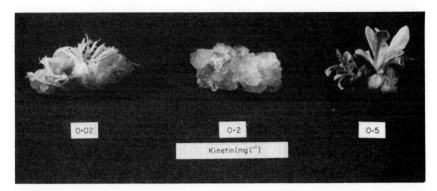

**Fig. 11.1**  Cultures of tobacco callus. The culture medium in each case contains IAA (2 mg l⁻¹). The culture receiving 0.2 mg l⁻¹ kinetin (centre) continues growth as a callus; with a lower kinetin addition (0.02 mg l⁻¹) it initiates roots and with a higher kinetin addition (0.5 mg l⁻¹) it initiates shoots. (Photograph by G. Asquith.)

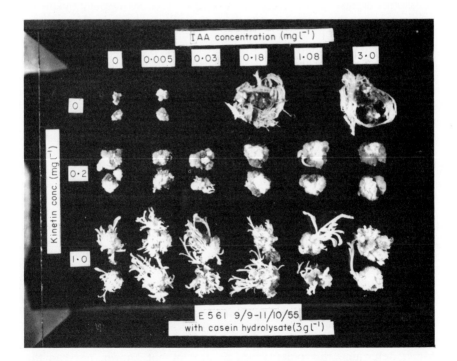

**Fig. 11.2**  Organogenesis in tobacco callus. Effects of increasing IAA concentration at different levels of kinetin and in the presence of casein hydrolysate (3 g l⁻¹) on growth and organ formation. Age of cultures 62 days. Note root formation in absence of kinetin and in the presence of 0.18 and 3.0 mg l⁻¹ IAA, and shoot formation in the presence of 1.0 mg l⁻¹ kinetin, particularly with IAA concentrations in the range 0.005–0.18 mg l⁻¹ IAA. (From Skoog and Miller, 1957, *Symposium of the Society for Experimental Biology,* **11**, 118–31.)

callus cultures points to cytokinins and auxin being generally involved in meristem initiation; some other callus cultures react like tobacco and cases are known where addition of auxin alone or of a cytokinin alone results in bud initiation. Other growth factors than auxin and cytokinin are also clearly involved. Application of the natural plant growth-inhibitor, abscisic acid, but not cytokinin, is essential for bud formation in cultures of potato and sweet potato and gibberellic acid will induce bud formation in *Chrysanthemum* callus. However, the number of callus cultures where it appears that it is just the endogenous level of known growth factors which prevents spontaneous bud initiation is limited. There are many callus cultures in which organogenesis cannot be induced by application of known growth factors or of natural extracts like coconut milk. In all such cases it may be that meristem initiation is prevented by a depletion or accumulation, during growth in culture, of as yet unknown growth regulating factors.

There is, however, another possible reason why no root or bud initiation has ever been observed in some callus cultures. Callus cultures normally arise by activation of division in partially or completely differentiated tissue cells in the organ explant in response to its excision from the plant (an injury response) and the chemical stimuli in the culture medium. Even when meristematic cells in the explant are involved they may already be committed to the formation of particular tissue cells (e.g. cambial cells to the formation of vascular tissue cells). The cells in the explant are induced to divide but the extent to which they undergo dedifferentiation during callus induction may depend upon the species, the nature of the organ explant and the cultural conditions operating during the induction process. Thus activation of the mechanisms involved in cell division and cell growth may occur without a sufficiently profound change in cellular physiology to permit the cells to express their totipotency via root and shoot initiation (see also Chapter 12, Embryo development, p. 250).

Prior to the formation of root and shoot primordia in a callus tissue there can be distinguished groups of tightly packed meristematic cells characterized by their small size, dense cytoplasm and prominent nuclei (these groups have been termed meristemoids). These groups apparently respond to directional organizing stimuli within the callus to form an organized organ primordium. The formation of these meristemoids may involve the operation of a contagion phenomenon whereby a cell embarking upon division entrains adjacent cells to divide and a 'sink' effect whereby a localized centre of cell division draws upon nutrients and metabolites from adjacent cells thereby inhibiting their division. Once the meristemoid has formed it must develop an axis of polarity (see p. 175) to give rise to a primordium. The nature of the physico–chemical gradients within the callus which determine this polarity are at present not known, nor can the induction of this polarity be recognized in terms of the development of an axis of symmetry in the structure of the initiating cells by either light or electron microscopy.

# Leaf initiation and growth

Leaves are arranged on the shoot in a regular manner as a consequence of the spatial regularity with which primordia arise at the shoot apex. This arrangement of the leaf primordia – *phyllotaxis* – depends upon the species and the stage of development and rate of growth of the individual plant (see Chapter 8, p. 153). Changes in the time interval separating the initiation of successive primordia (the *plastochron*) and changes in phyllotaxy occur during development and are correlated with changes in the growth activity of the apical meristem and with the shape of the apical dome of meristematic cells. New primordia arise in succession so that, at any moment in time, there is a free surface area of the apex above the last initiated primordium; it is in this 'next available space' that the next primordium will arise. This space must be able to accommodate the primordium as it develops and it must arise in the correct relationship to existing primordia if the phyllotaxy is to persist. Its point of origin appears to be controlled by the interaction of influences emanating both from the main axis meristem and the adjacent primordia. The chemical (or physical) nature of these influences awaits identification. A physiological interpretation of orderly leaf arrangement has yet to emerge.

The development of leaves involves the functioning in sequence and for a limited time of a number of zones of meristematic activity and an associated programme of cell expansion and differentiation. This developmental anatomy has now been worked out in some detail for the leaves of a very limited number of species. Such studies only serve to emphasize how far we are from understanding what controls such highly integrated developmental programmes. One possible experimental approach to this is to see how far such development is dependent upon the culture medium supplying particular nutrients and growth factors.

This experimental approach has been followed in studies of the growth and development in culture of excised leaf primordia of ferns (species of *Osmunda, Leptopteris* and *Dryopteris*). The culture medium solidified with agar has contained sucrose, inorganic ions and various combinations of vitamins and other growth factors. Although the most recently initiated primordia can be succcessfully cultured such primordia develop into shoots and their growth becomes indeterminate. Primordia, however, show, with passage of time from inception to excision, an increasing tendency to develop into leaves in culture and, in *Osmunda cinnamomea*, primordia which are about 800 $\mu$m long (there are 8 to 10 younger primordia at the apex) always give rise to leaves. The cultured primordia follow a pattern of growth essentially similar to that of attached fronds except that their growth is completed precociously and the resulting leaves are smaller than normal due to reduced cell number per leaf (rather than to any reduction in mean cell size) (Fig. 11.3). The cultured leaves remain healthy for many weeks after completing their development. Addition to the basal medium of vitamins, auxin, amino acids or substances such as coconut milk does not profoundly modify development although they may increase final leaf

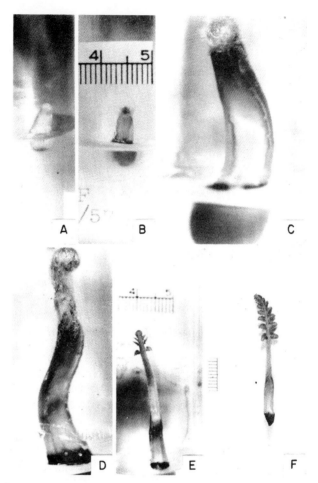

**Fig. 11.3** Culture of excised frond primordia of *Osmunda cinnamomea*.
(A) Primordium at time of excision (9.7 mm long).
(B) After 2 weeks' culture – crozier formation stage.
(C) After 5 weeks' culture – full crozier stage.
(D) After 7 weeks' culture – completion of crozier elevation.
(E) Crozier uncoiling.
(F) After 11 weeks' culture – development complete (average length 46 mm, average number of pairs of pinnae 9.7). (From Caponettio and Steeves, 1963, *Canadian Journal of Botany,* **41,** 545–56.)

weight. Crozier uncurling does not occur unless the cultures are exposed to light. These observations indicate: (1) that some influence (possibly a leaf-forming substance) moving from the shoot apex programmes the primordium to ensure its subsequent dorsiventral and determinate growth into a leaf; (2) that once the primordium is so determined it can complete its development to give a miniature leaf similar in form to the naturally developed frond without needing special metabolites or growth factors; (3) that leaf size is controlled by some influence of the shoot which either prevents cell expansion occurring precociously or ensures that cell divisions proceed in the leaf meristems at a sufficient rate and for an appropriate time.

## Apical dominance

Apical dominance is the phenomenon of suppression of lateral root emergence or lateral bud outgrowth by an actively growing main root or main shoot apex. Evidence has been obtained in work with roots that this suppression of lateral development is due to the release of inhibitor(s) (not however yet isolated in pure form) whose action cannot be reproduced by auxin applied at a growth-inhibitory concentration. The natural inhibitor, abscisic acid, known to occur in roots, is an inhibitor of root growth but not a specific inhibitor of lateral initiation and extension. A number of workers, using cultured excised roots, have reported that light can suppress lateral root initiation, that in this respect red light is much more active than light from other regions of the visible spectrum and that this action of red light is reversed by immediate subsequent exposure of the roots to far-red light. As discussed later in this chapter – photomorphogenesis (p. 225) – red light is now known to have a number of morphogenic effects which are reversible by far-red light.

Apical dominance in shoots is usually quoted as the most clear-cut example of a growth correlation in plants. An actively growing main shoot apex inhibits the growth of lateral buds nearest to the apex and the distance over which the suppression operates is a measure of the intensity of the apical dominance. If the apex is excised, punctured or arrested in its growth the lateral buds begin to develop and elongate and in many instances this growth can be arrested by applying a lanoline paste containing an auxin like IAA at the site of the removed apex. Apical dominance has therefore been explained on the basis that the growing shoot apex is a centre of auxin synthesis, that this auxin is transported basipetally from the meristem and that it accumulates at the loci of the axillary buds to a concentration which inhibits their growth. In some instances spraying intact shoots with gibberellins has increased their apical dominance and in other cases simultaneous application of IAA and gibberellin to decapitated shoots has been more effective than IAA alone in maintaining lateral bud inhibition. The observation that, when [14]C-labelled IAA has been applied to decapitated shoots of certain species, the accumulation of [14]C at the level of the inhibited buds has been enhanced by the simultaneous application of unlabelled gibberellic acid, has led to the view that gibberellin may act by promoting endogenous auxin synthesis or release or by enhancing its basipetal transport.

The concept that apical dominance results from a direct suppression of lateral bud outgrowth must be now modified in the light of additional experimental work. The actual level of auxin in lateral buds is often too low to be inhibitory and in certain instances lateral buds can be induced to grow by direct application of auxin to them. Although lanoline paste containing a high level of IAA does in many instances suppress lateral bud growth when applied to the decapitated shoot apex, this suppression usually persists for only a limited period and in some species (e.g. *Coleus* sp.) IAA application is quite ineffective in suppressing lateral bud growth. Apical dominance in the intact shoot or in decapitated shoots supplied with auxin can in some species be broken by the application of the cytokinin kinetin to the lateral buds (Fig. 11.4). With intact shoots this release is usually transient but if auxin is also applied to the buds their growth continues.

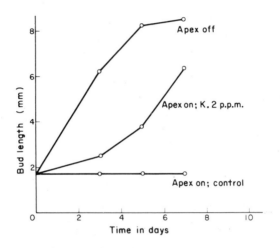

**Fig. 11.4**   Promotion of lateral bud growth of intact pea shoots resulting from application to the buds of kinetin (K. 2 ppm). (From Wickson and Thimann, 1958, *Physiologia Plantarum,* **11**, 62–74.)

Clearly an actively growing apical meristem is a 'sink' for the metabolites needed for the synthesis of protein and other cell constituents and some authors have attempted to explain apical dominance in terms of competition for such food materials; the operational 'sink' diverts food material from potential 'sinks'. This concept is supported by the observation that apical dominance in the shoot is intensified under conditions of nutrient deficiency. Evidence for 'auxin-directed' transport of nutrients has come from experiments where it has been demonstrated that the application of auxin to decapitated shoots causes rapid accumulation at the cut surface of radioactive $^{32}$P-phosphate applied to the base of the shoot or of $^{14}$C-sucrose applied to the leaves. No such accumulation occurs in the absence of applied auxin. Such 'auxin-directed' transport could result in nutrient competition between the main shoot meristem as a site of auxin release and the quiescent lateral buds which do not release auxin. The mechanism of the 'auxin-directed' transport is not fully understood but one effect of the auxin application to decapitated shoots in certain species has been inhibition of the development of vascular connections between the lateral buds and the main vascular supplies of the shoot. Cytokinin, apparently essential to bud growth, is synthesized in the root and transported to the shoot; 'auxin-directed' transport of this cytokinin to the main apex may starve the lateral buds of an essential growth factor.

Apical dominance is expressed not only in the inhibition of lateral roots and axillary buds but controls the orientation of the lateral roots and branches (affects their gravitropic sensitivity, see Chapter 10). In the potato the stolons which arise from a basal node of the stem normally grow horizontally at or below the soil surface and show elongated internodes, and rudimentary leaves. If the aerial shoot is decapitated the stolons turn upwards and become typical leafy shoots. If lanoline paste containing auxin and gibberellin is applied to the decapitated shoot apex the stolons retain their characteristic morphology and direction of growth; if kinetin is applied directly to the stolons this antagonizes the action of the auxin + gibberellin and a leafy shoot develops.

These studies of apical dominance, still obviously far from revealing the complete story, indicate that growth correlations are probably determined by the distribution patterns of growth regulators and nutrients and that growth regulators may play a critical role in determining these distribution patterns.

## Seasonal bud dormancy

Dormancy is defined as lack of growth in a potentially growing system. If the dormancy results from unfavourable external conditions such as low temperature or lack of water, growth is resumed as soon as conditions permit, and this type of dormancy is often referred to as quiescence or *enforced dormancy*. However dormancy is frequently internally imposed when no obviously unfavourable external conditions are operative, a state known as *innate dormancy*, true dormancy or rest. The phenomenon of apical dominance involves the imposition of dormancy on lateral buds by an active apical meristem. Trees and shrubs of temperate climates form resting buds which remain dormant during the winter months. These buds are in a state of innate dormancy, but the onset and breaking of dormancy are controlled partly by environmental stimuli.

In many woody species bud dormancy is under photoperiodic control. *Photoperiodism* is the physiological reaction of plants to daylength; in this instance the shortening of days in the autumn induces the apical buds to become dormant. When daylength is maintained above a critical value, growth can be extended for at least 18 months in some species, e.g. birch (*Betula* spp.), *Robinia pseudacacia*, *Liriodendron tulipifera*, whereas the imposition of short days brings about the cessation of growth in about two weeks. In many species the photoperiodic stimulus is perceived by the mature leaves, implying the transmission of an internal stimulus from the leaves to the growing apices, but in birch the apical bud itself is sensitive.

Endogenous control of the onset of dormancy also occurs, for not all species can be kept growing for prolonged periods in long days. In many temperate zone trees, resting buds are formed early in the summer, from June onwards; removal of the leaves will then make the buds grow out again, showing that the dormancy is imposed by the leaves. However, once the trees have experienced the short days of autumn, the buds are found to have become deeply dormant and are no longer affected by defoliation. The development of cold hardiness which is associated with winter dormancy is also partly a response to short photoperiods, but the falling temperatures of autumn are the main stimulus in the induction of cold hardiness. Experimentally, dormancy and hardiness can be induced independently of each other.

The breaking of dormancy in the spring may again be under photoperiodic control, with the buds themselves perceiving the stimulus; this is the case for birch, pine (*Pinus* spp.), beech (*Fagus sylvatica*) and larch (*Larix decidua*); in other species the buds respond to the rising temperatures of spring. In many species, however, e.g. poplar (*Populus* spp.), sycamore (*Acer pseudoplatanus*), the buds must first be chilled during the winter before they are capable of outgrowth: this is an adaptation which prevents premature outgrowth during a freak warm spell in the autumn. The chilling temperature must be below 10°C, with an optimum usually at 5–7°C; up to several months of low temperature treatment may be required.

Even in species native to humid tropics, perennials still show alternating periods of activity and dormancy, or at least of fast and slow growth. Beech transplanted to the tropics still shows a dormancy rhythm, but no longer with a 12-month periodicity. Plants appear to have an innate tendency to rhythmic growth; in climates with an unfavourable season, selection processes have operated to entrain the rhythm to an annual cycle, and to impose photoperiodic control, which ensures the onset of bud dormancy before the unfavourable conditions set in.

There is strong evidence for the involvement of ABA in dormancy induction. In buds of beech, apple (*Pyrus Malus*) and ash (*Fraxinus excelsior*), ABA concentration becomes high as the buds enter winter dormancy, and falls in the spring; short day treatment causes the concentration of ABA to rise in leaves of sycamore. Resting bud formation has been artificially elicited by the application of ABA in birch, sycamore and blackcurrant (*Ribes nigrum*). There is also evidence that gibberellins are active in the breaking of bud dormancy; increases in gibberellin content of buds during chilling treatment have been recorded, and dormancy can be broken by application of gibberellins, which can substitute for both long days and for low temperature. The hormonal relationships of bud dormancy control are, however, by no means clear-cut. Instances have been noted where short days fail to increase ABA levels, or where ABA application has failed to induce dormancy. The ratio between the contents of ABA and gibberellins (and perhaps also between the contents of ABA and cytokinins) may be more important than the absolute levels of either hormone.

# Leaf senescence and leaf abscission

Although plant development is a continuous process there can occur sharp qualitative changes such as the transition from vegetative to reproductive development. Similarly in some plants there occurs a transition from a juvenile to an adult form as reflected by leaf form and growth habit. This transition is known as *heteroblastic* development. Only the adult form is capable of flowering.

Maturity is sooner or later followed by ageing (senescence) and death. Senescence can be defined as the complex of catabolic events which, if not reversed, lead to death of the cell, organ or organism. Many plants (and particularly annuals and biennials) flower once during their life history (are monocarpic) and senescence leading to death of the whole plant follows their flowering and fruiting. In such plants the onset of senescence is clearly part of an inherited programme of development. The situation is less clear in perennial plants, particularly in woody perennials, where death due to disease or catastrophes (such as flooding, drought, wind damage, mineral deficiency) makes it difficult to establish a clear onset of senescence in the whole organism. However it is clear that as such plants increase in stature they show a fall in relative growth rate and that their leaf area to plant weight index falls. Such plants may therefore reach a stage when they are increasingly susceptible to unfavourable environmental influences. Such loss of vigour may be related to an increasing failure of the organism to nurture its regions of growth, to establish any steady state relationship between its meristems and its mature tissues.

At the level of individual plant organs it is difficult to distinguish between an inherited programme leading to senescence and an ageing of the organ imposed by the whole plant of which it is a part. If different organs enter into competition with one another for food materials the growth and development of one organ may result in the senescence and death of another. Thus developing leaves, flowers and fruits may initiate the onset of senescence in mature leaves.

There is evidence that under certain circumstances there can occur a progressive accumulation of damage to the genetic information of cells. This can apparently occur during the storage of dry seeds. The significance of this phenomenon in the growing plant is, however, very doubtful since the chances are that such impaired cells are eliminated at the growing points of the plant. It seems more pertinent to enquire whether the individual differentiated cells are programmed to senesce and die. Clearly certain lines of differentiation rapidly and without interruption lead to cell death (as for instance in the development of vessels, tracheids and fibres). In other cases however the differentiated cell reaches a metabolic 'plateau' which can persist (the pith cells of perennials can remain alive for many years). Are such cells intrinsically limited in life span? We refer to differentiation as involving sequential activation and repression of genes but we do not understand the timing mechanism in this sequence; perhaps when this is understood it will be possible to answer whether the timing sequence involves a determination of senescence and ensuing death.

As indicated above, senescence in annual plants follows more or less rapidly upon fruit development and the index of this senescence is the yellowing and withering of the foliage leaves. In herbaceous perennials we observe a similar senescence of the aerial shoot. In deciduous perennials there is an annual senescence and abscission of leaves and an associated induction of bud dormancy. Active abscission (falling) is not confined to leaves; a similar phenomenon is seen in flower and fruit fall. Leaf senescence is characterized by yellowing, export of amino acids from the leaf blade, decline in leaf protein content and drying out. Since these changes in annual plants are associated with fruit development it has been suggested that senescence is a consequence of leaf starvation. However in dioecious species senescence follows flowering in both male and female plants so that it is not apparently dependent upon competition between the leaves and the 'food sinks' represented by developing fruits. However senescence in leaves is affected by other parts of the shoot system and reversals of senescence following removal of the younger leaves and apical buds and supplying a source of nitrogen such as ammonium sulphate have been reported from experiments with barley and tobacco.

It has long been known that leaf excision promotes leaf yellowing. When leaves are excised and their petioles placed in water, amino acids move out of the blade and protein breakdown occurs well in advance of decay of the protein synthesis potential of the leaf cells as judged by incorporation of amino acids into protein by leaf discs and by subcellular fractions. If roots are initiated at the base of the petiole, yellowing of the leaves is immediately reversed and as long ago as 1939 it was suggested by A. C. Chibnall that some hormone of root origin might exert a controlling influence on protein metabolism in the leaf. This hormone may be a natural cytokinin for we know that cytokinins are synthesized in roots and transported from the root to the shoot; also that application of kinetin to detached leaves either via the petiole or directly to the leaf blade has been found in some species to arrest their senescence (Fig. 11. 5). Where kinetin is applied to the blade its effect is often strictly localized, regions painted with the kinetin remaining green while adjacent areas show yellowing. The areas receiving kinetin do not lose amino acids and amino acids migrate to these regions from the untreated areas (Fig. 11.6). The treated areas show enhanced protein and RNA synthesis compared with the untreated areas. The induction of senescence by fruit development and the promotion of senescence in older leaves by the presence of growing leaves and buds may on this basis be explained either by diversion of the root hormone to these regions or, and this would apply particularly to flowers and developing fruits, by these regions being themselves centres of cytokinin synthesis.

The above hypothesis is, however, not totally adequate, for senescence in some leaves cannot be arrested by kinetin application. In a few plants, gibberellins are known to be effective in delaying senescence, e.g. in the dandelion (*Taraxacum officinale*), and nasturtium (*Tropaeolum majus*). In yet other instances, particularly in leaves of woody perennials, senescence is

**Fig. 11.5** Detached leaves of *Xanthium* 10 days after detachment and culture with petioles in water (top row) and in 5 mg kinetin per litre (bottom row). Note retention of green colour in kinetin-treated leaves and yellowing in leaves cultured in water. (From Richmond and Lang, 1957, *Science, N.Y.*, **125**, 650–1.)

prevented by auxin application. The complexity of the hormonal regulation of leaf metabolism is illustrated by experiments with *Euonymus japonicus*. In this plant auxin application promotes leaf senescence. However, if the auxin (2,4-D) is applied to discrete areas of the leaf blade the treated areas of the leaf blade do not senesce (they retain their colour, show active respiration and protein synthesis and accumulation of $^{14}$C-labelled photosynthate and amino acids) but the surrounding areas show accelerated

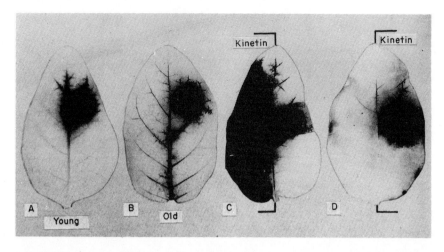

**Fig. 11.6** Radioautographs of leaves of *Nicotiana rustica* treated with a spot of a radioactive amino acid (DL-aminoisobutyric acid-$^{14}$C). A: a young leaf revealing the position on the leaf where the amino acid solution was applied; it has spread very little from the point of application; B: an adult leaf similarly treated — the radioactivity has spread into the vein system and moved from the leaf lamina into the petiole (not shown); C: adult leaf, the left hand side of which has been painted with kinetin solution – note the spread of amino acid applied to the other half only into this kinetin-treated half; D: adult leaf, the right half of which has been painted with kinetin and amino acid – the amino acid has spread very little so that its autoradiograph resembles that of the young leaf. (From Mothes, in *Régulateurs Naturels de la Croissance Végétale*, 1964, 131–40. Centre National de la Recherche Scientifique, Paris.)

senescence and the leaves absciss prematurely. The simultaneous senescence of leaves of deciduous perennials in autumn is under environmental control, being promoted by the shortening of daylengths and falling of temperatures in the autumn, but the environmental signals may act via hormone supply, inducing decreases in auxin and gibberellin contents in the leaves, while the ABA content often rises.

Leaf fall is the consequence of the development in the petiole of a *separation layer*, formed within an *abscission zone* where the cells are smaller than in neighbouring areas. Sometimes a row of petiole cells divides transversely to form a few tiers of thin-walled cells; sometimes the separation layer (1–5 tiers) differentiates without cell divisions. The separation layer forms a sharp physiological barrier across which senescence does not spread: the distal (leaf side) cells yellow and die, the proximal (axis side) cells remain healthy. In the separation layer itself, the cells secrete wall-digesting enzymes; the middle lamella is particularly affected so that cell separation occurs there, aided by cell expansion. Conducting cells in the separation region become blocked, and a layer of suberized or lignified cells may form in the proximal region as protective scar tissue. A similar process occurs in the peduncles of abscissing flowers and fruits.

The formation of the abscission zone is correlated with senescence in the leaf. As senescence develops the output of auxin to the petiole falls. In many experiments IAA applied to petiole stumps has retarded abscission, and IAA application to the adjacent stem tissue has speeded it up. On the basis of such observations the hypothesis was advanced that it is the auxin gradient across the separation layer which suppresses or initiates cytolysis, i.e. if the gradient is steep from high on the distal side to lower on the proximal side abscission is suppressed; if the gradient becomes less steep or is reversed, by fall in auxin output of the leaf blade or fruit or by rise in the stem auxin level, then abscission occurs. However, the response to auxin of explants (stem segment-petiole stump units) depends upon the time interval between excision and auxin application. Auxin applied immediately acts as expected from the gradient hypothesis. If however auxin application is delayed, then auxin applied either to the petiole stump or the stem tissue promotes abscission. This has led to a two-stage hypothesis; that auxin reaching the separation layer by diffusion from the leaf inhibits abscission but that once abscission has been triggered then auxin applied either distal or proximal to the layer promotes abscission. Several other hormones are known to be involved in abscission. In 1955, D. J. Osborne reported that diffusates from yellowing leaves contained a senescence factor (SF) which accelerated abscission. Around the same period, two abscission-promoting substances were demonstrated to be released from the peduncle of the cotton fruit; one of these substances has been identified as ABA, but the SF has not been chemically characterized. Finally, ethylene is believed to play a very important role; this gas is produced in large amounts in senescing leaves and one of its earliest noted effects in plants was the promotion of leaf abscission. The synthesis of hydrolases that produce cell wall dissolution in the separation layer is strongly stimulated by ethylene. The responsiveness of the petioles to ethylene nevertheless depends on the age of the leaf: cells must be already aged to some extent to respond. The precision of the spatial location of the events of abscission is striking: whatever the (hormonal) stimulus, only a few cells in a strictly fixed position respond by the formation of a separation layer.

## Photomorphogenesis

Light exerts profound effects on plant growth, differentiation and morphological development, the control of these processes by light being termed *photomorphogenesis*. A very simple yet informative experiment is to grow duplicate samples of some plant in the light and in darkness. The two sets of plants may then be scarcely recognizable as belonging to the same species. The light-grown individuals will be green, sturdy, with expanded leaves and of moderate height. The specimens grown in the dark will be *etiolated* – unpigmented, very tall but thin and weak, and with rudimentary or folded leaves. This simple experiment shows that the everyday appearance of plants, that we are accustomed to regard as normal, is not the outcome of a totally internally programmed developmental sequence, but results from the interaction of light with the genome of the plant. Light inhibits internode elongation, promotes leaf expansion (dicotyledons) or leaf unrolling (monocotyledons), promotes chlorophyll synthesis and chloroplast development, and stimulates the synthesis of secondary products such as anthocyanin pigments. As a generalization light can be stated to stimulate differentiation in contrast to rather undifferentiated elongation growth; every kind of character from ultrastructure to overall plant size is affected. While for sustained normal growth photosynthesis is needed to provide the substrate, the formative effects of light do not directly depend on photosynthesis. They can be evoked by light energies far too low for photosynthesis, and are exhibited in etiolated, chlorophyll-free plants. Light acts in morphogenesis as a specific environmental stimulus rather than as an energy source. The main photoreceptor pigment for photomorphogenic reactions is *phytochrome*. Currently two types of photomorphogenic response are recognized, the low energy reaction and the high energy reaction.

### The low-energy phytochrome response

The existence of the pigment called phytochrome was first deduced from studies of what now would be considered a typical low energy reaction (or induction–reversion response). The seeds of certain varieties of lettuce need light for germination. They are induced to germinate by exposure for a few minutes to *red* light (peak activity at 660 nm) of low irradiance level. A similar dose of *far red* light (peak activity at 730 nm) is inhibitory to germination, and the effect of a red light exposure is negated by an exposure to far red light. Alternating exposures of red and far red light can be given for many cycles and whichever wavelength is given last determines the response. These observations led Hendricks and Borthwick working at Beltsville, Maryland, in the 1950s to postulate the existence of a photoreceptor, named phytochrome in 1959, existing in two photoreversible forms. One form, $P_r$, absorbs red light at 660 nm, and is thereby converted to the second form, $P_{fr}$, active in promoting germination in the lettuce seeds. $P_{fr}$ shows peak absorption in the far red, and under far red illumination it is converted to $P_r$ again:

$$P_r \underset{730 \text{ nm}}{\overset{660 \text{ nm}}{\rightleftharpoons}} P_{fr}$$

Subsequently phytochrome was isolated from plant tissues; it is a blue-green chromoprotein and in solution it shows photoreversibility between the $P_r$ and $P_{fr}$ forms, with respective absorption maxima at 660 and 730 nm as predicted (Fig. 11.7). It is universal in green plants. With specially designed spectrophotometers, absorption by phytochrome is directly observable in etiolated tissues where it is not masked by chlorophyll. The active form $P_{fr}$ may be unstable in plant tissues. In etiolated plants in the dark $P_{fr}$ undergoes breakdown in monocotyledons, but in dicotyledons it reverts to $P_r$ either wholly, or partially with some breakdown also taking place, over a period of

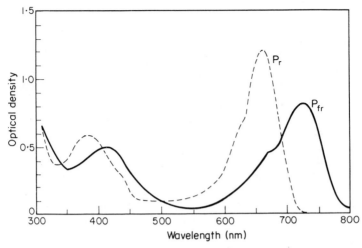

**Fig. 11.7** Absorption spectra of $P_r$ and $P_{fr}$ for a solution of oat phytochrome. (From Hartmann, 1966, *Photochemistry and Photobiology*, **5**, 349–66.)

a few hours. In green plants the loss of $P_{fr}$ is slower.

A multitude of photomorphogenic processes can be controlled by phytochrome in a manner similar to the germination of lettuce seed, i.e. they are promoted by low doses of red light, and far red light reverses the red light effect. The processes include inhibition of internode elongation, promotion of leaf expansion, straightening of seedling epicotyl and hypocotyl hooks, control of leaf turgor movements, stimulation of chlorophyll synthesis and anthocyanin synthesis, and many other reactions. The light energy (total dosage) for saturation of these reactions usually varies from 1–1000 J m$^{-2}$. The reciprocity law (p. 194) holds, so that, taking full sunlight as equivalent to 42 000 J m$^{-2}$ min$^{-1}$, it can be seen that full response would be achieved with a few seconds of sunlight at the most.

The low energy responses show that plant morphogenesis is sensitive to very small amounts of light, and their study has provided much information about phytochrome. The conditions under which the functioning of the low energy phytochrome reaction has been demonstrated are, however, artificial in the extreme. In nature plants do not grow in darkness interspersed with odd flashes of monochromatic red and far red light: prolonged periods of illumination with a mixed spectrum are the rule. While exposure of dark-grown plants to brief treatments with low energy red light alleviates some of the symptoms of etiolation, the appearance of the plants still remains far removed from that of plants grown under natural daylight conditions.

### The high energy reaction of photomorphogenesis

The low energy reactions promoted by brief illumination with red light are characteristically inhibited by a subsequent short exposure to far red. But when the far red illumination is prolonged, a quite different result may be obtained. This effect was first recorded by H. Mohr working with seedlings of mustard (*Sinapis alba*). Cotyledon expansion in mustard is elicited by short red illumination and the expansion is prevented by a *brief* far red exposure; but if far red light is applied for longer than 2–3 hours, the cotyledons proceed to expand even better than under the red light treatment. Many (though not all) of the other developmental processes, too, that can be stimulated by red light via the low energy phytochrome reaction can be stimulated even more strongly by prolonged exposure to far red light. Seedlings germinated in continuous far red illumination assume an appearance more or less the same as that of seedlings grown in white light. Mohr called this response the *high energy reaction* (HER) of photomorphogenesis; now it is sometimes also called the high irradiance response, HIR, or prolonged light reactions, PLR. For the HER the reciprocity law does not hold; the length of the period of illumination is critical rather than the light dose. The reaction is proportional to irradiance over a wide range of irradiance levels. The action spectrum of the HER shows a peak at 720–730 nm, where $P_{fr}$ absorbs, and there is also a triple peak in the blue and one in the near ultraviolet (Fig. 11.8). Two pigments are apparently involved: phytochrome, and a blue-absorbing pigment which is probably the same as the unidentified photoreceptor involved in phototropism, i.e. a carotenoid or a flavin. Studies to date have concentrated on the role of the phytochrome.

Investigations of the low energy reaction have indicated that numerous developmental processes require phytochrome in the $P_{fr}$ form. Studies of the

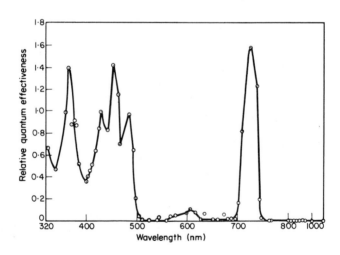

**Fig. 11.8**   Action spectrum for the inhibition of hypocotyl growth in *Lactuca sativa* L. (From Hartmann, 1967, *Zeitschrift für Naturforschung, B,* **22**, 1172–5.)

HER show that some of the *same* developmental processes are promoted by far red light, which converts phytochrome to the $P_r$ form. How may these two observations be reconciled? One possibility is that the HER also is dependent on phytochrome in the $P_{fr}$ form. The photoconversion of phytochrome in either direction is not 100% complete, there being some overlap between the absorption spectra of $P_r$ and $P_{fr}$ (Fig. 11.7). A *photostationary state* is set up on prolonged illumination, a dynamic equilibrium between $P_r$ and $P_{fr}$, equal amounts being converted in each direction per unit time. With monochromatic illumination of 660 and 730 nm, the proportions of $P_r$ and $P_{fr}$ at photoequilibrium are approximately as indicated below:

$$20\% \ P_r \xrightarrow{\ \ 660 \ nm \ \ } P_{fr} \ 80\%$$

$$97\% \ P_r \xleftarrow{\ \ 730 \ nm \ \ } P_{fr} \ 3\%$$

Even under pure far red illumination, a small fraction of the phytochrome is maintained in the active $P_{fr}$ form, and it is postulated that the HER depends on the presence of this small amount of $P_{fr}$ over a long period of time. If $P_{fr}$ is unstable, there may be less breakdown of phytochrome with only a small proportion in the $P_{fr}$ form at any one moment. An alternative proposal is that the HER depends on the continuous cycling of phytochrome between the $P_r$ and $P_{fr}$ – hence the need of prolonged periods of illumination. The dependence of the HER on irradiance levels would be explicable on the cycling hypothesis: the higher the irradiance, the faster the rate of cycling.

### The function of phytochrome in the field

Much less is known about phytochrome activity in normal green plants exposed to white light than about its reactions in etiolated plants treated with narrow wavelength bands of illumination. Phytochrome obviously signals to plants the presence or absence of light, to which the plant system reacts according to its competence. Seeds of some species are adapted for surface germination, and react to the light signal by becoming active; other seeds are adapted for germination under the soil and these are inhibited from outgrowth by the light signal. There is physiological advantage for a seedling which germinates underground to adopt the etiolated mode of growth: rapid internode elongation will bring it quickly into daylight, while the hook protects the plumule and there are no expanded leaves which might get damaged by passage through the soil. When light is reached, conversion of phytochrome largely to $P_{fr}$ channels growth into the 'normal' pattern. But the phytochrome system also enables a plant to sense the spectral composition of light, and this is thought to be of considerable significance. Under illumination by a mixed spectrum, phytochrome assumes a photostationary state in which the ratio of $P_r$ to $P_{fr}$ is determined by the proportions of different wavelengths in the light. During most of the daylight hours the phytochrome of plant organs exposed to direct daylight would contain *c.*

50% $P_{fr}$. The situation is, however, very different under a leaf canopy. The chlorophyll of leaves effectively filters out the red wavelengths but most of the far red above 700 nm is transmitted (Fig. 11.9). Plants shaded by others find themselves in an environment enriched with far red and they have less than 10% of their phytochrome as $P_{fr}$; many species respond to this by rapid internode elongation, which helps to bring them nearer to the top of the canopy. (Species which normally grow in the shade show little reaction, however.) Not only is total irradiance higher at the top, but the proportion of wavelengths strongly absorbed by photosynthetic pigments is higher.

The role of phytochrome in photoperiodic responses is discussed later in relation to the control of flowering (p. 237), and also its role in seed dormancy breaking (p. 263).

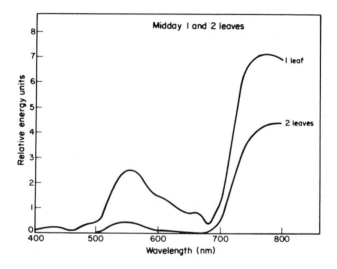

**Fig. 11.9**   Calculated spectral distribution of midday daylight transmitted through one or two leaves of sugar beet. (From Smith, 1973, in *Seed Ecology*, ed. by Heydecker, 219–31. Buttersworths, London.)

### The biochemistry of phytochrome action

In trying to explain the action of phytochrome at the cellular level the same difficulty is encountered as with plant hormones: this one photoreceptor is able to control a very large number of processes, even as plant hormones all exhibit a multiplicity of effects. Action of phytochrome ($P_{fr}$) as an enzyme has been considered, but seems highly improbable in view of the large number and diverse nature of the processes that lie under phytochrome control. Phytochrome action at the genetic level as gene repressor or activator has also been suggested. Yet some phytochrome-controlled events occur within 1 minute of a light treatment. Again, it is hard to visualize how many genes could be controlled by one single agent. The idea currently most favoured is that phytochrome acts on membranes. There is evidence of phytochrome association with membranes within cells, $P_{fr}$ being more firmly membrane-bound than $P_r$. The $P_r \rightleftharpoons P_{fr}$ transition is suggested to affect membrane permeability and membrane transport. What solutes get transported, and the ultimate effect on the cell, would depend on cell type and its state of development, i.e. on the competence of the reacting tissue.

## FURTHER READING

GEORGE, E. C. (ed) (1979). *Differentiation and the Control of Development in Plants – Potential for Chemical Modification*. British Plant Growth Regulation Group Monograph No. 3. Wessex Press, Wantage, Oxon.

KENDRICK, R. E. and FRANKLAND, B. (1983). *Phytochrome and Plant Growth*, 2nd edition. Studies in Biology, No. 68, Edward Arnold, London.

MOHR, H. (1972). *Lectures on Photomorphogenesis*. Springer-Verlag, Berlin.

SMITH, H. (1975). *Phytochrome and Photomorphogenesis*. McGraw-Hill, London.

SMITH, H. (1982). Light quality, photoperception, and plant strategy. *Annual Review of Plant Physiology*, 33, 481–518.

STREET, H. E. (1966). Growth, differentiation and organogenesis in plant tissue and organ cultures. *Cells and Tissues in Culture*, ed. Willmer, E. N., Vol. 3. Chap. 10, 631–89. Academic Press, New York.

STREET, H. E. (1969). Growth in organized and unorganized systems: knowledge gained by culture of organs and tissue explants. In *Plant Physiology*, ed. Steward, F. C., Vol. 5, Chap. 6, 3–224. Academic Press, New York and London.

THIMANN, K. V. (ed.) (1980). *Senescence in Plants*. CRC Press Inc., Roca Baton, Florida.

TORREY, J. G. and CLARKSON, D. T. (1975). *The Development and Function of Roots*. Academic Press, London.

TRAN THANH VAN, K. M. (1981). Control of morphogenesis in *in vitro* culture. *Annual Review of Plant Physiology*, 32, 291–311.

WAREING, P. F. and PHILLIPS, I. D. J. (1981). *Growth and Differentiation in Plants*, 3rd edition, Pergamon Press, Oxford.

WOOLHOUSE, H. W. (ed.) (1967). Aspects of the Biology of Ageing. *Symposium of the Society for Experimental Biology*, 21, Cambridge University Press, London.

## SELECTED REFERENCES

DHINDSA, R. S., PLUMB-DINSA, P. and THORPE, T. A. (1981). Leaf senescence: correlated with increased levels of membrane permeability and lipid peroxidation, and decreased levels of superoxide dismutase and catalase. *Journal of Experimental Botany*, 32, 93–101.

LIEBERMAN, S. J., VALDOVINOS, J. G. and JENSEN, T. E. (1982). Ultrastructural localization of cellulase in abscission cells of tobacco flower pedicels. *Botanical Gazette*, 143, 32–40.

SEXTON, R. and ROBERTS, J. A. (1982). Cell biology of abscission. *Annual Review of Plant Physiology*, 33, 133–62.

THOMAS, H. and STODDART, J. L. (1980). Leaf senescence. *Annual Review of Plant Physiology*, 31, 83–111.

TORREY, J. G. (1976). Root hormones and plant growth. *Annual Review of Plant Physiology*, 27, 435–59.

# 12

# Reproductive Development

Vegetative growth sooner or later leads to a transition to reproductive development signalled by the initiation of flower primordia. Plants will not flower, nor in many cases respond to the environmental stimuli which ensure subsequent flowering, until they have completed a part of their vegetative development – until they have reached a 'ripeness to flower'. The sharpness of this transition and the extent to which the reproductive phase occurs along with continuing vegetative development or is marked by a virtual cessation of further vegetative growth vary very widely between species. The shoot meristems usually undergo a major shift in morphology associated with the switch from initiating leaves and vegetative axillary buds to the formation of flowers and their subtending bracts. This morphological change is the basis for early detection of flower induction.

Reproductive development will here be interpreted to encompass not only flower initiation but flower development, the formation of the haploid sex cells, fertilization and the growth and ripening of the fruit including development of the embryo and seed structures from the fertilized ovule. We are therefore here concerned to survey the physiological aspects of all the changes which intervene from the initiation of reproductive development to the formation of viable seed capable by its germination of giving rise to the next generation of plants.

## Flowering

Our knowledge of flowering derives from studies with a selected number of species from amongst those cases where appropriate conditions of temperature and/or daylength are essential to flowering or can greatly hasten or delay the onset and abundance of this process. It should however be emphasized that many species are not exacting and will flower under almost any conditions compatible with continuing growth. Any general theory must therefore be able to explain flowering in plants differing profoundly in the sensitivity of this process to environmental stimuli.

## Vernalization

Some plants require a cold treatment *(vernalization)* before they will flower and in certain of these cases it has been shown that it is the stem apex which is the sensitive region and that this has to reach the required maturity before the cold treatment is effective (biennial *Hyoscyamus niger*, and *Beta vulgaris*). In photoperiodic plants (those whose flowering depends upon exposure to an appropriate daylength) it is the leaves which must reach the appropriate maturity; photoperiodic induction is effective only after a particular number of nodes have been formed. Winter cereals have a cold requirement and a long-day requirement; the cold requirement can be achieved by the embryo of the immature seed or in the moistened mature seed but the daylength requirement can be met only when the young plant has grown to the appropriate stature.

Vernalization may be essential to flowering or merely hasten the onset of flowering. Winter rye has no absolute cold requirement but when vernalized the winter varieties flower as rapidly as the spring varieties. By contrast the biennial variety of *Hyoscyamus niger* (henbane) has an absolute cold requirement and if overwintered at too high a temperature will remain vegetative indefinitely. Both the above species are long-day plants (see discussion of photoperiodism below) but cold requirements for flowering are also found in a short-day variety of chrysanthemum and in a number of day-neutral species (various spp. of *Geum, Lychnis* and *Erysimum*, and in *Senecio jacobaea, Pyrethrum cinerariaefolium, Saxifraga rotundifolia*).

The temperature most effective in vernalization is at or near to 6°C, and the duration of treatment for maximum acceleration of flowering can range according to species from 4 days to 3 months. Vernalization can be reversed by an immediately following period of high temperature. This suggests that there is a neutral temperature, below which vernalization occurs and above which vernalization tends to be reversed. It has been shown that in *Hyoscyamus* this neutral temperature is at about 20°C.

The German botanist G. Melchers first demonstrated that vernalization could be transmitted from a vernalized to an unvernalized *Hyoscyamus* plant through a graft union. To explain this he postulated the formation during vernalization of an effector substance and named this **vernalin**. To explain the accumulation of such a substance at low temperature it has been postulated that some destroying system is preferentially reduced in activity at low temperature compared with a system controlling the synthesis or partial synthesis of vernalin. The reversibility of vernalization by high temperature applied immediately would then be by activation of the destroying system and stabilization of vernalization, unless high temperature immediately follows the cold treatment, would imply that the unstable substance accumulating at low temperature is subsequently converted to the stable vernalin. These relationships can be expressed diagrammatically thus:

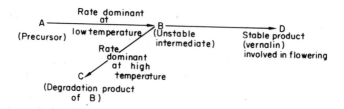

Application of gibberellins can substitute for the cold treatment for vernalization (Fig. 12.1). Further it has been shown that the vernalized plants have enhanced endogenous levels of substances with gibberellin activity as compared with unvernalized control plants of the same species. Although vernalin has not been identified, it is possible that it is a particular gibberellin or mixture of gibberellins. An objection of this view is that in some instances the pattern of growth and flowering following gibberellin treatment is not exactly that which follows cold treatment. Also the vernalized state is transmitted in the meristem from cell to cell through many cell generations; it is difficult to visualize how a certain hormone level could be maintained in such circumstances. (See also p. 265).

## Photoperiodism

Experimental study of the role of light in controlling flowering dates from the work by W. W. Garner and H. A. Allard first published in 1920. These workers noticed that although the Maryland Mammoth variety of tobacco, despite its vigorous vegetative growth, failed to flower and set seed during the summer, nevertheless root stocks transferred to the greenhouse gave rise readily during winter to small flowering plants. They also found that successive spring sowings of a variety of soybean (*Glycine max (soja)*) all came into flower at the same time and that flowering occurred quickly and in quite small plants grown in the greenhouse in winter. No significant promotive effect on the flowering of these plants could be traced to the moisture conditions, temperature and irradiance within the greenhouse. Garner and Allard therefore examined the effect of daylength by extending the natural day with artificial light and shortening the day by placing the plants in light-proof cabinets at the appropriate time. The outcome of these studies was a demonstration that these varieties of tobacco and soybean required a period of exposure to short days in order to flower and that after an appropriate period of short days (***photoperiodic induction***) they would flower irrespective of the subsequent daylength. Garner and Allard then examined further plants for their response to daylength and concluded that some plants had no special daylength requirements (day-neutral plants), others were short-day plants (SDP) and still others long-day plants (LDP). The phenomenon of flower induction by appropriate daylength treatment is therefore a further instance of photoperiodism (see bud dormancy, p. 219).

**Fig. 12.1** Carrot plants (var. Early French Forcing). *Left:* control; *centre:* maintained at 17°C but supplied 10 mg of gibberellin daily for 4 weeks; *right:* plant given vernalizing cold treatment (6 weeks). All photographed 8 weeks after completion of cold treatment. (From Lang, 1957, *Proceedings of the National Academy for Science, U.S.A.,* **43**, 709–17.)

The situation exposed by the pioneer studies of Garner and Allard has proved, on further study, to be much more complex. There are, for example, plants with an absolute requirement for short-day induction, but others whose flowering is only hastened by short days. Some plants require a combination of daylengths, e.g. short days followed by long days. There are complex interactions between temperature and daylength, e.g. an absolute daylength requirement at one temperature and no daylength requirement at another. These are examples of the many response types which will one day have to be fitted into a comprehensive theory of flowering. We shall however confine our attention mainly to a consideration of a plant with an absolute requirement for short-day induction like *Xanthium pennsylvanicum* (syn. *X. strumarium*) and an obligate long-day plant such as the annual strain of *Hyoscyamus niger*.

When grown under normal conditions all plants have a light requirement for flowering since they need photosynthetic products to grow and develop. However plants with no daylength requirement can be brought to flowering in darkness if supplied with sugar. The need for a 'high irradiance' light reaction for flowering is therefore a photosynthetic requirement not specifically related to flower induction. With this background it is possible now to discuss an important discovery regarding photoperiodism, the discovery that it is the length of the *dark* period that is normally determinative in both short- and long-day plants. A SDP is induced by a dark period or succession of dark periods *exceeding* a critical length (in *Xanthium* the dark period must be 8.5 h or longer), a LDP flowers provided the dark periods are *less than* a critical length (in *Hyoscyamus* the dark period must not exceed 13 h). Recognition of the importance of the dark period followed from the demonstration that a SDP like *Xanthium* placed under inductive conditions will flower if the light period is interrupted by a short dark period but will not flower if the dark period is interrupted by a short light period. Similarly a LDP will flower in short days provided the dark period is appropriately interrupted.

Photoperiodic induction occurs in the leaves and the flowering stimulus moves out from the leaves to the meristems where flowers are to be initiated. Not only can a partially defoliated *Xanthium* plant be caused to flower by appropriate induction of the remaining single leaf but a single induced leaf can cause flowering when the other leaves on the same plant are maintained under non-inductive conditions. These observations together with the failure of applied metabolites (such as sugars and amino acids) to cause flower initiation indicate that it is not the general synthetic output of the leaves which is involved but their production under inductive conditions of a hormonal stimulus to flower initiation (a flowering hormone or *florigen*). The concept of a flowering hormone, probably identical in all photoperiodic plants, is supported by the observation that the flowering stimulus can be transmitted from an induced to a non-induced plant of the same species across a graft union and by several demonstrations that this can also be achieved between different species with different photoperiodic requirements, e.g. the induction of flowering in a SDP (non-induced) by transmission from a LDP (induced), and between day-neutral and day-sensitive plants.

Contemporary with their study of the light stimulation of lettuce seed germination, the Beltsville group (see Photomorphogenesis, Chapter 11, p. 226) also examined the action spectrum of the suppression of flowering in *Xanthium* by a short break (5 min) of low irradiance light given during the effective dark period of not less than 8.5 hours. These studies, first published in 1954, showed that red light (peak activity between 640–680 nm) was most effective (Fig. 12.2) and that the effect of red light could be nullified by a succeeding exposure to far-red (710–740 nm). This implicated phytochrome in photoperiodic induction.

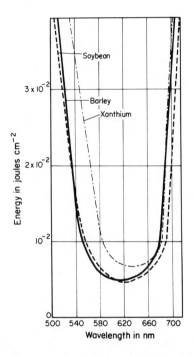

**Fig. 12.2** Action spectrum for light breaks in otherwise inductive or non-inductive dark periods. The light breaks in the inductive dark periods of the short-day plants (*Xanthium* and Biloxi soybean) suppress flower initiation; the light breaks in the non-inductive dark periods of the long-day barley plants promote spike formation and stem elongation (after Borthwick, Hendricks and Parker, 1948, *Botanical Gazette*, **110**, 103–18). The wavelength region where the energy requirement is lowest indicates the most effective wavelength for respectively the suppression or promotion of flowering.

The discovery of phytochrome raised the whole question of the processes which proceed in the critical dark period. These are not understood although there is experimental evidence strongly suggesting that a sequence of events is involved and that one of these is a timing reaction (Fig. 12.3). If red light suppresses flowering in SDP it must be assumed that phytochrome

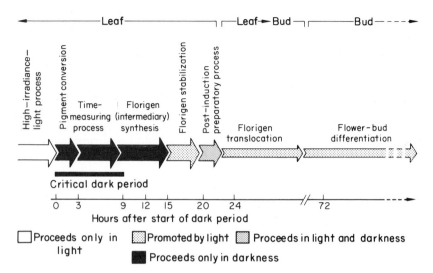

**Fig. 12.3** Partial processes involved in flower induction in *Xanthium*. Scheme advanced by Lang in *Encyclopedia of Plant Physiology*, ed. Ruhland, 1965, **15**(1), 1380–1536. Springer-Verlag, Berlin.

in the dark period is normally in the $P_r$ form and when all the phytochrome is in this form the dose of red light required for maximum inhibition should reach a fixed value (before that less red light would be required to reverse the existing $P_r$ to $P_{fr}$). Experimental studies based on this reasoning show that early in the critical dark period much of the phytochrome is in the $P_r$ form (in less than the first 4 h of the critical 8.5 h in *Xanthium*). Further points for consideration here are (*i*) that plants can measure the duration of the dark period accurately ($\pm$ 15 minutes in *Xanthium*), (*ii*) that irradiation with far-red (thereby completing rapidly the $P_{fr}$–$P_r$ change which occurs more slowly in darkness) early in the dark period does not in general reduce the length of the dark period required for induction and (*iii*) that cobalt ions, which do not delay $P_r$ formation, nevertheless increase the critical dark period. Thus the time-measuring process is not the conversion of $P_{fr}$ to $P_r$ nor apparently is this process triggered off by a critical level of $P_r$ relative to $P_{fr}$. The release of flowering stimulus from the leaves does not immediately follow the critical dark period and its release once initiated continues subsequently for many hours in *Xanthium,* and as discussed below for very long periods in some plants. Synthesis of the flowering stimulus therefore may be regarded as the terminal step in a multi-stage induction process. Although the biochemical nature of the processes involved is still very incompletely understood it is possible to express this multistage induction process by tentative schemes such as that shown in Fig. 12.3.

Although *Xanthium* is exceptional in that it can be induced to flower by a single critical night (often referred to as a single inductive cycle), it is found when using a population of plants that about 3 inductive cycles are needed to cause 100% of the plants to flower. Further inductive cycles increase the number of flowers formed and the speed with which flowering takes place. A similar situation appears where, as is usually the case, a number of inductive cycles are essential to flowering.

The change that takes place during photoperiodic induction can be very stable or decay very rapidly. Although in *Xanthium* export of the flowering stimulus from induced leaves does not persist, nevertheless, once the plant is induced, leafy shoots formed subsequently can transmit the flowering stimulus to non-induced plants by grafting and the induced plant will go on flowering for a long time in non-inductive conditions. In the flowering *Xanthium* plant the flowering hormone pervades the plant and is not diluted out; persistent production of hormone seems to be localized in the stem apices. A rather different situation is seen in the SDP, *Perilla ocymoides*. Here also there is persistence of the induced state but it remains localized in the leaves submitted to the inductive cycles. As these particular leaves age and die the plant becomes vegetative again. Biloxi soybean plants require continuous inductive conditions for flowering and formation of flowers ceases very rapidly when the plants are transferred to non-inductive conditions.

The discussion developed above has been in terms of the formation of a specific positive stimulus to flowering, of a floral hormone (florigen). The most telling evidence for this hypothesis is the transmission of the flowering stimulus through grafts (Fig. 12.4). If a dark period in excess of some critical value is needed to produce such a morphogenetic factor or its precursor in SDP, we might conclude that in LDP a dark period in excess of the critical value acts in the reverse sense and suppresses the formation of the same agent. Such a paradox suggested that the critical dark period may not be directly concerned in the synthesis of the flowering hormone. LDP will flower in continuous light, i.e. they have no need of a dark period to develop the flowering stimulus. Again there are a number of plants which are photoperiodic at one temperature and day-neutral at another temperature. Further, if the influence of dark periods in photoperiodic plants was to alter the level of other cell constituents which in turn controlled florigen synthesis then darkness could act in the same sense in both SDP or LDP. The dark reaction could tend to raise the level of such a factor – a night in excess of the critical length being needed to achieve the appropriate concentration in SDP whereas in LDP the factor could reach an inhibitory level if the critical period of darkness was exceeded. Alternatively the dark period could be one in which the level of some factor decreased, a dark period in excess of the critical length being needed in SDP to lower a previously inhibitory concentration.

Do other growth regulating substances exert an influence on flowering? First there is the observation that in *Chenopodium* inductive photoperiods lead to marked increases in the growth rate of the leaves and that photoperiodic induction of *Xanthium* is associated with a marked stimulation of cell division in the apical meristems. We have already referred to the possible identity of vernalin with gibberellins and there are many

**Fig. 12.4** Transmission of the flowering stimulus from a short-day plant (*Nicotiana tabacum* 'Maryland Mammoth') to a non-induced long-day plant (*Hyoscyamus niger,* annual variety). In each case the *Hyoscyamus* graft is on the left. A: entire graft on short days; short-day partner induced, long-day partner flowering (flowers on donor removed). B: short-day partner on long days, long-day partner on short days – no flower formation. (From Khudairi and Lang, via Lang, *Encyclopedia of Plant Physiology,* ed. Ruhland, 1965, **15**(1), 1380–1536. Springer-Verlag, Berlin.)

recorded instances of the promotion of flowering in LDP from application of gibberellins. Auxin applications are used commercially to induce flowering in the pineapple (*Ananas comosus*) and in the litchi tree (*Litchi chinensis*). In pineapple the auxin seems to act by inducing ethylene synthesis. More recently auxin-induced flowering in two LDP, winter barley and *Hyoscyamus niger,* has been reported. There are many instances of the inhibition of flowering from auxin application. The present weakness in the hypothesis that critical dark periods are concerned with regulating levels of auxins or gibberellins or possibly of other regulators to levels compatible with florigen synthesis is that significant changes in the endogenous levels of auxins, gibberellins or cytokinins following photoperiodic induction have not been demonstrated. However in view of the limited specificity of bioassays for such growth regulators and uncertainty about the quantitative nature of extraction techniques this evidence does not warrant rejection of the above hypothesis. Here again as in the other morphogenetic phenomena in plants we may have complex interactions between growth regulating substances leading to certain cells being programmed to synthesize a flowering stimulus.

### The timing reaction in photoperiodism

Reference was made earlier (Chapter 8, p. 151) to diurnal rhythms of growth and to their endogenous control by a 'biological clock'. Such rhythms are often referred to as *circadian rhythms* (from *circa* about, and *diem* day) because their period approximates to 24 h. Such rhythms are involved in leaf movements such as can be observed in the runner bean, in the opening and closing of flowers (e.g. in *Kalachoë blossfeldiana*), in root pressure (see Chapter 4) and in aspects of metabolism (such as the dark-fixation of $CO_2$ in succulents like *Bryophyllum fedschenkoi*). The hypothesis that such rhythms are implicated in photoperiodism was first advanced by Erwin Bünning in the early 1930s and defended by him in subsequent publications. He postulated the continuous alternation of two phases, a photophile (light-loving) phase and a scotophile (dark-loving) phase (also called photophobe – light-fearing) and that light received during the former was promotive of flowering. while in the latter light inhibited (darkness promoted) flowering. A precisely regulated circadian rhythm of such a kind involving quantitative and qualitative changes in responsiveness to light could form the basis of the timing mechanism in photoperiodism. Evidence in support of the involvement of the 'biological clock' in photoperiodism comes mainly from studies where SDP have been submitted to very long dark periods (thereby involving cycles in excess of 24 h). For instance when soybean plants received 8 h of light, followed by dark periods ranging from 8 to 62 h, flowering was obtained when the light+dark period was approximately 24 h or a *multiple* of 24 h. Plants remained vegetative when the 'artificial days' had intermediate values (e.g. 36 h). When plants of the SDP *Chenopodium rubrum* (inducible by a single photoperiod) were submitted to 72 h dark periods interrupted at various times by 2 min 'light-breaks' of red light, the light-breaks effectively inhibited flowering when given 6, 33 or 60 h from the beginning of the dark period, i.e. at the times when the plant, on an approximately 24-h inductive cycle, would have been in the dark. Light-breaks were quite ineffective at or near 18 and 46 h into the dark period (Fig. 12.5), when the plant would normally have been in the light. Such results are consistent with the operation of a rhythm with a period of 27–30 h divided into two phases of reversed light sensitivity. The elusiveness of florigen, now possibly linked in its synthesis to a 'biological clock' of unknown nature, emphasizes the incompleteness of our knowledge of the control of flowering.

### Flower development

The above discussion has concentrated upon the events which determine

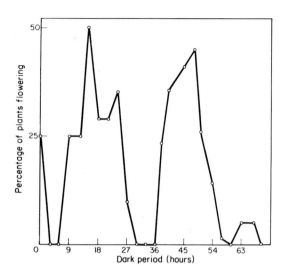

**Fig. 12.5**   The effect on the flowering of *Chenopodium rubrum* of red 'light-breaks' of 2 min duration when applied at random times ( o ) during a period of continuous darkness of 72 h. After the single dark period the plants were kept in continuous light and scored for flowering. (Modified from Cumming, Hendricks and Borthwick, 1965, *Canadian Journal of Botany*, **43**, 825–53.)

that flowering will occur in photoperiodic plants. It has not considered flowering in 'day-neutral' plants where no specific environmental control of flowering operates, nor factors affecting the profusion of flowering nor the processes involved in the transformation of a vegetative apex into a reproductive apex and the factors controlling the development of functional flowers. Heteroblastic development and the transition from the juvenile to the adult vegetative form (p. 221) are aspects of an endogenously controlled development (sometimes referred to as phasic development). In day-neutral plants transition to flowering is similarly endogenously controlled. It may nevertheless involve a similar sequence of changes (leading to synthesis and release of florigen) as those which operate in photoperiodic plants. This is supported by the existence of quantitative or facultative SDP and LDP which will flower ultimately under any daylength but flower more rapidly under appropriate photoperiods. Furthermore flowering has been induced in a long-day tobacco variety by grafting onto a flowering day-neutral variety.

The abundance of flowering can be controlled by nutritional factors and, in photoperiodic plants, by the number of successive photoperiodic cycles. Often the speed of flowering (interval between induction and flower appearance) and the number of flowers formed increase as the number of consecutive inductive photoperiods increases. However the photoperiodic conditions essential for flower induction may not be optimal for flower development. A high nitrogen status in some cases promotes vigorous vegetative growth but reduces the profusion of flowering and fruiting (this is the case in some fruit trees); in other cases (e.g. *Xanthium*) high nitrogen promotes vigorous vegetative growth under non-flowering conditions and increases the abundance of flowers under inductive conditions. A number of synthetic growth-retarding substances are now widely used in ornamental horticulture to produce 'dwarf' plants; some of these, applied to particular species, also speed up flowering and increase flower number. Some of these retardants are considered to act by inhibiting the synthesis of endogenous gibberellins (high levels of gibberellins are inhibitory to flowering in some species) and are therefore termed anti-gibberellins (e.g. AMO 1618, CCC, phosphon D). A number of these observations point to an antagonism between vegetative growth and reproductive development; examples of reproductive development as the trigger for senescence of vegetative structures have already been considered (see p. 221).

Most studies on flower development have been in terms of morphology and developmental anatomy (for an excellent account of such work the reader is referred to *Plant Anatomy. Part 2. Organs* by Elizabeth G. Cutter, Edward Arnold, London, 1971). The science-based bulb industry of the Netherlands has however made very detailed studies of the influence of conditions of storage, particularly temperature, upon flower development. The development of flowers in *Tulipa gesneriana* involves three main stages, each with a different temperature requirement: (*i*) differentiation of the embryonic flower bud, a stage involving active cell division and requiring high temperature (20°C); (*ii*) an apparent rest which needs about 13 weeks at an optimum temperature of 8–9°C; (*iii*) a period of rapid growth of the floral organs and elongation of the flower stalk which takes about 9 weeks at an optimum of 13–15°C.

The role of growth-regulating substances in the growth of floral organs is very incompletely understood. The apparently controlling role of auxin in the growth and orientation of the flower stalk of *Fritillaria meleagris* has however been demonstrated (Fig. 12.6). The level of diffusible auxin in the stalk closely follows its pattern of growth. By careful excision experiments evidence was obtained that the source of this auxin moving into the flower stalk was the ovules. The drooping stage (b in Fig. 12.6), which corresponds with the first auxin peak, occurred at a time when there was intense cell division in the ovules and formation of the embryo sac. The subsequent fall in 'auxin activity' was shown in part to be due to the release of an inhibitor from the young anthers, the output of this inhibitor falling sharply as the stage of anthesis approached. Other examples of growth regulators from one floral organ affecting the growth of another part of the flower are known. Thus removal of young stamens from the flower of *Glechoma hederacea* causes a marked reduction in growth of the corolla. The growth can be restored by gibberellin application but not by auxin.

Some evidence has been obtained of the role of growth regulators in sex expression in species which produce unisexual flowers. In a number of species auxin application has promoted development of female as against male flowers, and gibberellic acid of male as against female flowers. An

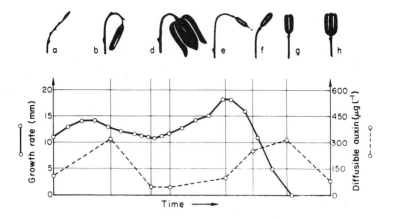

**Fig. 12.6** Stages (a–h) in the development of the flower of *Fritillaria meleagris* showing the reversal of the gravitropic response of the flower stalk during the development of the flower and fruit. Graphs: top curve, growth rate of the flower stalk in mm per day; lower curve, amount of auxin diffusing from the flower and through the flower stalk. (From Kaldewey, 1957, *Planta*, **49**, 300–44.)

interesting series of experiments with excised cultured flower primordia of cucumber (*Cucumis sativus*) confirmed this 'triggering' of development towards female or male flowers in monoecious strains by IAA and gibberellic acid. However with primordia excised from hermaphrodite plants (producing bisexual flowers) these hormones failed to produce unisexual flowers.

## Fertilization

The physiological processes involved in the production of the male (germinated pollen grains) and female (8-nucleate embryo sac) gametophytes and within these of the male nuclei and egg cell are very incompletely understood. Central to this aspect of reproductive development is the induction of meiosis in the pollen (microspore) mother cells and in the megaspore mother cell. Attempts to study the development of the microspores by culturing excised anthers have emphasized that further development can be achieved in culture only after meiosis has been initiated in the spore mother cells and that successful completion of meiosis is normally achieved only when the spore mother cells at the time of anther excision are already advanced in meiosis to the pachytene stage. Work with anthers of onion excised at or near this stage of meiosis has demonstrated the beneficial effect on tetrad formation and the later stage of uninucleate microspore development of supplementation of the sugar–mineral salt medium with cytokinin, gibberellin and RNA or RNA nucleotides. These studies however shed no light on what factors programme cells to embark upon meiosis – the critical step in the alternation of generations and in genetic segregation.

Studies on pollen grain development in attached anthers have shown that the spore mother cells lose most of their cytoplasmic RNA and protein at the time of microspore formation. This may be an essential preparation for the change in cell metabolism associated with the switch to the gametophytic pathway which leads to the formation of the mature pollen grains. The production of such grains involves new synthesis of RNA and protein (the precursors for which are probably transported to the newly formed tetrad from the tapetal layer of the anther wall), temporary isolation of the tetrad of spores by enclosure within an impermeable polysaccharide (callose) wall, and synthesis of the several-layered pollen grain wall. The latter includes the ektexine layer which becomes impregnated with the highly resistant sporopollenin, probably derived from material released by the breakdown of the callose wall of the tetrad and the walls of the tapetal cells. Finally the microspore undergoes an asymmetric mitosis to give rise to a smaller generative cell with a condensed usually lens-shaped nucleus (from which by a further mitosis arise the two male nuclei) and a vegetative cell with a larger and more diffusely staining spherical nucleus (tube nucleus).

If the anthers of a number of Solanaceous plants (*Nicotiana* spp. including tobacco, *N. tabacum, Datura* spp., and *Atropa belladonna*) are excised immediately prior to or at the time of this pollen grain mitosis and cultured in a simple sugar–mineral salt medium a large proportion of the microspores become non-viable, a proportion mature into pollen grains and a proportion are directed from the gametophytic pathway and embark upon a continuing sequence of cell divisions to give rise to haploid plantlets (pollen embryoids) (Fig. 12.7). The proportion of microspores which embark upon this sporophytic pathway can be increased by submitting the anthers to a low temperature shock (4°C) for 48 hours before transferring them to the culture medium. This phenomenon has now been induced in anthers of a wider range of species, and is of considerable practical importance as a source of haploid plants which spontaneously or following colchicine treatment can give rise to homozygous fertile diploid lines. It shows that sporophyte versus gametophyte development is not controlled by ploidy level and that isolated cells can spontaneously develop the necessary polarity required for the early emergence in embryology of a shoot and a root pole. To achieve this switch to sporophyte development it is necessary to remove, by anther excision, an influence arising from attachment of the stamen to the parent sporophyte and to do this while the microspores are still uninucleate or have only just embarked upon the microspore mitosis – before they are fully committed to the gametophytic pathway. A similar increasing determination to a fixed pathway of development was noted in relation to the culture of excised leaf primordia (see p. 215).

The act of fertilization requires the pollen grains released from the mature anther to reach the stigma and to germinate there to give rise to a pollen tube which will grow into the stigma, down the style into the ovary and usually ultimately to the micropyle of the ovule. Pollen grains can however in most cases be germinated *in vitro*. The pollen of some species will show some

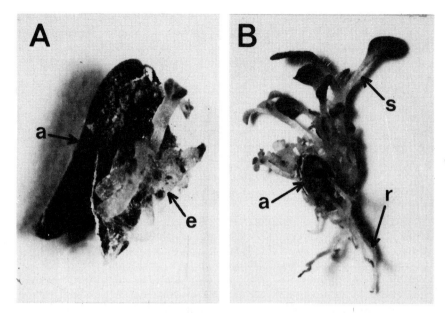

**Fig. 12.7** Excised anther cultures of *Nicotiana tabacum* showing embryogenesis from microspores. A: stage of dehiscence of the cultured anthers showing embryoids (e) within the anthers (a). B: seedlings developing from the embryoids, s, Shoot with leaves; r, roots. (Photographs by M. Horner, Botanical Laboratories, University of Leicester.)

germination in water but the percentage germination is low, pollen tube growth slow and restricted, and pollen tube bursting common. The percentage germination and the extent of pollen tube growth can usually be increased by using a sugar solution (sucrose often being the most effective sugar). The sugar acts both as a nutrient and to make the osmotic potential of the solution more favourable. Further improvement in pollen tube growth can often be achieved by including, in the culture solution, calcium and other cations and boric acid (within the concentration range 0.001–0.01%). In a culture solution supplying these constituents at appropriate concentrations percentage germination and pollen tube growth similar in rate and extent respectively to those occurring during normal fertilization can be achieved with some species. Further there is evidence that all these factors are supplied naturally to the pollen tubes by the stigma and style tissues. In other cases additional factors are required in culture (gibberellin, particular amino acids) and in still other cases it is not possible to achieve fully effective germination and pollen tube growth *in vitro*. Pollen tubes growing on the stigma and within the style secrete enzymes which soften the cutin of the stigma and the middle lamella of the cell walls (thereby facilitating pollen tube penetration) and also produce auxin (or a factor promoting auxin synthesis) involved in the initiation of fruit development. The initial penetration of the pollen tube into the stigma involves its positive hydrotropic and negative aerotropic responses. In the final stages of its growth the pollen tube shows a positive chemotropism to a substance released at the micropyle of the ovule. Effective pollination requires the pollen to land on a compatible stigma. Successful pollination normally occurs only with pollen of the same species and where self-incompatibility occurs only with pollen from a different plant of the same species. Incompatibility occurs because of failure of the pollen tube to penetrate the stigma (e.g. failure to receive an appropriate stimulus to synthesize and release the necessary cutinase) or owing to inhibition of pollen tube growth within the stigma and style as a result of a genetically determined interaction between a pollen tube and a stigma-style product. The stigma-secreted S-factors are probably glycoproteins, which react with cell wall components of incompatible pollen tubes and thereby inhibit their growth.

The generative nucleus in the pollen grain undergoes mitosis to form the two male nuclei; this may occur before the pollen grains are shed or during pollen tube growth. The pollen tube entering the ovule via the micropyle comes into contact with the embryo sac at the site of the synergids. Often one of the synergids breaks down and the tip of the pollen tube enters this degenerating cell; the male nuclei are released through a sub-apical pore in the pollen tube. The two male nuclei migrate into the embryo sac (it is not clear how this movement is effected) and one fuses with the egg nucleus to give the diploid zygote from which the embryo develops and the other fuses with the diploid nucleus of the central cell to give the triploid endosperm nucleus.

## Fruit development

The development of fruit is normally a consequence of pollination; unpollinated flowers fall, pollinated flowers show fruit set. Germinating pollen is a rich source of auxin and there is evidence that pollen tube growth not only supplies auxin but that pollination may activate auxin synthesis by the tissues of the gynaecium, particularly by stylar and/or locular tissue. Recognition the importance of this auxin for fruit set followed from the demonstration by Gustafson in 1936 that in some species application of auxin to unpollinated flowers led to the development of seedless (partheno-carpic) fruit. However about 80% of horticultural species cannot be set by auxin application. Applications of gibberellins can similarly induce parthenocarpic fruit development in some species, some of which will not set fruit by auxin application. Singly or in combination, applications of these growth regulators however succeed only with a very few species, and these are plants in which some degree of natural parthenocarpic fruit development occurs. Where natural parthenocarpy occurs or is induced by growth regulator application it seems that for a time ovule development including growth of the seed coat and nucellus proceeds and that the seedless nature of the mature fruit follows from abortion of the pseudo-seeds. In most species it is the double fertilization which leads to active growth of the nucellus and seed coat and to an associated rapid increase in fruit growth. The importance of seed development to fruit growth is well illustrated by studies in the strawberry where fruit size and form are determined by the number of fertile achenes (Figs 12.8 and 12.9). In this case the promotion of receptacle growth

**Fig. 12.8** Effects of developing achenes on the growth of the strawberry receptacle. 1: Unpollinated flower: no development of receptacle. 2: One pollinated achene: growth of the receptacle around it. 3: Several pollinated achenes: several areas of receptacle growth. 4: Many pollinated achenes. (From Nitsch, 1965, *Encyclopedia of Plant Physiology,* ed. Ruhland, **15**(1), 1537–1697. Springer-Verlag, Berlin.)

can be achieved, following removal of all achenes, by an appropriate application of the synthetic auxin, 2-naphthoxyacetic acid (2-NOA) (Fig. 12.10), and it can be shown that the developing achenes are a rich source of diffusible auxin. Many other developing seeds have been shown to be centres of auxin synthesis. Young seeds are also rich sources of natural gibberellins and the first successful isolation (in 1958) of a gibberellin from a higher plant was the isolation of gibberellin $A_8$ from *immature* seeds of *Phaseolus multiflorus*. Endosperm in the free nuclear stage (the outstanding example being coconut milk) and immature seeds and fruitlets are rich sources of cytokinin activity; the first natural cytokinin to be isolated (1964) was from immature maize kernels and was hence termed zeatin (Fig. 9.3, p. 160).

Fruit development following pollination involves growth of the seeds to

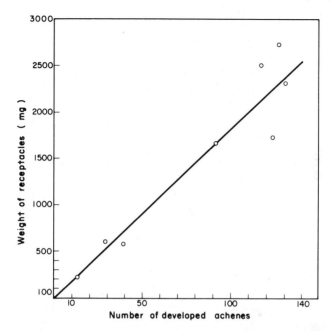

**Fig. 12.9** Proportionality between number of developed achenes and weight of the receptacle in strawberry. (From Nitsch, 1950, *American Journal of Botany,* **37**, 211–15.)

maturity and enlargement of the ovary or receptacle. Cell division, cell expansion and accumulation of food reserves are involved in seed development (see Chapter 8). Growth of the fruit may, subsequent to pollination, involve only cell expansion but in many cases there is also a short phase of cell division. The development of fleshy fruits involves considerable accumulation of organic metabolites (organic acids, sugars) into the succulent pericarp and associated tissues. The importance of seeds to fruit development suggests a controlling influence of auxins, gibberellins and cytokinins liberated by the seeds. These growth regulators thus make the developing fruit a 'sink' for food materials synthesized in the leaves and fruit development is normally associated with a marked check to vegetative growth and in annuals with senescence of the whole plant. Removal of fruits immediately halts leaf senescence. In *Phaseolus vulgaris* a similar diversion of $^{32}P$ and of the carbon of $^{14}CO_2$ from the leaves to the peduncles can be achieved either by leaving the pods attached or applying a mixture of IAA, GA and kinetin in lanoline paste to the decapitated peduncles.

The concept that the growth-regulating substances essential to fruit development are synthesized in the fertilized ovules is supported by experiments involving the culture of excised flowers. Fruit development to maturity has been demonstrated in a number of species using a simple medium containing only inorganic salts and sucrose always provided that the flowers had been pollinated two or more days before they were separated from the mother plant (Fig. 12.11). Studies with unpollinated tomato

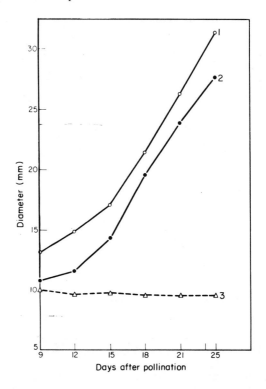

**Fig. 12.10** Growth curves of three strawberries (variety Marshall) which were pollinated on the same day. 1: Control. 2: Fruit which had all its achenes removed on the ninth day and replaced with a lanoline paste containing 100 mg $1^{-1}$ of the synthetic auxin, 2-naphthoxyacetic acid. 3: Fruit which had all its achenes removed on the ninth day and replaced with plain lanoline. (From Nitsch, 1950, *American Journal of Botany,* **37**, 211–15.)

flowers showed that parthenocarpic fruit development could be obtained by incorporating 2-NOA into the culture medium. Addition into the medium of tomato juice (which contains a cytokinin) gave larger fruits. Similar cultures of pollinated flowers of a number of species have shown the importance of the sepals in nitrate assimilation, and indicated that the sepals (and sometimes also the petals) synthesize auxin, gibberellin and cytokinin and export these hormones to the developing fruit.

## Embryo development

The fertilized ovum develops into the embryo plant within the seed and in the immediate vicinity of the endosperm. To begin with the zygote is to begin at the beginning of morphogenesis. The first division of the zygote leads to a predictable cleavage; the plane of division and the extent to which the two daughter cells are unequal are the first features of an embryology

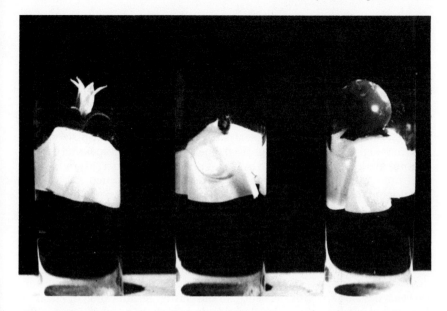

**Fig. 12.11**　Culture of a tomato fruit from a pollinated excised flower. Flower trimmed down to ovary (middle picture). *Right:* small fruit developed in culture (photograph supplied by Dr. J. P. Nitsch).

characteristic of the species. Thus for instance in some pteridophytes the daughter cell adjacent to the neck of the archegonium is the first cell of the embryonic shoot, whereas in others the shoot axis is directed away from the neck. Polarity is early recognizable in the zygote of flowering plants often due to the location of the large vacuole towards the micropylar end. The first division of the zygote is transverse and often unequal to give a smaller terminal cell rich in mitochondria and plastids and a large vacuolated basal cell (at the micropylar pole). The basal cell by further divisions gives rise to the suspensor, the terminal cell to the pro-embryo proper. The globular stages of the pro-embryo include quadrant, octant, 16- and 32-celled stages. In dicotyledonous embryos the transition from the globular to the heart-shaped stage (resulting from the formation of the primordia of the cotyledons) clearly identifies the axis of the embryo with the incipient plumular meristem directed away from the micropyle and a provascular strand directed to the root pole where the root meristem and root cap will be organized (this often involving cells arising from the adjacent cell (hypophysis) of the suspensor).

In certain instances it has been possible to show gradients in the zygote cytoplasm with respect to the density of membranes of the endoplasmic reticulum, ribosomes, mitochondria and other cytoplasmic structures. However the protoplasmic pattern can be disturbed by centrifuging and yet the normal polarity can persist or be rapidly re-established. It has therefore been suggested that the polarity lies in the plasmalemma or in an outer immobile (gelled) cytoplasmic layer which is not displaced during centrifugation. Thus aspects of the heterogeneous distribution associated with polarity can be described (albeit imperfectly) but the determining forces in polarity are as yet quite unknown.

Zygotes in which the influence of environmental factors on polarity has been studied are the fertilized ova of brown algae, and particularly of *Fucus* spp. The fertilized ovum, originally spherical, becomes pear-shaped as a rhizoid begins to grow out from the basal pole and a wall is formed cutting off the first rhizoidal cell from an apical cell. Further divisions take place in both the rhizoidal and apical cell to give a filament with a distinct apex and base. If zygotes are kept in the dark and as free as possible from external gradients of environmental factors polarity is nevertheless manifested by rhizoid development . If the zygotes are submitted to various environmental gradients these can however determine the axis of polarity. Unilateral illumination results in rhizoid development on the shaded side, an electrical field in its development towards the positive pole, a pH gradient (6.0–8.0) in its development on the acid side, an auxin gradient in its development towards the higher concentration and a gentle temperature gradient in its development towards the warmer side. If zygotes are allowed to cluster, rhizoids develop towards the centre of the cluster. Stratification of *Fucus* or *Sargassum* ova by centrifuging does not determine the point of origin of the rhizoidal outgrowth. These ova appear to have an incipient polarity either endogenously initiated or developed during oogenesis which is labile and can in consequence be orientated by certain environmental gradients. Such observations however do not indicate what physico-chemical processes are involved in the initiation and fixation of the polarity.

The maintenance of polarity in an embryo appears to depend upon the association of its cells and it seems that both physical forces developing as the embryo enlarges and polarization of its biochemical activities together determine the growth and associated segmentation pattern. Physical forces, such as those of surface tension, will tend towards maximum reductions in the extent of the surfaces created by cell divisions. Cells will thus tend to divide by walls of minimal area, so that the new walls are perpendicular to the previous direction of growth and at right angles to those already present. Modifying the basic pattern of segmentation will be the chemical gradients and physical restraints imposed upon the embryo as it develops within the ovule.

Significant progress has recently been made in our understanding of the factors which induce cells to embark upon embryogenesis and of the nutritive requirements of such cells. This progress arises from studies on natural polyembryony and on the formation of 'adventive embryos' or embryoids from cultured callus tissues and cell suspension cultures.

Polyembryony (presence of more than one embryo within each seed) can arise in several ways. For instance it may involve the young embryo splitting into two and each fragment then generating an embryo. This phenomenon shows the plasticity of the young embryo; it also shows that the earlier concept of the role of the daughter cells of the zygote being strictly determined (prescribed as to their contribution to the separate regions of the embryo) right from the first few divisions of the zygote is untenable. Polyembryony can also arise by the development of embryos from cells of the nucellus of the ovule. This shows that the act of fertilization is not essential to embryogenesis, and that cells of the sporophyte can be embryogenic. Further the segmentations observed during the early development of such nucellar embryos rarely correspond closely with those during embryo development from the zygote. This again shows that a strict (and genetically controlled) pattern in the early divisions is not essential for the formation of a perfect embryo (an embryo which achieves the characteristic morphology of the species as it matures). One of the species showing natural polyembryony from cells of the nucellus is *Citrus macrocarpa*. A callus developed in culture from the nucellus of this species or from *Citrus reticulata* will give rise on its upper surface to numerous plantlets and it is these that have been named 'adventive embryos' or embryoids. Similar embryoids arise profusely from the surface of a callus culture derived from young flower buds of *Ranunculus sceleratus* (Fig. 12.12) and seedlings derived from these embryoids spontaneously initiate new embryos all along

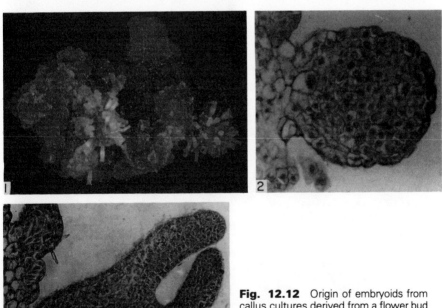

**Fig. 12.12** Origin of embryoids from callus cultures derived from a flower bud of *Ranunculus sceleratus*. 1: Embryoids growing out of the surface of the callus culture. 2: Late globular embryoid on the surface of the callus. 3: Mature embryoid. (Photographs supplied by R. N. Konar and K. Nataraja.)

their stem surfaces starting at the base of the hypocotyl. These new embryos arise from single epidermal cells of the stem and pass through the 4-celled, 8-celled, globular, cordate and torpedo stages typical of the embryology of the species (Fig. 12.13). Callus or cell suspension cultures derived from carrot embryos and embryos of *Cichorium endivia* also show a prolific spontaneous capacity to give rise to embryos apparently by segmentation in single cells.

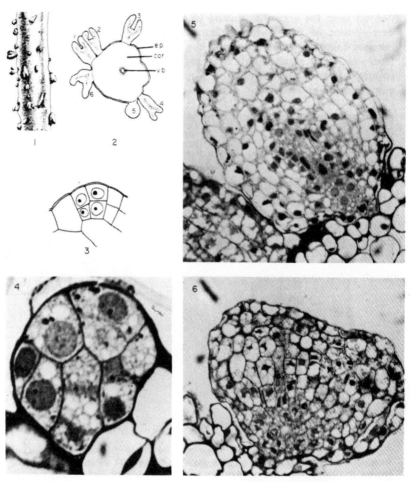

**Fig. 12.13** Embryo development from the epidermis of the hypocotyl of embryoid-derived seedlings of *Ranunculus sceleratus*. 1: Portion of the hypocotyl bearing accessory embryos. 2: Transverse section of the hypocotyl showing six developing embryos; ep, epidermis; cor, cortex; vb, vascular bundle. 3: 4-celled stage of epidermal embryo. 4: More advanced stage. 5: Full globular stage. 6: Heart-shaped stage. (1, 2, 3 from Konar and Nataraja, 1965, *Phytomorphology*, **15**, 132–7; 4, 5, 6 from Konar, Thomas and Street, 1972, *Journal of Cell Science*, **11**, 77–93.)

Perhaps even more striking than the cases quoted above are the instances when embryogenesis has been induced in callus or cell cultures derived from the organs of mature plants. The classical instance is the development of embryo-like structures from cell suspension cultures initiated from the young phloem of the storage root and from the living tissues of other vegetative parts of the carrot plant (Fig. 12.14). Suspension cultures of carrot can be grown in an undifferentiated state in a defined culture medium containing auxin. Such suspensions consist of small aggregates of highly cytoplasmic dividing cells and larger vacuolated free-floating cells released as the enlarging aggregates fragment in the moving culture solution (Fig. 12.14, 3ii, **a**). When such suspensions are transferred to a medium from which the auxin has been omitted many of the superficial cells of the cell aggregates give rise to embryoids which remain attached to the aggregates up to the late globular stage (Fig. 12.14, 4ii, **b**.) These embryoids then break away and continue their development as free-floating structures until they show typical seedling morphology (Fig. 12.14, 5, 6). At this stage they can be transferred to filter paper bridges (Fig. 12.14, 7) and grown to the stage where they can be planted out to give rise to normal carrot plants. The superficial cells of the aggregates which give rise to the embryoids undergo regular segmentations forming the pro-embryo and suspensor and produce normal globular embryos of single cell origin (Fig. 12.14 **c, d, e**). The segmentations however do not correspond to those described as typical of zygotic embryology in carrot. Similar embryogenesis has now been described in suspension culture of *Atropa belladonna* and embryogenesis from superficial cells of callus cultures has been reported for a considerable number of different species. These studies fully justify the statement made in 1961 by F. C. Steward and H. Y. Mohan Ram that 'the capacity to produce the plant body does not, however, reside in the zygote alone— indeed in the light of recent work it may well persist, even though suppressed, in almost any living cell of the plant body ... cells which have passed through many cell generations in culture, may still retain a degree of totipotency which is comparable with that of the zygote'. *Totipotency* in the sense used in this quotation is the potentiality to embark upon any of the differentiation sequences observed in tissue development or to recapitulate the sequence of development which separates the zygote from the mature embryo of the species. The evidence that, under the appropriate conditions, somatic (body tissue) cells can initiate embryos also indicates that plant cells can undergo differentiation without loss of the initial genetic potentialities of the zygote from which they were ultimately derived (see also p. 177).

How far can we define the conditions which permit somatic cells to function like the zygote? These certainly include release of the cells from the limitations (physical and biochemical) imposed upon them by the plant body ('cells are as they are because of where they are'). This release is effected by their *in vitro* culture. The conditions of culture also provide the nutritional factors and growth factor stimuli to induce rapid division in previously non-dividing cells. Further this meristematic activity is expressed in the absence of a morphogenetic field imposing a pattern of tissue differentiation

256

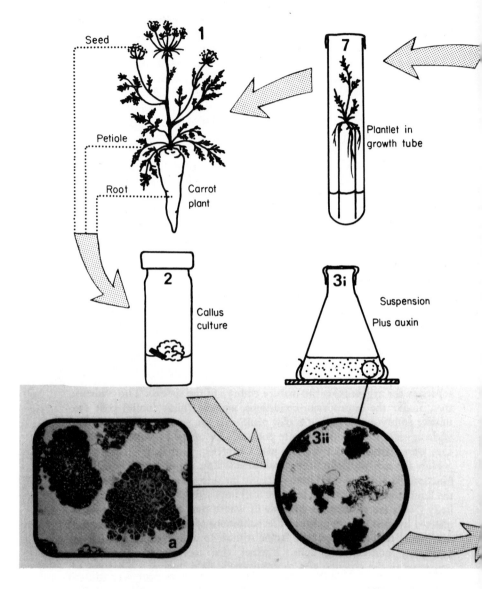

**Fig. 12.14** Propagation of carrot via somatic embryogenesis induced in suspension cultures. Sequence of steps involved indicated by broad arrows. 2: Callus culture derived from petiole and growing on medium solidified with agar. 3i and 4i show flasks containing the suspension cultures mounted on a platform shaker which imparts a rotary motion to the cultures. 3ii: Low power microscope view of the suspension cultured in presence of auxin showing the small-celled aggregates which will give rise to embryoids; **a**, section of such aggregates at higher magnification. 4ii: Low power microscope view of suspension after culture in medium lacking auxin shows the embryoids developing attached to the cell

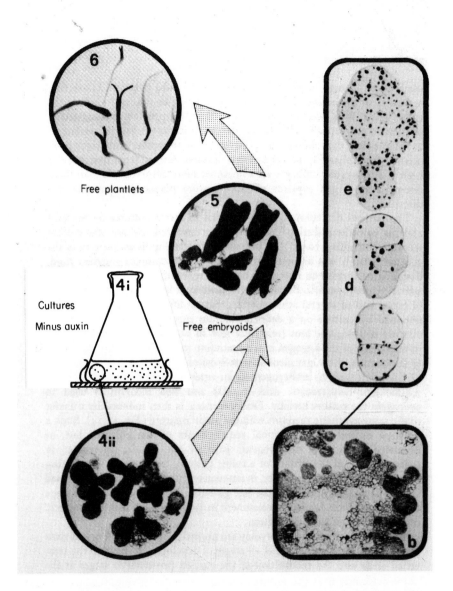

aggregates; **b**, section of such an aggregate with well developed globular embryoids; **c, d, e,** a sequence showing stages in the development of a globular embryoid at higher magnification. 7: Plantlet growing on a filter paper bridge which conducts culture solution to the root system. The plantlet developed in such a tube can be transplanted to a medium such as vermiculite. All stages conducted under aseptic conditions from sterlization of organ explant from the carrot plant to transplantation of plantlet from growth tube to vermiculite. For full description see text. (Figure prepared from research material by Dr. Lyndsey A. Withers, Botanical Laboratories, University of Leicester.)

such as operates upon dividing cells within a plant organ. These two conditions alone, however, do not necessarily lead to expression of embryogenic potential. Many actively growing callus and cell suspension cultures cannot at present be induced to generate embryoids. We do not know why this is but a clue to this is probably contained in the observation that in species where embryogenesis can occur in culture whether or not this actually occurs can depend upon the nutritive and environmental conditions during the stage of the original callus induction from the organ explant and even on the nature and age of the organ explant used. Some change in physiology (gene expression) over and above that necessary to render the cell actively dividing may be needed to induce the embryogenic state (see also Chapter 11 where this problem is discussed in relation to the capacity of cultured tissues to initiate organ primordia, p. 214). The facts that embryoids readily give rise to further embryoids from their surface cells (Fig. 12.13) and that cultures derived from immature or even mature zygotic embryos are sometimes more embryogenic than those derived from organ explants of more mature plants also point in this direction.

The pioneer detection of embryogenesis in carrot cultures by Steward and his coworkers during the early 1960s involved the use of a culture medium containing 10–20% of coconut milk (the liquid endosperm of the coconut) and it was argued that the use of this complex nutrient fluid, designed to nurture an immature embryo, was critical to the induction of somatic embryogenesis. Subsequently however it was shown, with cultures of carrot and of several other species, that prolific embryoid yield can be obtained by culture on a defined medium supplying sugar (sucrose), a mixture of inorganic ions (relatively rich in potassium and nitrogen and in which part of the nitrogen is as ammonium ions or glutamine), a mixture of B vitamins, the sugar alcohol *meso*-inositol and a cytokinin. For active proliferation without embryogenesis an auxin is also added ($0.5$–$2.0$ mg $1^{-1}$ 2,4-dichlorophenoxyacetic acid – 2,4-D) and this medium is used to propagate the culture serially. Embryogenesis is then induced by transfer to the same synthetic medium with the auxin omitted (Fig. 12.14). Such a medium defines the nutritional requirements of the cultures but, as discussed below, the embryogenic cells of the cellular aggregates in suspension, or at the surface of a callus culture, may have more exacting requirements. Clearly, however, in suspension cultures, embryoids released as free-floating structures at the late globular or early heart-shaped stage can continue their further development in isolation from the meristematic cell aggregates in such a medium.

The evidence that the embryoids are normally of single-cell origin rests primarily on the observation of all stages of development down to the two-celled stage and the recognition of the earliest pro-embryo stages at the surface of the aggregates in suspension (Fig. 12.14 **a, b, c**) and at the surface of embryogenic callus. It is also clear that the embryoids show a consistent polarity – the root pole and suspensor developing towards the centre of the aggregate or callus tissue and the plumular pole outwards from it. The polarity is induced by the associated cells. There is still some uncertainty as to whether cytoplasmic discontinuity (breakage of plasmodesmata) is established between the cellular aggregate and the embryogenic cell before the segmentations of the embryogenesis occur. The balance of evidence is however that this is not so although at a very early stage (in *Ranunculus sceleratus* and carrot at least by the 4-cell stage) discontinuity is clearly visible and the pro-embryo appears to have a continuous external delimiting cuticle.

These studies of somatic embryogenesis indicate that the segmentations giving rise to the globular embryoids are more uniform between species than is the case for their zygotic embryos. The apparently specific (and often regarded as phylogenetically significant) early segmentation patterns of zygotic embryos probably reflect the influence on segmentation pattern of chemical gradients and physical forces imposed on the embryo by the ovule in which it is developing.

A rather different approach to the experimental study of embryo nutrition is that of the aseptic culture of immature embryos dissected out from fertilized ovules. Such studies have shown that the growth requirements of the embryo become less exacting as embryology progresses. Successful completion of embryology can be achieved only with embryos which have at least completed the globular stage before isolation though this limitation may be due to the difficulty of removing younger embryos without injury and the susceptibility of very young embryos to osmotic shock. Successful culture of young embryos has required the use of culture media containing not only sucrose, salts and a form of reduced nitrogen but also an appropriate mixture of growth-regulating substances. Thus globular embryos (50–80 $\mu$m in length) of *Capsella bursa-pastoris* were successfully cultured by using a medium containing IAA, kinetin and adenine. Culture of small embryos of *Hordeum sativum* was achieved in a medium containing glutamine (or a complex mixture of amino acids) and coconut milk. Coconut milk and liquid endosperms obtained from several species, yeast extracts and protein hydrolysates have proved to be effective supplements to basal media for the culture of immature embryos of a number of species, particularly of *Datura* spp., *Cucurbita maxima* and *Cocos nucifera*. Liquid endosperms are the natural nutritive environment of immature embryos and their activity appears to be due to their content of organic nitrogen compounds, of sugar alcohols and of growth-regulating substances. A particular balance of growth-regulating substances seems to be of great importance in early embryology and to prevent premature expansion of the embryonic cells. Some experiments have pointed to the importance of a high osmotic value of the bathing medium in early embryology and later of a lower osmotic value to permit germination of more mature embryos. The significance of the osmotic value in regulating embryo development is however uncertain. Thus in experiments with immature *Capsella* embryos a high osmotic value (achieved by 12–18% sucrose) was only promotive of development in the absence of an effective addition of growth-regulating substances and was less effective than the latter.

Since no one has achieved successful culture of the isolated zygote it is possible that the really exacting nutritional requirements for embryogenesis have already been met by the stage at which embryos can be successfully excised and cultured. The view is supported by the great difficulty which is encountered when attempts are made to raise in culture a callus tissue from a single cell. The development of such single-cell clones at present requires the nurturing of the single cell by placing it in direct contact with or very close to an actively growing multicellular colony (Fig. 12.15). Some progress has been made towards replacing the 'nurse' colony by using more complex culture media and carefully regulating the carbon dioxide concentration of the culture atmosphere but not to the point of inducing completely isolated single cells to divide. The reasons for the origin of embryoids at or in the surface of cultured cell aggregates may not only be that this location promotes the establishment of polarity but that it is one where the initiating cell is nurtured by the associated growing multicellular mass. Further, the ability of defined media to support the growth of the released embryoids

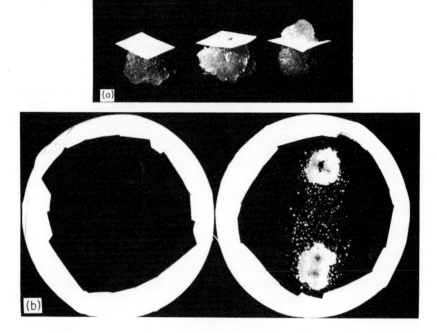

**Fig. 12.15** **(a)** Growth of single isolated cells on the upper surface of a piece of filter paper in contact with a 'nurse' piece of callus. (*Left:* at time of placing the single cell in position; *right:* when a callus has developed from the single cell. (From Muir, 1953, Ph.D. thesis, University of Wisconsin.) **(b)** Colony formation on an agar plate seeded with a suspension of cultured cells derived from *Acer pseudoplatanus* and incubated for 28 days. Both dishes were seeded with the cell suspension but the dish on the right was also seeded with two pieces of callus of the same culture. Note cells have divided and given rise to visible colonies adjacent to and in particular between the two pieces of callus whereas no colonies have developed on the other petri dish not seeded with growing callus.

may be due to its having been conditioned ('enriched') by the growing culture as a whole.

The endosperm nucleus divides very soon after the double fertilization, often well in advance of division in the zygote. When both begin division at about the same time the rate of mitosis in the developing endosperm is much more rapid. The growth of the endosperm is at the expense of the nucellus and it normally does not slow down until the nucellus has almost disappeared. Throughout endosperm development embryo growth is slow and there is usually during this period no visible disorganization of the endosperm cells adjacent to the embryo. It may well be therefore that the embryo, at least during its early stages of development, absorbs nutrients directly from the nucellus via the suspensor and this is supported by the existence of well developed plasmodesmata throughout the suspensor and pro-embryo and by the presence of internal wall projections in the basal cells of the suspensor closely resembling those that occur in transfer cells (see p. 105). The endosperm however may function as a nutritive tissue during the latter stage of embryo development even in those cases where the ripe seed is endospermic; clearly there is a massive transfer of material from the endosperm to the cotyledons in non-endospermic seeds. During the development of the embryo and endosperm, the integuments of the ovule become the testa of the seed.

## Seed dormancy

The next generation of sporophytes is established by the germination of the seed. When shed, some seeds are immediately capable of germination if provided with water and oxygen at an appropriate temperature; this applies to many cultivated species, which have been selected for ease of germination. The seeds of many species, however, show innate dormancy (p. 219) when shed. A dormant seed frequently shows great hardiness towards unfavourable environmental conditions, and for many species the seed is the survival stage in which the unfavourable season is endured. Regulatory mechanisms in innately dormant seeds ensure that the dormancy is broken in circumstances which offer a reasonable chance for the seedling's survival. The more extreme the habitat, the more elaborate are the mechanisms of dormancy control.

A simple dormancy mechanism is the possession of seed coats which are almost impermeable to water (e.g. many Leguminosae) or less often to oxygen–cocklebur (*Xanthium*), wild oat (*Avena fatua*). Sometimes the coat is water-permeable, but so hard that the embryo is unable to take up water and expand; examples are water plantain (*Alisma plantago-aquatica*), pigweed (*Amaranthus retroflexus*) and the shepherd's purse (*Capsella bursa-pastoris*). A hard coat is not necessarily thick; the entire seed of *Capsella* has a diameter of only about 1 mm. In temperate climates such coats are gradually softened by microbial decay. A considerable time may elapse before the seed can germinate, so that seeds produced in a season may not be softened till the next spring, and thus they overwinter in the dormant state. The rate of decay will moreover vary between individual seeds dispersed at about the same time, so that germination will be spread out over a long span of time, and the probability is that at least some of the seeds will chance to grow out under good climatic conditions. In hot arid regions subject to periodic fires, seeds may need to be cracked open by fire (e.g. in the sumac, *Rhus ovata*); seedlings will then grow when the mature vegetation has been destroyed. The hard seeds of some desert shrubs are abraded when a sporadic rainstorm floods the normally dried-up watercourses, and torrents of water roll the seeds along the stream beds. This occurs with seeds of ironwood (*Olneya*) and paloverde (*Cercidium*); germination then takes place at the rare times when the water supply is plentiful.

Many seeds however exhibit a physiologically imposed dormancy, in which germination is not prevented by mechanical constraints but by a physiological block which is built up during seed ripening, and the depth of dormancy can be influenced by the conditions to which the parent plant is exposed during the period. The basis of physiologically controlled dormancy is often much more obscure. A clear-cut case is that of seeds shed with rudimentary embryos; the embryos then obviously must grow to fill the seed before outgrowth can take place. Seeds can be blocked from germination by the presence of water-leachable inhibitors. In certain desert ephemerals, only prolonged leaching will remove the inhibitor, so that the seeds can germinate only after heavy rain; since intermittent light showers fail to elicit germination, the inhibitor must be replenished between leachings if it is only partially removed. In other instances the breaking of dormancy involves complex biochemical changes within the seed tissues; although there is much descriptive knowledge about the conditions necessary for overcoming dormancy, understanding of the underlying physiological mechanisms is still incomplete.

### Photoblastic seeds

Light-sensitive seeds are termed *photoblastic*; they are *positively photoblastic* if (white) light promotes germination, *negatively photoblastic* if germination is inhibited by light, and there are also *photoperiodic* seeds. To react to light the seeds must be at least partly imbibed. Positively photoblastic seeds include those of plants of open habitats – willowherbs (*Epilobium* spp.), mistletoe (*Viscum album*) and foxglove (*Digitalis purpurea*), as well as the lettuce varieties intensively investigated in the laboratory. Many members of the Liliaceae have negatively photoblastic seeds; two other well-known examples are *Nigella damascena* (Ranunculaceae) and *Phacelia tanacetifolia* (Hydrophyllaceae). Birch seeds are photoperiodic, requiring long days for germination. The light stimulus is in all cases perceived by the phytochrome system. In positively photoblastic seeds, white and red light promote germination but far red inhibits, whether applied briefly or for prolonged periods. The light-promoted seeds need a high proportion of their phytochrome as $P_{fr}$ for germination and they are inhibited from outgrowth in the shade of other plants, where the illumination is enriched in far red (Fig. 11.9). Less is known about the situation in negatively photoblastic seeds. It is believed that they require a very high proportion of phytochrome to be in the $P_{fr}$ form, but contrary to most other plant systems they seem to synthesize $P_{fr}$ in the dark. Inhibition by white light is then attributable to its action in causing partial conversion of the $P_{fr}$ to $P_r$.

### After-ripening

Many seeds require an *after-ripening* period before they will germinate. Numerous cereal and other grass seeds need a period of storage in the dry state, from about a month to several years. A very common requirement for seeds of temperate zone plants is for after-ripening in an *imbibed* state at *low* temperature. Seeds of trees and shrubs show this requirement, e.g. species of pine (*Pinus*), spruce (*Picea*), apple (*Malus*) and rose (*Rosa*). In agriculture and horticulture this treatment has been applied by spreading seeds between layers of moist substrate and cold moist after-ripening is therefore often called *stratification*. The temperature range for stratification is 1–10°C, with the optimum mostly around 5°C; adequate aeration must be available. The time for completion ranges from 1–6 months, with a progressively greater proportion of seeds becoming able to germinate as time proceeds. Seeds of some species can germinate at the stratification temperature, but in most cases germination demands a higher temperature level. As for buds, the chilling requirement inhibits germination till the winter is past.

During stratification metabolism is active and many biochemical changes have been noted in seed tissues. The rate of respiration frequently rises, and increases in the content of ATP, and of total adenosine nucleotides take place. This can be taken as a sign of a general activation of metabolism. The activities of a number of enzymes increase; rises in catalase and peroxidase activity have been reported, and in the activities of several hydrolases, with concomitant hydrolysis of reserves. Protein digestion and rises in the level of soluble nitrogenous compounds are especially widespread; reserve lipids may disappear. Many metabolic changes are obviously associated with the breaking of dormancy during after-ripening; the problem is to identify the critical controlling factor(s), and to discover the biochemical significance of the low temperature requirement as opposed to its ecological significance.

### Hormones and membranes

Many studies have been conducted on the hormonal relationships of seed dormancy control. During dormancy break there is a trend for an increase in gibberellin and cytokinin levels, and/or a decrease in the concentration of

ABA and other inhibitors. The ratio of growth promoters to growth inhibitors accordingly rises (compare bud dormancy, p. 220). Increases in gibberellin content during chilling have been recorded in numerous species; e.g. beech (*Fagus sylvatica*) and oak (*Quercus robur*); in seed of sugar maple (*Acer saccharum*) there is first a rise in cytokinins, then in gibberellins, while at the same time ABA content goes down (Fig. 12.16). Decreases in ABA levels on chilling have been observed also in American ash (*Fraxinus americana*), peach (*Prunus persica*), and walnut (*Juglans nigra*). Gibberellin application may substitute for low temperature treatment in many cases, $GA_4$ and $GA_7$ being most effective; cytokinins and ethylene can also sometimes break dormancy. Nevertheless not all observations accord well with the idea that gibberellins are primary control agents in the lifting of seed

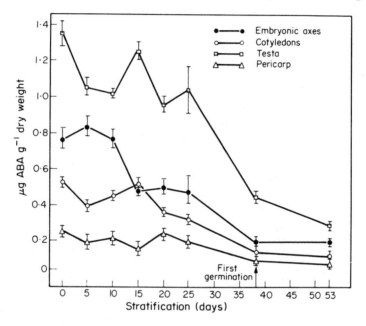

**Fig. 12.16**  Effect of stratification at 5°C on ABA concentration in various parts of the fruits of sugar maple, *Acer saccharum;* values as means ± S.E. The ABA content was estimated in extracts by gas-liquid chromatography. (Adapted from Enu-Kwesi and Dumbroff, 1978, *Zeitschrift für Pflanzenphysiologie*, **86**, 371–7.)

dormancy by stratification. Apple seeds need chilling, but fail to respond to exogenously applied gibberellic acid. In the seed of hazelnut (*Corylus avellana*) gibberellin content increases slightly on chilling, but most of the rise in gibberellins comes after dormancy has been broken: the chilling treatment seems to promote *subsequent* gibberellin synthesis. Lack of comprehensive analyses on one species make interpretation of data in this field difficult; gibberellin content has been followed in seeds of some species, ABA levels in others, and so on. Where multiple controlling factors are involved, it may be futile to try to interpret data relating to one factor by itself. Most of the estimates of hormone contents in seeds are based on classical bioassays and should be checked by more modern methods. It may also be that, as discussed in Chapters 9 and 10, hormones are not such specific primary controlling factors in plant development as was originally thought.

Three instances have now been given where treatment at low temperature is needed to permit a developmental process to proceed: breaking of seed dormancy, breaking of bud dormancy, and vernalization. Dormant bulbs, corms and tubers may also require chilling before they can grow out. This low temperature requirement has been interpreted in terms of regulation of balances between alternative metabolic pathways of varying temperature sensitivity (vernalization p. 234). In seeds where coats, bulkiness and compactness of tissue may limit oxygen supply, it has been suggested that, at higher temperatures, respiration may be stimulated to such an extent that oxygen shortage results in fermentation; the products of fermentation then inhibit metabolism. More recently attention has been directed to the effects of temperature on membranes. Temperature affects the fluidity and consequently the permeability of biological membranes. A lowering of temperature could therefore have profound effects on the traffic of solutes between subcellular compartments and hence on cellular metabolism. Phytochrome and hormones have been postulated to act at the membrane level. Perhaps the primary act of dormancy breaking is to be sought at the level of membrane organization rather than in some individual metabolic reaction.

FURTHER READING

BEWLEY, J. D. and BLACK, M. (1982). *Physiology and Biochemistry of Seeds in Relation to Germination. Viability, Dormancy and Environmental Control.* Springer-Verlag, Berlin.

STREET, H. E. (1969). Growth in organized and unorganized systems: knowledge gained by culture of organs and tissue explants. In *Plant Physiology*, ed. Steward, F. C., Vol. 5B, Chap. 6, 3–224. Academic Press, New York and London.

STREET, H. E. (1975). Experimental embryogenesis – the totipotency of cultured cells. In *Textbook of Developmental Biology*, eds. Graham, C. F. and Wareing, P. F., Blackwell Scientific Publications, Oxford.

TAYLORSON, R. B. and HENDRICKS, S. B. (1977). Dormancy in seeds. *Annual Review of Plant Physiology*, **28**, 331–54.

VINCE-PRUE, D. (1975). *Photoperiodism in Plants.* McGraw-Hill, London.

ZEEVAART, J. A. D. (1976). Physiology of flower formation. *Annual Review of Plant Physiology*, **27**, 321–48.

ZEEVAART, J. A. D. (1978). Phytohormones and flower formation. In *Phytohormones and Related Compounds – A Comprehensive Treatise*, ed. Letham, D. S., Goodwin, P. B. and Higgins, P. B., Vol. II, Chap. 6, 291–327.

SELECTED REFERENCES

BÜNNING, E. (1973). *The Physiological Clock*. Academic Press, New York.
IMAMURA, J., OKABE, E., KYO, M. and HARADA, H. (1982). Embryogenesis and plantlet formation through direct culture of isolated pollen of *Nicotiana tabacum* cv. Samsum and *Nicotiana rustica* cv. Rustica. *Plant and Cell Physiology*, **23**, 713–16.
NITSCH, J. P. (1965). Physiology of flower and fruit development. In *Encyclopedia of Plant Physiology*, ed. Ruhland, W., **15**, (1), 1537–1647. Springer-Verlag, Berlin.
OHTANI, T. and KUMAGAI, T. (1980). Action spectra for the light inhibition of flowering and its reversal in *Lemna paucicostata T-101*. *Planta*, **149**, 332–5.
POLOWICK, P. L. and GREYSON, R. I. (1982). Anther development, meiosis and pollen formation in *Zea* tassels cultured in defined liquid medium. *Plant Science Letters*, **26**, 139–45.
SUNDERLAND, N. (1973). Pollen and anther culture. In *Plant Tissue and Cell Culture*. ed. Street, H. E., 205–39. Blackwell Scientific Publications, Oxford.

# Units of Measurement – Conversion Table

The system of SI units (Système International d'Unités) was introduced in 1960. In this system, the basic units of mass, length and time are the kilogramme, (kg), metre (m) and second (s); a number of common units, e.g. litre, hour, are abandoned. Older units however still abound even in current scientific literature, and a few are retained in this text. A conversion table is given opposite.

Further information on SI units and Symbols may be obtained from:

INCOLL, I. D., LONG, S. P. and ASHMORE, M. R. (1981). SI units in publications in plant science. In: *Commentaries in Plant Science*, ed. Smith, H., Vol. **2**, 83–96. Pergamon Press, Oxford.
PENNYCUICK, C. J. (1974). *Handy Matrices*. Edward Arnold, London.
*Symbols, Signs and Abbreviations*. Symbols Committee of the Royal Society, Royal Society, London 1969.
*The Use of SI Units*. British Standards Institution Publication, PD 5686, London 1972.

| SI units | Previously employed units |
|---|---|

### Length
Basic unit metre, m
$1\ \mu m$, micrometre $= 10^{-6}$ m
$1$ nm, nanometre $= 10^{-9}$ m

Basic unit metre, m
$1\mu$, micron $= 10^{-6}$ m
$1\ m\mu$ millimicron $= 10^{-9}$ m

### Volume
Basic unit cubic metre, $m^3$
$10^{-6}\ m^3 = 1\ cm^3$
$10^{-9}\ m^3 = 1\ mm^3$

Basic unit litre, $I = 10^{-3}\ m^3 = dm^3$
$1$ ml, millilitre $= 10^{-3}$ I
$1\ \mu l$, microlitre $= 10^{-6}$ I

### Energy
Basic unit joule, $J = 1$ kg $m^2\ s^{-2}$

Basic unit erg $= 10^{-7}$ J
$1$ cal, calorie $= 4.18$ J

### Temperature
Degrees Kelvin, °K
$1°K = 1°C$, but the Kelvin scale
starts at absolute zero so that
$0°K \simeq -273°C$, and $0°C \simeq 273°K$

Degrees centigrade, (Celsius), °C

### Pressure
Basic unit Newton per square metre,
the Pascal (Pa), N $m^{-2}$
$(1N = 1$ kg m $s^{-2})$
KPa, kilopascal $= 10^3$ Pa
MPa, megapascal $= 10^6$ Pa

Bar or atmosphere
$1$ bar $= 10^5$ Pa
$1$ atm $= 1.01325 \times 10^5$ Pa

### Amount of substance and concentration
Mole (symbol mol), amount of
substance of a system which contains
as many elementary units as there
are carbon atoms in 0.012 kg of
carbon-12
A solution which has a concentration
of $10^3$ mol $m^{-3}$

Gram-molecule (g-mole), the quantity
of a compound whose mass in g is
equal to its molecular weight;
numerically equals mole

$1$ M, a solution which has a concentration
of $1$ g-mole $I^{-1}$

---

Planck's constant (h)
$6.6256 \times 10^{-34}$ J s

$6.6256 \times 10^{-27}$ erg s

---

Avogadro's number $6.02252 \times 10^{23}$ mol$^{-1}$

# Index

**Bold** page numbers refer to major entries. *Italicized* page numbers refer to illustrations.

ABA (abscisic acid), 34, 115, 119, *160*, **164**, 186, 202, 214, 217, 220, *264*
abscisins, **164**, 206
abscission, 148, 161, 163, 164, 165, **221**
abscission zone, 224
absolute growth rate, 144
absorption, of ions, 76, **77**
 of water by roots, **53**
 spectrum, *28*, *205*, *227*
acclimation, 110
accumulation, of ions, **77**
*Acer platanoides*, 143
 *pseudoplatanus*, 220, *260*
 *saccharinum*, 113
 *saccharum*, 91, *263*
achenes, 248
acid-growth theory, **172**
actinomycin-D, 14, 18, 170, 174
action spectrum,
 of light breaks in photoperiodism, *237*
 of photomorphogenesis, *228*
 of photosynthesis, *29*
 of phototropism, *205*
activated diffusion, 103
adenine, 259
adenosine diphosphate (*see* ADP)
adenosine triphosphate (*see* ATP),
*Adenostoma fasciculatum*, *54*
ADP (adenosine diphosphate), 26, 36, 83
adsorption, 77
adventive embryos, **252**
*Aegopodium podagraria*, 208
aeration, of soil, 65, 67, 77
aerial roots, 52
aerobic respiration, 12
aerotropism, 246
*Aesculus hippocastanum*, 9
after-ripening, 263
agar blocks, 158, *159*, 190, *191*, 196

agaves, 97
age of cells, 139
ageing, 221
aggregation of proteins, 118, 125
agravitropism, 190
agricultural yield, 110, 120, 127, 132
*Agrostis stolonifera*, 131
*Agrostis tenuis*, 132
alcohol, 13
alcohol dehydrogenase, 83
aldehyde oxidase, 85
aleurone grains, 6
 layer, 18, 19, 21, 22, 23
alfalfa, 53, 127
algae, 1, *78*, 85, 88, 114, 143, 169, 252
*Alisma plantago-aquatica*, 142, 262
alkaloids, 92
*Allium cepa*, *197*
allometric growth, 153, *155*
allometry formula, 154
allosteric effectors, 185
aluminium, 75
amides, 91
amino acids, 6, 14, 16, 20, 23, 45, 91, 118, 131, 212, 215, 222, *224*
aminoisobutyric acid, *224*
ammonium, 76, 77, 84, 91, 258
 nitrate, 76
 sulphate, 222
Amo 1618, 243
*Amoeba*, 169
amylase, 16, 83
α-amylase, 16, 17, 18, **21**, 184
β-amylase, 16, 17, 18
amyloplast, 6, 198
anabolism, 18
anaerobic conditions, 9, 101, 104
anaerobic respiration, 12, 13
*Ananas comosus*, 240
anions, 76, 79, 84
anthers, 84, **244**

culture of excised, **244**, *246*
anthesis in the anther, 246
anthocyanin production, 226
anti-auxin, 108
antibodies, 166
anti-gibberellins, 243
aphids, 90, 97
aphototropism, 191
apical cell, 252
 dome, 153, 212, 215
 dominance, 153, 160, 164, **217**
 growth of cells, 178
 meristem, 89, 138, 153, 201, 215, 218, 239
apomixis, 4
apoplast, *80*, 81, 82, 105
apple, 53, 86, *91*, 126, 220, 263, 264
aquatic plants, 49, 52, 65, 76, 117, 127, 142
*Arachis hypogaea*, 6, 9, 192
arginine, 212
*Arisaema triphyllum*, 144, *179*
arithmetic growth, 146
artichoke, 170
*Arum maculatum*, *140*
 spadix of, 139
ash, 220, 264
aspartate, 43
*Aspergillus niger*, 143
*Astragalus*, 75
atmospheric conditions, influence on water uptake and movement, **68**
ATP (adenosine triphosphate), 11, 12, 26, *27*, 36, 41, 45, 81, 83, 92, 101, 116, 119, 173
ATPase, 83, 173
*Atropa belladonna*, 245, 255, 263
autocatalytic reaction, 148, 150
autonasty, 209
autoradiographs, 92, *224*
autotrophic, 24

autotropism, 191
auxanometer, 136, *137*
auxin, 86, 94, **157**, 165, 169, 170, **171**, 178, 181, 201, 208, 212, 215, 223, 240, 243, 246, 247, 255
  bound, 159
  destroying enzymes, 206
  diffusible, 159, 190, 243
  directed transport, 94, 218
  diversion, *191*, 201, 206
  extractable, 159, 202
  gradient in abscission, 225
  hydrolysable, 159
  mechanism of action, 159, **172**, 185
  transport of, **107**, 165, 175, 187, 201, 217
  relation of structure to activity, 161
  synthesis, 187, 201, 207, 247, 248
  synthetic, 106, *160*, 248
*Avena*, 17, 33, *35*, *173*, *174*, 190, 201, 261
*Avicennia germinans*, 130
Avogadro's number (*N*), 37, 267

bacteria, 143, 145, 185
  psychrophilic, 143
  thermophilic, 143
bamboos, 144, 145
*Bambusa*, 145
bark, 90, *123*, 126
barley, 15, *17*, 19, 21, *22*, 81, 184, 185, 222, 240
bean, 4, *8*, *9*, *11*, 12, 13, 16, 18, 21, 59, 86, 139, *143*, *186*
beet, 86, 94
benzyladenine, *186*
*Bertholletia excelsa*, 6
*Beta vulgaris*, 40, 233
*Betula*, 91, 113, 219
  *pubescens*, 72
bicarbonate ions, 77
bidirectional transport, 102, 103
biennial, 221
bimolecular lipid layer, 125
bioassay, of hormones, 159, 162, 165, 187, 202, 206
biological clock, 151, 241
birch, *72*, 91, 113, 219, 220, 262
  forest, 39
biosynthesis of macro-molecules, 26, 44, 45
blackcurrant, 220
black locust tree, *123*, 126, 127

bleeding sap, 58, 59
blue-green algae, 143
blue-light, in photomorpho-genesis, 228
borate, diphenol complexes, 85
  sugar complexes, 85
boric acid, 84, 246
boron, 75, **84**
*Borya nitida*, 114
*Botrytis cinerea*, 145
bound auxin, 159
  water, 114, 119, 125
boundary layer, 31, 33, 68, 71
*Brassica*, 16, 86
  *napus*, 127
  *oleracea* var. *acephala*, 140
brazil nut, 6
brown-heart, 86
*Bryonia*, 146
*Bryophyllum fedtschenkoi*, 241
bryophytes, 114, 132
bud, 120, 124, 126, 153, 175, **211**, 217
bud dormancy, **219**
bulbs, 94, 121, 243, 265
bundle sheath, 43, *89*

C₃ cycle, 41
C₃ plants, **42**
C₄ plants, **42**, 84
cabbage, 16
cacti, 44, 113, 127, 133
calcium, 74, 75, 76, 77, *78*, 79, 81, 83, 86, 246
*Callitriche intermedia*, 154
callose, 99, 106, 145
callus colonies, *260*
  cultures, 169, 178, 180, 212, 253
Calvin cycle, 41
CAM, **42**
cambial activity, control of, 181, 182, *183*
cambium, 89, 90, 128, 139
*Camellia sinensis*, 75
*Canavalia ensiformis*, 6
canavanine, 6
capillarity, 57
*Capsella bursa-pastoris*, 208, 259, 262
carbohydrate, 6, 12, 19, 42, 85, 90, 125
carbon dioxide, 12, 26, *27*, 28, 30, 93, 208, 212, 260
  diffusion, **31**
  fixation, 13, **36**, 44, 116
carbon dioxide, radioactive (¹⁴CO₂) use of, 41, 93, 100, 105, 249

carbon gains, 39
carbon turnover, **26**
carbonic anhydrase, 83
carboxylating enzymes, **41**
α-carotene, *205*
carotenoids, in photofropism, 204, 228
carrot, *155*, *235*, 255, *256*, *257*
casein hydrolysate, 212, *213*
Casparian strips, 56, *57*, *80*, 81, 82
castor bean, *5*, 6, 9, 12, 13, 14, 18, 19, 20
catabolism, 18, 221
catalase, 85, 263
cation exchange, 76, 77, 79
cations, 76, 77, 79, 83, 246
cavitation, 61, *62*, 116
CCC, 243
celery, 86
cell, counts, 137
  cultures, 163, 169, 171, **255**
  cycle, 145, **168**
  division, 10, 138, **139**, 145, 154, 161, 163, **168**, 175, 214, 224, 239, 249, 250, 252
  elongation, 10, 154, 158, 163, 164, 184, 185, 195
  expansion, 138, **139**, 145, 158, **171**, 185, 192, 248
  growth, 26, 115, **157**, 168, 192
  lineage, 177
  multiplication, 139
  number, 137, *140*, 141
  size, 137, *140*, *141*, 169, 171, 179
  suspension cultures, **254**
  wall, 6, *8*, 16, *32*, 34, 49, 50, 56, 60, 77, 79, 83, 105, 116, 118, 131, 132, 171, 172, 178, 179
  wall colloids, 49, 77
  wall-plasticity, 171
  wall synthesis, 115, 138, 171, 172
cellobiase, 16
cellobiose, 16
cellular differentiation, 138, 144, **157**, 171, **175**, 221
cellulase, 16
cellulose, 16, 172, 173, 179
  microfibrils, 173, 179
central cell, 247
centrifugal force, 194, *196*
centrifuge, 194, *196*
*Cercidium*, 262
cereals, 4, *17*, 18, 19, 83, 159, 187, 193, 202, 263

*Chamaegigas intrepidus*, 114
chamise, *54*
chelates, 91
chemical potential of water,
  $\mu_w$, 48
chemotropism, 190, 247
*Chenopodium*, 239, 241, *242*
chilling, requirement, 220,
  233, 263, 265
  resistance, 101, 120
*Chlorella*, 1
chloride, 35, 76, *78*, 84, 118,
  130, 131
chlorine, 74, 75, 84, 86
chlorophyll, *28*, 37, 40, 83,
  226, 227, 230
chloroplasts, 18, 34, 41, *89*,
  117, 124, 207, 226
CHOLODONY—WENT theory
  of tropisms, 190, **201**, **206**
chromatography, 165, *166*
chromoprotein, 226
chromosome aberrations,
  178
chromosomes, 168, 177
chrysanthemum, 214, 233
*Cichorium endivia*, 254
circadian rhythms, 241
*Circaea*, 192
circumnutation, 151
*cis*-polyhydroxyl
  configuration, 85
9–9′ *cis*-β-carotene, *206*
*Citrullus vulgaris*, 139, 141
*Citrus*, 69, 70, 86, 164
  *grandis*, 70
  *limonia*, *69*, 70
  *macrocarpa*, 253
  *reticulata*, 253
clay soil, 65, *66*
climacteric, 187
climbing plants, 151
$CO_2$ (*see* carbon dioxide)
cobalt, 74, 75, 84, 86
cocklebur, 261
cocoa, 86
coconut, 5, 19, 258
  milk, 169, 214, 215, 248,
  258, 259
*Cocos nucifera*, 5, 169, 259
cohesion theory, **59**
colchicine, 245
cold requirement, 233, 220,
  263, 265
cold resistance, (*see* freezing
  resistance)
coleoptile, *5*, 114, *158*, *159*,
  *161*, *167*, 172, *173*, *174*,
  187, 189, 190, *191*, 193,
  195, 201, 202, *203*
coleorhiza, 5

*Coleus*, 94, 107, 218
colloidal imbibition, 9, 49
colloids, 49, 50, 77, 125
columns of water, 59
*Commelina benghalensis*, 84
Commelinaceae, 192
companion cells, 90, 92, 98,
  104, **105**, 106, 177
compensation point, *30*, 31
competition for food
  materials, 218
Compositae, 6
conditioned medium, 261
conduction of tropic
  stimulus, 197
conifers, 56
constant, exponential
  growth, 146
linear growth, 146
logarithmic growth, 146
contact exchange, 77
conversion, lipid to
  carbohydrate, 19
conversion table of units of
  measurement, 267
cooling effect of
  transpiration, 129
copper, 75, 83, 132
corky-core, 86
correlation, growth, 217
cotton, 21, 61, *155*, 164, 225
cotyledons, 4, *5*, 6, 9, 11, 12,
  16, 18, 23, 105, 139, 228,
  251
Crassulacean acid
  metabolism (*see* CAM)
cress, 195
criteria of essentiality – of
  mineral elements, 74
criteria, of stress resistance,
  112, 121
  of survival, 112, 121
critical dark period, 236
crop efficiency, **38**
crop yields, 39
crozier, *216*
cucumber, 23, 140, *148*, *149*,
  244
*Cucumis anguria*, 141
  *sativus*, 23, *140*, 143, *148*,
  *149*
cucurbit fruits, 139, 141
*Cucurbita*, 101, 103, 104, 146
  *maxima*, 23, 259
  *pepo*, 95, 141, 143
Cucurbitaceae, 6, 98, 154
culture, of embryos, **259**
cultured plant cells, **255**
cultures of flower primordia,
  244, 249
cuticle, 31, 33, 34, 70, 113,

259
cuticular resistance, 33
cuticular transpiration, 70
cutin, 246
cutinase of pollen, 247
cyanide, 101
cycloheximide, 174
*Cymbalaria muralis*, 149
cytokinesis, **168**
cytokinins, 23, 86, 106, 107,
  *160*, **163**, 170, 171, 178,
  *183*, 222, 244, 248, 258, 263
cytolysis, 225
cytoplasmic
  strands – in phloem, 100,
  103
  streaming – in phloem, 103

dandelion, *176*, 198, 222
dark, acidification, 44
  $CO_2$ fixation, 13, 44, 241
  period, 236, 238
  respiration, 41, 45
date palm, 6, 7, 18, 19, 144
*Datura*, 245, 259
daylength, 219, 233, 234
day-neutral species, 233,
  234, 242
death, 11, 117, 211, 221
decapitation, of coleoptiles,
  158, 159, 175, 190, 196
  of roots, 195
dedifferentiation, 214
definition of growth, 135, 139
defoliation, 90, 94, 236
dehydration, 112, 114, 117,
  118, 119, 121, 122, 123,
  124, 126
dehydrogenases, 83
denaturation, 118, 125, 128
dendrograph, 59
deoxyribonucleic acid (*see*
  DNA)
depression of freezing point,
  121
desert plants, 61, 113, 127,
  262
desiccation, 47, **112**, 124,
  126, 129
  of seeds, 113
desiccation resistance, (*see*
  water stress resistance)
deuterium hydroxide, 47
deuterium oxide, 47
development, 138, 184, **211**,
  **232**
  progress of growth during,
  150
  reproductive, **232**
  sequence of processes
  involved in, *138*

vegetative, **211**
dextrin, 16
dextrinase, 16
*Dianthus*, 192
diatropism, 189, **207**
2, 4-dichlorophenoxyacetic
  acid (2, 4-D), 106, *160*,
  169, 183, 223, 258
differential growth, 153, 191
differentiation, 89, **138**, 144,
  **157**, 171, **175**, 215, 221, 226
  patterns of, **180**
diffuse-porous, 55
diffusible auxin, 159, 190,
  243
diffusion, 19, 43, 88, 95, 145
  of $CO_2$, **31**, 40, 43
  of water vapour, **68**
diffusion resistance,
  of cuticle, 33
  of stomata, 33
  of testa, 9
*Digitalis purpurea*, 10, 262
2, 4-dinitrophenol (DNP),
  81, 101
disulphide (SS) bridges, 117,
  118, 126
diurnal changes, in direction
  of translocation, 106
diurnal fluctuations, in
  moisture content, **53**
  in sap composition, **91**
diurnal leaf movements, 241
diurnal rhythm, 34, 58, 120,
  150, 241
diurnal rhythms of growth,
  150, 151
DNA (deoxyribonucleic
  acid), 14, 168, **169**, 177,
  178
Donnan free space (DFS), 79
dormancy, 122, 163, 164,
  **219, 261**
  of buds, 164, **219**, 222, 265
  of seeds, **261**
dose-response curve,
  hormones, *161*, 187
  phototropism, *203*
double fertilization, 4, 247,
  261
drought, 110, 111, 114, 120,
  127, 221
drought hardiness (*see* water
  stress resistance)
dry weight, 133
dry weight changes in seeds,
  *20*
*Dryopteris*, 216
dwarf pea assay, 161, *163*,
  165
dye injection, 59, 90

*Echinocystis*, 165
edge diffusion, 33
egg cell, 244
egg nucleus, 4, 247
Einstein (*E*), 37
ektexine, 245
elasticity of cell wall, 172
electrical field, 252
electrochemical potential
  gradient, 78
electromagnetic radiation,
  26
electron transport, 129
electro-osmosis, 100, 104
electrophoresis, 15
*Eleocharis*, 193
elongation growth, 138
embryo, 4, 10, 14, 15, 21,
  114, 180, 233, 247, 250
embryo culture, **259**
  sac, 4, 243, 244, 247
embryogenesis, **246, 250**
embryoids, 245, *246*, *253*,
  255, *256*, 258
embryology, 169, 245, 250,
  255
  stages in, 251, 254, 255
embryos, immature, **259**, 262
endodermis, 56, *57*, 79, *80*,
  81, 82, 197
endogenous rhythms, 151
endogenous rhythms in
  photoperiodism, 241
endomitosis, 170
endoplasmic reticulum, 15,
  98, 99, 171, 198, 199, 200,
  251
endosperm, 4, 5, 6, 9, 11, 12,
  16, 18, 19, 21, 184, 248,
  250, 259, 261
energy, free (*see* free energy)
  conservation, 45
  conversion, 28, 37
  flow, **26**
  release in respiration, 45
  transduction, 12, 27, 39,
  40, 83
entropy, 46
enzymes, 10, **14**, 92, 117,
  124, 125, 129, 130, 131,
  138, 171, 173, 185, 263
enzyme proteins, 15, 17, 117,
  138
  synthesis, **15**, 170, 184
epicotyl, *5*, 202, *203*, 227
epidermis of leaf, 32, 71, 177
epidermis of root, 55, 56, *57*,
  79, 80, 177
epigeal, 13
*Epilobium*, 262
epinasty, 208, 209

epiphytes, 52
*Erysimum*, 233
essential elements, **74**
esterase, 16, 18
ethylene, 86, 106, 107, **164**,
  166, 186, 202, 209, 225,
  240, 264
etiolation, 144, *155*, 192, 202,
  203, 225, 227
*Euonymus japonicus*, 223
Euphorbiaceae, 6
evaporation, 65, *72*
exanthema, 86
exchangeable ions, 76, 77
expansion growth, 138
extracellular freezing, 124

Fagaceae, 6
*Fagus sylvatica*, 220, 264
far-red light, reversal of
  effects of red light, 217,
  **226**, 237
fatty acids, 16
feedback control, 23, 34
female gametophyte, 244
fermentation, 13, 265
fertilization, 4, 192, **244**, 253,
  261
fertilized ovum, 177, 211,
  250, 252, 259
fibres, 98, 106, 178, 221
fibrils – in sieve tubes, 99,
  *100*, 101
field capacity of soils, **65**
fixation of carbon dioxide,
  13, **36**, 116
flavins, in phototropism, 204,
  205, 207
florigen, **236**, 242
flower cultures, **249**
flower development, **241**
flower stalk – growth of, 192,
  243
flowering, 163, 221, **232**
flowering hormone, 236, 239
flowering stimulus, 163
flowers, 94, 120, 126, 192,
  **193**, 247
fluorescein, 90
fluorescent dyes, 90, 95
foxglove, 10, 262
*Fragaria vesca*, 140, 192
*Frankenia grandiflora*, 131
*Fraxinus americana*, 264
  *excelsior*, 220
free energy (*G*), 31, **36**, 44,
  45, 47, 48, 49, 78
  change (Δ*G*), 36, 45
free nuclear stage of endosperm, 248
freezing point of plants, 126

freezing stress, **120**
  avoidance, 120, *121*, **125**
  damage, **124**
  measurement, 121, 122
  resistance, **120**
French bean, 107
frequency of radiation (v), 37
fresh weight, 135
*Fritillaria meleagris*, 243, *244*
frond of ferns, 215
frost, 110, 120, 127, 133
  hardiness, 120
fruit cultures, 249, *251*
  development, 141, 221,
    **247**
  diameter, 141
  growth, 135, 139, 141
  set, 164, 247
  shape, 152, 154, 247
  size, 141, 247
fruits, 139, 192
*Fucus*, 255
fungal hyphae, growth rate
  of, 47, 145
fungal sporangiophores, 198
6-furfurylaminopurine, 169
fusicoccin, 173, *174*

g, **194**
G₁ and G₂ phases of cell
  cycle, 168
galactose, 16
α-galactosidase, 16
galvanotropism, 190
gametophyte, 175, 245
  female, 244
  male, 244
gas–liquid chromatography
  (GLC), 166
Geiger counter, 92
gel electrophoresis, 15, 171
gene, activity, 185
  bank, 120
  expression, 185
general resistance, 129
generative cell, 245
genetic, damage, 221
  engineering, 133
  potentialities, 178, 255
  repression, 184, 221, 230
geoelectric effect, 201
geotropism (*see*
  gravitropism)
geranium, 93
germination, seed, **4**, 105,
  114, 158, 164, 165, 172, 262
*Geum*, 233
giant chromosomes, 177
gibberellic acid, (GA₃), 21,
  185, *186*, 211, 214, 244
gibberellins, 21, 106, 107,

160, **161**, 165, *166*, 171,
  184, 198, 202, 217, 220,
  222, 234, *235*, 240, 243,
  247, 263
girdling experiments, 90
glands, 105, 131
*Glechoma hederacea*, 43
glucose, 16, 19, 45
glucose-1-phosphate, 16
α-glucosidase, 16
glutamic acid, 13
  decarboxylase, 13
glutamine, 258, 259
glycerol, 16
glycophyte, 130, 131
*Glycine max (soja)*, 6, 234
α-glycosyldisaccharides, 181
glyoxylate, 19
  cycle, 19
glyoxysome, 20
Golgi apparatus, 15
*Gossypium barbadense*, 21,
  61, 67
gradients, in meristems and
  organs, 175
grafts, 233, 236, 239
Gramineae, 4, 5, 6
grand period of growth, 147,
  *149*, 150, 152
grapevine, 61, 106
gravitropism, 84, 189, **193**,
  207, 219
gravity, action of, 190, 192,
  **193**, 207, 208
greenhouse crops, 40
grey speck, 86
growth, 44, 115, 116, **135**,
  185
growth, allometric, 153, *155*
  arithmetic, 146
  by cell expansion, 138
  centre, 214
  conditions necessary for,
    47, **142**
  constant exponential, **146**,
    *148*
  constant linear, **146**, *148*
  constant logarithmic, **146**
  correlation, 178
  definition of, 135, 139
  exponential, 146
  grand period (curve) of,
    147, *149*
  heterogonic, 153, 155
  localization, **139**
  measurement of, **135**
  minimum and maximum
    temperatures for, 142
  movements, **189**
  regulating substances, 157,
    177, 178, 180, 181, 184,

190, 212, 219, 243, 247,
  249, 259
  rhythms, 150
  rooms, 3
  growth rate, 180, 191, 193,
    *194*, 201, 202, 206
  growth rate, absolute, **144**
    relative, **144**, 221
  growth retardants, 243
  guard cells, 32, 34, 35, 71, 83,
    117, 177
  guttation, 58

H⁺ ions, 77, 104, **173**
H⁺ pump, **173**
haemocytometer, 137
halophytes, 61, 84, 126, **130**,
  131, 133
haploid plants, 245
haptotropism, 190
hardening, 110, 118, 122,
  125, 127, 129, 220
hardiness (*see* stress
  resistance)
Hatch–Slack cycle, 43
hazelnut, 264
heart-rot, 86
heat, release of by plants, 45,
  46
heat resistance, **127**
heavy metals, 132
heavy water, 47
*Hedera helix*, 61, 122, *123*,
  126
*Helianthus annuus*, 6, 21, 140
hemicellulose, 6, 16, 173
henbane, 233
HER (high energy reaction
  of morphogenesis), 226,
  **227**
*Heracleum sphondylium*, 102
herbaceous plants, 53, 56
herring sperm DNA, 169
heteroblastic development,
  221
heterogonic growth, 153
heterotrophic, 24
*Hieracium*, 208
high energy reaction of
  photomorphogenesis
  (HER), 226, **227**
high performance liquid
  chromatography (HPLC),
  *166*
high photosynthesis plants,
  **40**
high temperature stress, **127**
  damage, 128
  resistance, **127**
  tolerance, **129**
hilum, 9

HIR (*see* HER)
histones, 169
hop, 146
*Hordeum sativum*, 259
  *vulgare*, *17*, 21
hormones, 85, 88, 142, 144,
  **157**, **184**, 189, 202, 206,
  211, 222, 263
  assays for, **165**
  modes of action, 174, **184**
  transport of, 88, **106**
horse-chestnut, 9
*Humulus*, 146
hydration, 9
  phase of germination, **8**
  shells, 119
hydrogen bonds, 173
hydrogen ions (H +), 76, 77,
  104, **173**
hydrolases, **15**, 18, **21**, 105,
  116, 224, 263
hydrolysis of food reserves,
  11, **15**, **21**, 47, 118
hydrolytic enzymes, (*see*
  hydrolases)
hydrophilic bonds, 126
hydrophilic colloids, 125
hydrophobic bonds, 125, 126
hydrostatic pressure
  gradient, 97
hydrotropism, 55, 190, 246
α-hydroxy-sulphonates, 42
hydroxyl ions, 76, 77
*Hyoscyamus niger*, 233, 236,
  240
hypocotyl, 6, 185, *186*, 187,
  191, 198, 203, *254*
  hook 107, 185, 227, 229
hypogeal, 13
hyponasty, 209
hypophysis, 251

IAA (indol-3yl-acetic acid),
  159, *160*, *161*, 169, *176*,
  180, 182, 185, 201, 206,
  217, 225, 244, 249, 259
  dose response curves of,
  *161*, 187
  methyl ester, 185
ice formation, 121, 124, 126
  extracellular, 122, 124, 127
  intracellular, 121, 124
*Ilex verticilliata*, 63
imbibition, **8**, 12, 114
  forces of, 49
immobilization of ions, 93
*Impatiens balsamia*, 145
  *Hawkerii*, 145
indicator plants, 86
indol-3yl-acetic acid (*see* IAA)
induction, photoperiodic,

233
induction–reversion
  response, 226
inductive cycles, 236, 238
inhibitors of growth, 164,
  202, 212, 262
initiation of, buds, **211**
  leaves, 153
  meristems, 214
  organs, 153, 154, **211**
  roots, 153, **211**
injury, effect on growth, 142
  effect on metabolite
  transport, 94
  effect on respiration, 115,
  119
innate rhythm, 152, 220
inositol hexaphosphate, 5
integuments, 4
intercalary meristems, 139
intercellular air spaces, 31,
  69
interrelationship between
  growing seed and storage
  tissues, 21
invertase, 105
ion exchange capacity, 76
  pumping, 35, 79, 80, 104,
  133
  transport, **79**
  uptake, 77
*Ipomoea hederacea*, 142
iron, 74, 75, 76, 83
ironwood, 262
irradiance (*see* light
  irradiance)
irrigation, 40, 130, 133
isocitratase, 19
ivory nut palm, 6
ivy, 61, *123*, 126
ivy-leaved toadflax, 192

*Juglans nigra*, 264
juvenile stage, 221

*Kalanchoë blossfeldiana*, 241
kinetin, *160*, 163, 169, 211,
  *212*, *218*, 222, *223*, *224*, 259
klinostat, 191, 194, 208
Kranz anatomy, 43

*Laburnum anagyroides*, *193*
lactic acid, 13
*Lactuca sativa*, 4, 10
*Lagenaria vulgaris*, 141
lanthanum salts as tracers, 81
larch, 220
*Larix decidua*, 220
laser beam auxanometer, 136
latent heat of evaporation, 59

latent time, 194
lateral buds, 153, **211**
  roots, 55, 193, **211**, 217
  shoot, 153
*Laurus nobilis*, 70
LD₅₀, 112
LDP (*see* long-day plant)
leaching of metabolites, 9
lead, 81, 132
leaf, abscission, **221**
  area index, 39, 221
  cultures, 215, *216*
  fall, 224
  growth, **140**, *141*, 144, *148*,
  *149*, 215, 226, 227, 239
  as photosynthetic organ,
  28
  senescence, **221**
  shape, 152, *154*
  venation, *89*
leaf surface, external, 31, 69,
  70
  internal, 31, 70
leakage of salts, 80, 81
leaves, **28**, 59, 67, 89, 92, 93,
  94, 105, 106, 114, 118, 120,
  122, 124, 129, 153, 192,
  208, 212, 225, 236
Lecythidaceae, 6
Leguminosae, 6, 105, 161
lemon, *69*, *162*
*Lens culinaris*, 195, *203*
lentil, 195
*Lepidium sativum*, 143, 195,
  198, *199*, 200
*Leptopteris*, 215
lettuce, 4, 10, 226, 262
lianas, 55, 56, 63
lichens, 114, 132
life history, 139
life span, 140, 221
light, absorption, **28**, 204,
  *205*, 226, 230
  action of unilateral, **189**,
  *191*, **202**
  break effect, 236, 237, *242*
  compensation point, *30*, 31
  conduction, 204
  effect on stomata, 34
  influence on growth, 144,
  225
  irradiance, *30*, 31, *35*, 36,
  39, 112, 144, 203, 228
  requirement for
  germination of seeds,
  226
  saturation, morphogenesis,
  227
  saturation, photo-
  synthesis, *30*
light-sensitive seeds, 226, 262

lignin biosynthesis, 85
*Ligustrum lucidum*, 61, 67
Liliaceae, 5, 262
lime, 106
liminal direction, 192
limiting factors, 31, **35**, 39,
  43, 110, 127, 142, 144, 212
linear elongation growth,
  135, 136, 146
lipase, 16
lipids, 5, 12, 13, 16, 19, 20,
  118, 125, 126, 129, 263
  unsaturated, 127, 130
lipid reserve droplets, 6, 7
lipoprotein, 100
liquid endosperm, 169, 248,
  259
liquid ice, 47
*Liriodendron tulipifera*, 219
*Litchi chinensis*, 240
loading, of phloem, 105
long day plants (LDP), 233,
  **234**
low photosynthesis plants, **40**
low temperature stress, **120**
lupin, 16, 136, 140, *141*
*Lupinus albus*, 75, 140
  *angustifolius*, 136
lutein, *28*
*Lychnis*, 233
*Lycopersicon esculentum*,
  58, 61, 67
lysine, 212

M-phase of cell cycle, 168
macromolecules, 26, 45, 117
macronutrients, 75, 83
magnesium, 74, 75, 76, 78,
  79, 82, 83, 91
maize, 5, 6, 12, 19, *20*, 38, 39,
  40, 41, 52, 55, 86, 107, 139,
  142, 143, 150, *151*, 161,
  165, 190, *194*, 195, 202,
  205, 248
malate, 35, 43, 44
malate synthetase, 19
male gametophyte, 244
male nuclei, 4, 245, 247
maltose, 16
manganese, 74, 82, 83, 86
mangrove, 130
marrow, 146
marrowstem kale, 140
marsh spot, 86
Maryland mammoth
  tobacco, 235, *240*
mass flow, **95**
mass spectrometry, 166, *167*
mathematical analysis of
  growth, **146**
matric potential, **49**

mature stage, 221
maximum temperature for
  growth, 142, *143*
measurement of growth, **135**
mechanism of action of,
  hormones, 162
mechanism of phloem
  translocation, **95**
*Medicago sativa*, 53
megaspore mother cell, 244
meiosis, 244
membranes, 6, 56, 60, 79, 80,
  85
  ATPase in, 173, 175
  chilling effects, 120, 265
  freezing and, 125
  gravitropism and, 198
  heat stress and, 128
  hormone action on, 173,
    184, 185
  leakage, 9, 80, 125
  phytochrome effect, 231
  receptors, 173
  salinity and, 131
  in seeds, 6, 9, 15, 19, 265
  transport, 35, 45, 79, 88,
    105
  in vernalization, 265
  water stress and, 117
meristematic cells, 80, *138*,
  214
meristemoids, 214
meristems, 84, 94, 120, 138,
  139, 152, 157, 169, 171,
  175, 177, 180, 201, 214,
  215, 234
  ro role of boron in, 84
mesocotyl, *5*
*meso*-inositol, 258
mesophyll, 43, 60, 69, 89, 105
  resistance, 33, 34, 69
mesophytes, 49, 114, 115,
  117
messenger-RNA (m-RNA),
  **14**, 15, 18, 23, 170, 174, 184
metabolic disturbance – due
  to stress, 115, 118, 119,
  124, 125, 128
metabolite transport –
  control of direction, **93**,
  218
metal contamination, 132
  tolerance, 132
metallophyte, 132
metastable state, of water, 61
3-methyl-lumiflavin, *205*
microfibrils, 173, 179
micro-nutrients, 75, 83
  as activators of enzymes,
    83
micro-organisms, 132, 143,

145
  thermophilic, 143
micropore system, 63
micropyle, 9, 245, 247, 251
microspore mother cell, 244
microtobules, 179
middle lamella, 172, 222
millet, *137*
mineral deficiency, 39, **86**, 93
  symptoms of, 86
mineral ions, 44, 74, 82, 88,
  90, 91, 92
mineral nutrients in soil, **76**
mineral nutrition, **74**, 142
minimum temperature for
  growth, 143
*Minuartia verna*, 132
*Miscanthus sinensis*, 75
mistletoe, 262
mitochondria, 12, 15, 21, 41,
  79, 98, 105, 117, 124, 168,
  171, 251
mitosis, 145, **168**, 261
  asymmetric, 245
mitotic spindle, 168
mobilization of food
  reserves, **15**, 94
molecular films, 102
molybdenum, 74, 75, 83, 86
monocarpic plants, 221
morning glory, 142
morphogenesis, **152**, 161,
  **211**, 250
morphogenetic field, 255
*Morus*, 122
mottle-leaf, 86
*Mucor stolonifer*, 145
multidirectional movement
  of metabolites, 88, 93
multinucleate, 178
Münch hypothesis, **95**
mung bean, 16, *32*, 198
mustard, 228
mutants, non-tropic, 190,
  192, 199, 207
*myo*-inositol, *183*
*Myristica fragrans*, 5
*Myrothamnus flabellifolia*,
  114

NADP (nicotinamide
  adenine dinucleotide
  phosphate), 36, 42
naphthaleneacetic acid, *160*,
  169, 182, 183
naphthoxyacetic acids, 160,
  248, 250
nastic movements, **207**
nasturtium, 222
necrosis, 84
neutral red, 112

neutral temperature, 233
next available space concept, 215
nickel, 132
*Nicotiana*, 107
*Nicotiana rustica*, 224
*Nicotiana tabacum*, 10, 40, 169, 245, 246
nicotinamide adenine dinucleotide phosphate (NADP), 36, 42
nicotinic acid, 211
*Nigella damascena*, 262
*Nitella*, 78
nitrate, 76, 77, 84, 90
  assimilation, 84, 91, 250
  reductase, 84, 115, 116
nitrogen, 6, 74, 75, 83, 91, 126, 137, 222, 242, 258, 259
  fixation, 84
nodes, of grasses, 192, 197
nucellus, 4, 247, 253, 261
nuclear transplantation, 177
  volume, 178, *179*.
nucleus, 4, 98, 105, 168, 171, 177, *179*, 260
nucleus, generative, 245, 247
  tube, 245
number of cells, 137, 140, 142, 215
nurse colony, 260
nutation, 151
nutmeg, 5
*Nymphoides peltata*, 100

O₂ (*see* oxygen)
oak, 6, 146, 263
oat, 17, 18, 86, 159, 172, 195, 202
*Olneya*, 262
onion, *197*, 244
ontogeny, 211
oogenesis, 252
optimum temperature for growth, 142, *143*
  photosynthesis, 36
  stratification, 263
  vernalization, 233
Orchidaceae, 4
orchids, 4
organelles, 7, 9, 79, 98, 119, 168, 171, 197, 198
organic acids, 43, 44, 91, 249
  anions, 79
  substances – transport of, **88**
organogenesis, 153, 177, 212
ornithine, 212
orotic acid, 170
orthotropism, 189
*Oryza sativa*, 7, 12, 142

osmoregulation, 131, 133
osmosis, 9, 35, **49**, 58, 60, 65, 95, 97, 100
osmotic potential, $\psi_\pi$, **49**, 58, 59, 67, 246, 259
  pressure, 49, 51, 118, 119
osmotic shock, 259
*Osmunda cinnamomea*, 215, *216*
ovary, 4, 245, 249
ovule, 4, 243, 245, 247, 249, 252, 253
ovum, 175, 177, 211, 250
oxaloacetate, 43
oxidation, of lipids, 12
  of proteins, 12
oxidoreductase, 83
oxygen (O₂), 190, 261, 265
  concentration, 41, 67, 77, 265
  and growth, 142, 172
  uptake, *11*, 12, 41, *143*, *172*

palm sugar, 97, 99
Palmae, 6, 106
paloverde, 262
*Papaver*, *193*
parthenocarpic fruits, 163, 164, 247, 250
partial molal volume of water ($\overline{V}_w$), 48
partial processes in flower induction, *238*
pea, 4, *5*, 6, 12, 13, 15, 21, 23, 86, 145, *155*, 161, *163*, 165, 170, 177, 180, *181*, 190, 202
peach, 264
peanut, 6, 9, 12, 192
pectins, 173
*Pelargonium zonale*, 93
pentoses, 16
PEP (*see* phosphoenol-pyruvate)
peptidases, 16, 18
peptides, 16, 45
perception, 195, 197, 206, 208
perennating organs, 121
perennials, 106, 114, 121, 125, 126, 146, 152, 221
pericarp, 4, 249
pericycle, 56, *57*, 81, 153, 211
periderm, 55
*Perilla ocymoides*, 239
perisperm, 4, 5, 11
permanent wilting point, (PWP), **66**
permeability, 10, 35, 56, 64, 67, 83, 85, 117, 126, 127, 198
peroxidase, 85, 263

peroxisome, 41, 43
Pfr (*see* phytochrome)
pH gradient, 173, 252
pH, of phloem sap, 91
  of soils, 76
  of xylem sap, 91
*Phacelia tanacetifolia*, 262
phase of activated metabolism and growth, **10**
*Phaseolus*
  *aureus*, 32, *89*
  *coccineus*, 161
  *limensis*, 9
  *lunatus*, 13
  *multiflorus*, 248
  *vulgaris*, 8, *11*, 13, 14, 18, 32, 107, *143*, *185*, 249
phenolic compounds, 85
phenoxyacetic acids, 160
phenylaliphatic acids, 160
*Phleum pratense*, 75, 177
phloem, 18, 19, **88**, *89*, *100*, *181*, *183*
  structure of, **98**
phloem sap – composition of, 91
phloem translocation, 19, **88**
  mechanism of, **95**
*Phoenix dactylifera*, 6
phosphate, 16, 76, 78, 81, 83, 92, 93
phosphoenolpyruvate, (PEP), 43
  carboxylase, 43
3-phosphoglycerate, (PGA), 41, 42
phosphoglycolate, 41
phospholipase, 83
phospholipids, membrane, 83, 127
phosphon D, 243
phosphorus, 5, 74, 75, 77, 83, 91, 92, 93, 94
  radioactive (³²P), 92, 168, 170, 218, 249
phosphorylase, 16
photoblastic seeds, **262**
photochemical reactions, 35, 37
photomorphogenesis, **225**
photonasty, 209
photons, 37
photo-oxidation of auxin, 207
photoperiodic induction, **233**
  stimulus, 108, 219
photoperiodism, 219, 220, 233, **234**
photophile phase, 241
photophobe phase, 241

photoreceptor in
phototropism, 228
photorespiration, **40**
photostationary state, 229
photosynthesis, 26, 27, 28,
89, 115, 117, 127, 128, 226,
236
action spectrum, *29*
energy conversion in, **28**
photosynthetic efficiency,
26, 27, **36**
phosphorylation, 38
pigments, *28*, 42
productivity, **38**, 44
products, 26, 41, 43, 44
phototropic curvatures,
succeeding positive and
negative, *203*
phototropic stimulus, 189,
**202**
phototropism, **202**
phyllotaxis, 153, 215
phytase, 16, 18
*Phytelephas macrocarpa*, 6
phytin, 5, 16, 18
phytochrome, **226**, 237, 238,
262
phytoplankton, 39
*Picea*, 263
pigweed, 262
piliferous layer, 55, 177
pine, 13, 21, 263
pineapple, 165, 240
*Pinus* sp, 21, 220, 263
strobus, 122
Piperaceae, 5
*Pisum*, 105
*arvense*, 18
*sativum*, 5, 6, 12, 13, 122,
143, 195
pith cells, 90, 107, 139, 169
pith segments, 169
plagiotropism, 189, **207**
Planck's constant (*h*), 37, 267
plant communities, 39, 77
distribution, 77, 120
factors, effect on water
uptake and movement,
**69**
weight index, 221
*Plantago*, 208
plantlets, 246, 253
plasmalemma, 10, 19, 34, 49,
79, 80, 81, 83, 98, 99, 105,
117, 124, 125, 173, 174,
179, 180, 184, 198, 200, 251
plasmatic growth, 138
plasmodesmata, 43, 56, *80*,
81, 105, *200*, 259, 261
plasmolysis, 50, *123*, 124
plasticity – of cell wall, 171,

172, *173*
plastid, *7*, 79, 98, 104, 198,
207, 251
plastochron, 215
plumular hook, 107, 165, 229
plumular pole, 251, 258
plumule, 4, *5*, 10, 21, *22*, 229
polar nuclei, 4
polar transport, **107**, 160, 164
polarity, **107**, 175, 180, 189,
197, 199, 200, 214, 245, 258
pollen, embryoids, *246*
enzymes in, 246
germination, 84, 245
grain formation, 245
grains, 122, 244
tube growth, 135, 145, 245,
246, 247
wall, 245
pollination, 164, 267
polyembryony, 252
polygenic control, 44, 133,
*Polygonatum multiflorum*,
208
*Polygonum sachalinense*, 75
polyhydric alcohols, 85
polyphenol oxidase, 85
polyploidy, 178
polyribosomes, 14, 170
polysaccharides, 16, 26
polysomes (polyribosomes),
14, 171
*Polystictus versicolor*, 47
polytene chromosomes, 169
pond water, 78
poplar, 220
*Populus*, 220
potassium, 35, 74, 75, 76, 78,
81, 82, **83**, 86, 91, 92, 102,
104, 118, 258
potato, 126, 139, 214, 219
P-protein, **98**
Pr (*see* phytochrome)
preprophase band, 180
presentation time, 194, 198,
203
primordia, excision, 212
primordium, 140, 153, **211**
privet, 61
programmes of,
development, 216, 221,
225,
senescence, 221
proline, 115, 131
prosthetic groups, 83
protein, 5, 12, 13, 16, 21, 26,
83, 98, 116, 123, 124, 126,
128, 171, 177, 263
protein body, 6, 7, *8*, 16
protein synthesis, 45, 115,
170, 171, 174, 184, 222, 245

in germination, **14**, 18
proteinases, 16, 17, 18, 23
protein-N, estimation of,
137, *172*
proteolytic enzymes, 16
proton, gradient, 45,
pump, 173
protoplasmic colloids, 9, 50,
125
protoplasmic streaming, 26,
80, *100*, 103, 108, 179, 185
protoplast, 50, 60, 79, 116,
117, 118, 124, 126, 171
protoxylem arches, 180, *181*
*Prunus persica*, 264
psi($\psi$) (*see* water potential)
psychrophilic bacteria, 143
pumpkin, 23, 95
pyrazol-1-yl-alanine, 6
*Pyrethrum cinerariaefolium*,
233
*Pyrus Malus*, *91*, 220

$Q_{10}$ (temperature coefficient),
35, 41, 128
quantum, 37, 46
efficiency of photosynthesis,
**36**
magnitude of (*E'*), 37
quantum yield, 29
quaternary ammonium
compounds, 131
*Quercus ilex*, 122
robur, 6, 264

radicle, 4, *5*, 15, 21, *22*, 172
radicle meristem, 180, 181
radioactive, bromide ($^{82}Br^-$),
93
calcium ($^{45}Ca$), 81
compounds, movement of,
**92**, 95, 201, 202, 206, 224
ions, 81, 93
phosphorus ($^{32}P$), 81, 92,
93, 218
potassium ($^{42}K^+$), 93, 102
sodium, ($^{24}Na^+$), 93
radioimmunoassay, of
hormones, 166, 202
radish, 14, 184, *191*, 203
raffinose, 92
*Ranunculus repens*, *193*
sceleratus, *253*, *254*, 259
*Raphanus sativus*, *183*, *191*
raspberry, 193
reaction time, 194, 203
receptacle of strawberry,
growth of, *248*, *249*
reciprocity law, 194, 203,
227, 228
red light, effects on

flowering, 237
 morphogenetic effects,
  217, **226**
redistribution of auxin, *191*,
 201, 206
regulators of gene activity,
 174, 184, 225, 230
relative growth rate, 144
relative humidity (RH), **49**,
 69, 111
relative water content
 (RWC), 111, 114, 116
reproductive development,
 **232**
respiration, 26, *27*, 38, 39,
 41, **44**, 77, 101, 105, 201,
 223, 263, 265
 dark, 41, 45
 at extreme temperatures,
  128, 143
 during germination, **11**
 during water stress, 115,
  119
 in the light, **41**
 maintenance, 46
 quotient (RQ), **12**
 rate as growth criterion,
  137,
 respiratory enzymes, 12,
  16
 substrate, 12
resurrection plants, 113, 114,
 119, 133
reversal of phototropic
 curvatures, *203*
rheotropism, 190
rhizocaline, 211
rhizoid, 252
rhizomes, 208, 121
*Rhizophora mangle*, 130
*Rhus ovata*, 262
rhythms, circadian, 241
 diurnal, 150, 241
 growth, 150, 151
*Ribes nigrum*, 220
riboflavin, *205*, 206
ribonuclease, 16, 18
ribonucleic acid, (*see* RNA)
ribosomal RNA (r-RNA),
 15, 170
ribosomes, 14, 105, 170, 171,
 251
ribulose bisphosphate
 carboxylase, **41**
rice, *7*, 12, 14, 17, 142, 190,
 192
*Ricinus communis*, 5, 6, 9, *62*
ring-porus, 56
ringing experiments, 90, 126
ripeness to flower, 232
RNA (ribonucleic acid), 14,

16, 18, 168, 170, 171, 177,
 184, 244, 245
RNA synthesis, 15, 18, 170,
 184, 222, 245
*Robinia pseudacacia*, 122,
 *123*, 126, 219
root, aerial, 52
 cap, 107, 195, 197, 199,
  202, 251
 cultures, 55, 180, *181*, 182,
  *183*, 184, 212
 gravitropism of, 193, *194*,
  197, 202
 hairs, 53, *57*, 177
 initiation, 153, 178, **211**
 ion uptake by, **77**, 82
 meristems, 89, 139, 180,
  251
 pole, 175, 245, 251, 258
 pressure, **57**, 63, 65, 68, 95
 surface area, 53, *69*
 systems, 53, *54*
 tip, growth of, 55, 135,
  139, 145, 146, *172*, 182,
  191
 water uptake by, **53**, 63, 69
rooting of cuttings, 108, 161,
 175
roots, auxin control of
 growth, *161*
 cation exchange properties
  of, 77
 growth rate of, 55
 hormone transport in, 106
 lateral, 55, **211**, 217
*Rosa*, 213
rubidium, 75
*Rubus idaeus*, *193*
rye, 12, 14, 15, 53, 54, 233

SI units, 267
*Saccharum officinarum*, 40
*Salicornia europaea*, 130, 131
salinity stress, **130**
 damage, 130
 resistance, **131**
*Salix lasiandra*, 93
salt, glands, 131, 133
 marsh, 130
 uptake, **78**, 131
sand culture, 74, 76
sand dune, 52
sandy soils, 65, *66*
sapwood, 55
*Sargassum*, 252
*Saxifraga rotundifolia*, 233
scald, 86
sclerophylls, 56
scotophile phase, 241
scutellum, 5, 12, 18, 19, 21,
 22

sea water, 130, 133
*Secale cereale*, 12, 15, 53, 54,
 145
secondary thickening, in
 roots, 181, *183*, 184
secretion of enzymes, 18, 19,
 21, 23, 222, 246
 K+ ions in phloem, 104
 K+ ions in stomata, 35
 salts, 80, 81, 131
SDP (*see* short-day plant)
seed, coat, 4, *5*, 247, 261
 composition, 4
 dormancy, 12, **261**
 germination, 4, 226, 229,
  261
 metabolism, **4**
 reserves, **6**
 structure, **4**
seed survival, of cold, 122,
 125
 of desiccation, 113, 118
 of heat, 127
seeds, importance in fruit
 development, 248
*Selaginella* gametophyte, 135
selenium, 75
self-incompatibility, 247
semipermeability, 19, 49, 80,
 99, 112, 125
*Senecio jacobaea*, 52, 233
senescence, 19, 147, 148,
 161, 163, 164, **221**, 249
sepals, physiological
 functions of, 250
separation layer, 244
sex determination, 161, 243
S-factors in pollination, 247
SH (sulphydryl) groups, 117,
 118, 126
shade plants, *30*, 31, 127, 230
shoot, apex, *96*, 107, 108, 153
 bud, 178, 181, 212
 growth, 135
 lateral, 153, *193*, 208
 meristem, 212, 139
 pole, 175, 245, 251
short day plants (SDP), 233,
 **234**
sickle-leaf, 86
sieve plate pores, **98**, 106
sieve plates, 98
sieve tube sap, **90**, 97
sieve tubes, 85, 95, 97, **98**,
 177
silicon, 74, 75, 76
silver maple, 113
*Sinapis alba*, *30*, 143, 192,
 228
single cell clones, 260
sink, 23, 94, 97, 101, 222, 249

slime, in sieve tubes, 98
sodium, 75, 76, 82, 83, 84, 86,
    130, 131
soil, aeration, 65, 67, 77
    colloids, 77
    conditions and water
        uptake, 65
    ions, 76
    mineral nutrients in, 76
    serpentine, 132
    solution, 65, 76, 80
    temperature, 65, 121, 122,
        128
    water potential, 66
soil water, 65
    yielding capacity, 65
Solanaceae, 245
sorghum, 17
*Sorghum vulgare*, 17
source to sink transport, 94,
    96, 97, 101
soybean, 6, 19, 98, 234, 239,
    241
speckled yellows, 86
S-phase of cell cycle, 168
sporangiophores of fungi,
    198
sporophyte, 175, 245, 252,
    261
sporopollenin, 245
spruce, 263
squash, 101
SS (disulphide) bridges, 117,
    118, 126
stachyose, 92
stamen, 243
staminal filaments, growth
    of, 144
starch, 6, 16, 17, 19, 21, 47,
    118
starch grains, *8*, 16, *197*
starvation, 94, 197
statocytes, *197*, 199
statolith theory, **197**, 207
statoliths, **197**
stelar parenchyma, 80
stem-crack, 86
stem tip culture, 212
stigma, 245, 247
stimuli, transport of, 189,
    197, 219
stimulus quantity law, 195,
    203
stomata, *32*, **33**, 44, 70, 83,
    113, 180
stomatal aperture, 33, 40, *71*,
    72
stomatal closing, 34, 44, 71,
    164
stomatal frequency, 33, 70
stomatal opening, 34, 36, 44,
    69, 70, *115*, 117

stratification, 263
strawberry, 140, *248*, *250*
streaming of cytoplasm, 103,
    179
stress avoidance, **110**
stress resistance, **110**
    breeding for, **132**
    chilling, 120
    freezing, 110, **120**
    general, 129
    heavy metals, **132**
    high temperature, **127**
    low temperature, **120**
    salinity, **130**
    terminology, 100
    water, **111**
stress tolerance, **100**
strontium, 75
style, 245
*Suaeda maritima*, 130, 131
suberin, 81
sub-threshold stimulation,
    195, 203
succulence, 44, 84, 241
sucrose, 19, 92, 93, 101, 125,
    181, 246, 258
sucrose in phloem sap, **92**,
    101
sugar, alcohols, 126, 258, 259
    beet, 40
    cane, 39, 40
    maple, 91
    phosphates, 16
    transport, 43, 85, 94, **95**
sugars, 5, 13, 26, 43, 85, 91,
    95, 97, 115, 118, 125, 181,
    246, 249
sulphate, 77
sulphur, 74, 75, 83, 86
sulphydryl (SH) groups, 83,
    117, 118, 126
sumac, 262
sunflower, 6, 21, 86, 107,
    140, *141*, 142, 202
sunlight, 28, 38, 227, 230
supercooling, 121, 124, 126
superoxide, dismutase 83,
    118,
    radicals, 83, 118
surface tension and cell
    division, 252
survival, criteria of, 112, 122
suspensor, 251, 258, 261
swede, 86
sweet potato, 214
sycamore, 220
symplast, 56, *80*, 81, 82
synchrony in cell division,
    168
synergids, 247
synthetic auxins, *160*
*Syringa vulgaris*, *182*

Système International
    d'Unités, 267

table of units, 267
tactic movements, 191
*Talbotia elegans*, 114
*Tamarix aphylla*, 131
tapetum, 245
*Taraxacum officinale*, 175,
    *176*, 222
tea, 75
tea-yellows, 86
temperature, and growth,
    142, *143*, 168
    night and day alternation,
        142
    and photorespiration, 42
    and photosynthesis, 35,
        41, 127
    regulation in plants, 120,
        129
tension, 57, 59, 64, 82, 117
testa, 4, *5*, 9, 13
tetrad formation, 244, 245
tetrazolium test, 112, 122,
    *123*
thawing, rate of, 125
thermal death point, 127, 128
thermochemical reactions,
    30, 35
thermocouple psychrometer,
    51
thermonasty, *210*
thermotropism, 190
thiamine, 211
thigmotropism, 190
*Thlaspi alpestre*, 132
threshold stimulation, 194,
    195, 196, 202, 227
thymidine, 168
*Tilia*, 106
*Tillandsia usneoides*, 52
timing reaction in
    photoperiodism, 237, **241**
tissue culture, 3, 120, 137,
    170, 171
tobacco, 10, 40, 86, 94, 140,
    169, 212, *213*, 222, 242
tomato, 56, 61, 139, 151, 164,
    250
tonoplast, 49, 79, 99, 117
top-sickness, 86
totipotency, 177, 178, 214,
    255
toxic elements, 75, 132
*Tradescantia*, 192
    *discolor*, 127
transcellular strands, *100*,
    103
transcription, 14, 18, 175,
    185
transduction, of energy, 12,

27, 39, 83
of stimuli, 197, 198, 200, 206
transfer cells, 105
transfer RNA (t-RNA), 14, 15, 170
translation, 14, 171, 185
translocation, 11, 21, 43, 44, 85, **88**
transmission of flowering stimulus, 236, 239
transpiration, 40, 44, 52, 58, **59**, *68*, *71*, *72*, 82, 91, 95, 128, 129, 144
cohesion theory of, **59**
cuticular, 70
pull, 95
rate, 52, **56**, 59, 65, 69, 70, 71, 82
transpiration stream, 52, 81
transport of metabolites, 88
control of direction of, **93**
long distance, 88, 106
medium distance, 88, 106
short distance, 88, 197
trichoblasts, 177, 180
2, 3, 5-triiodobenzoic acid (TIBA), 107
triploid endosperm nucleus, 4, 247
*Triticum dicoccum*, *119*
*Triticum vulgare*, 6, 143, 145
tritium ($^3$H), 103, 168
*Tropaeolum*, 153, *154*, 208, 222
tropical grasses, 40, 42
tropical rain forest, 39
tropisms, **189**
tubers, 94, 119, 140, 265
*Tulipa gesneriana*, 243
turbidity of cell suspensions, 137
turgor, 47, 84, 111, 119, 142, 171, 189
turgor pressure, 47, **49**, 97, 98, 104, *116*, 172
turnip, 155
two-stage hypothesis of abscission, 225
tyloses, 56
tyramine methylpherase, 170
tyrosine, 212

uncoupling of respiration, 46, 116
unicellular organisms, 88, 169, 191
unilateral stimuli, 189, 191, 203, 206
unisexual flowers, 161, 243
units of measurement, 267
unstirred layer, 33
uranium salts as tracers, 81

vacuolar sap, 78
vacuoles, 44, 47, 49, 50, 56, 79, 80, 131, 132, 171, 251
van Niel equation, 36
vapour pressure, 48, 49, 51, 111, 124
vascular bundles, 18, 43, *89*, 93, 94, 99
vascular cambium, 89, 90, 139
vascular tissue, *5*, 89, 90, 107, **180**
vascularization, in callus, 180, 181, *182*
vegetative cell, of pollen, 245
vegetative development, **211**
velocity of light (*c*), 37
velocity of transport, in phloem, 95
in xylem, 56
of hormones, 107
venation, 89
verbascose, 92
*Verbena ciliata*, *71*
vernalin, 233
vernalization, **233**, 265
vessels, 55, *61*, 79, *80*, 179
viability, 112, 113, 118
*Vicia faba*, 9, 84, 145
vigour, loss of, 221
*Viola calaminaria*, 132
viruses, 92, 103
viscosity of protoplasm, 119
*Viscum album*, 262
vital stains, 112, 122
vitamins, 84, 128, 169, 211, 215, 258
*Vitis*, 106
vitrification, 124
volume, of a cell mass, 137
of cells, 137
wall ingrowths, 105, 261
wall pressure, 49
wall thickenings, 179
walnut, 264
water, absorption from soil, 52
bound, 114, 119, 125
columns, 59
content of plants, 47, 52, *53*, 111, 113
of seeds, 7, **8**
culture, 74
free space (WFS), 79
lily, 100
melon, 139
movement – motive forces in, **57**
route of movement, **55**, 90
plantain, 142, 262
potential ($\psi$), 9, **48**, 58, 66, 70, 78, 84, 95, 111, 112, 113, 114, 130

measurement of, **51**, 112
properties of, 47
relations, **47**
of whole plants, **52**
saturation deficit, (WSD), 111
supply and growth, 142
– tritium labelled, 102
uptake and movement, factors controlling, **65**
vapour pressure, 48, 69, 111, 124

water stress, 34, 39, 40, 71, **111**, 127, 130, 131
avoidance, 112, 113
measurement, 111
metabolic disturbances, **115**, 118, 119
resistance, **111**, 121
structural damage, **117**
tolerance, **118**, 121
Went curvature test, 159
wheat, 6, 14, 15, 17, 18, 40, 58, 94, 114, 127, 142, 250
whiptail, 86
white bud, 86
willow, 93, 175
willowherb, 262
wilting, 34, 50, 61, 62, **66**, 71
wind, 34, 69
winter cereals, 233
wounding, 142, 177, 195

*Xanthium pennsylvanicum*, 236, 238, 239, 243, 261
*Xanthium strumarium*, 236
xanthophyll, *28*
xanthoxin, 164, 206
xeromorphic character, 113
*Xerophyta* spp., 114
xerophytes, 113, 117
X-ray, analysis, 113
treatment, 192
xylem, 18, **55**, **61**, 79, 88, *89*, 90, 96, 179, 211
diameter of vessels, 55
sap, 55, 57, 58, 59, **90**, 106
vessels, 55, 59, *80*, 81, 221

yeast extract, 169, 259
yellowing of leaves, 222, *223*
yield, agricultural, 110, 120, 127, 132

*Zea mays*, *5*, 6, **20**, 38, 143, 167, 195, 201
zeatin, 160
*Zebrina pendula*, 33
zinc, 75, 83, 86, 132
zygote, 4, 247, 250, 255, 259
polarity of, 251
zymogen, 17